Fault Diagnosis in Dynamic Systems

Theory and Applications

Prentice Hall International Series
in Systems and Control Engineering

M. J. Grimble, Series Editor

BANKS, S. P., *Control Systems Entineering: Modelling and Simulation, Control Theory and Microprocessor Implementation*

BANKS, S. P., *Mathematical Theories of Nonlinear Systems*

BENNETT, S., *Real-time Computer Control: An Introduction*

CEGRELL, T., *Power Systems Control*

COOK, P. A., *Nonlinear Dynamical Systems*

LUNZE, J., *Robust Multivariable Feedback Control*

SÖDERSTRÖM, T., and STOICA, P., *System Identification*

WARWICK, K., *Control Systems: An Introduction*

PATTON, R., CLARK, R. N., FRANK, P. M. (editors), *Fault Diagnosis in Dynamic Systems*

Fault Diagnosis in Dynamic Systems

Theory and Applications

Edited by

RON PATTON, PAUL FRANK and ROBERT CLARK

PRENTICE HALL

New York London Toronto Sydney Tokyo

First published 1989 by
Prentice Hall International (UK) Ltd,
66 Wood Lane End, Hemel Hempstead,
Hertfordshire, HP2 4RG
A division of
Simon & Schuster International Group

Printed and bound in Great Britain at the
University Press, Cambridge

Library of Congress Cataloging-in-Publication Data

Fault diagnosis in dynamic systems.
    bibliography: p.
    Includes index.
    1. Automatic control—Reliability. 2. Process control
  —Reliability. 3. Fault location (Engineering)
  I. Patton, R. J., 1949– . II. Clark, R. N.,
  1925– . III. Frank, P. M., 1934– .
  TJ213.F33 1989   629.8   89-4002

  ISBN 0-13-308263-6

British Library Cataloguing in Publication Data

Patton, R. J. (Ron J.) *1949–*
  Fault diagnosis in dynamic systems theory
  and application.
  1. Dynamical engineering systems. Faults
  I. Title II. Clark, R. N. (Robert N.) *1925–*
  II. Frank, P. M. (Paul M.) *1934–*
  620.7'3

  ISBN 0-13-308263-6

---

1 2 3 4 5 92 91 90 89

ISBN 0-13-308263-6

Contents

PREFACE

This text presents an introduction to a range of topics which fall within the domain of Fault Diagnosis in Dynamic Systems including both theoretical and application aspects. It is the first multi-authored book on this subject and, as such, the main aim has been to provide a wide coverage of the state of the art in Fault Diagnosis from a number of points of view. It is of particular value to have contributions from experts throughout the world working in industry, government establishments and academic institutions. The book is one of a number of texts published in the Prentice Hall Series on Control Engineering.

Fault Diagnosis has become an issue of primary importance in modern process automation as it provides the pre-requisites for fault tolerance, reliability or security, which constitute fundamental design features in any complex engineering system. The book focuses to a large extent on *fault detection and isolation* (FDI) techniques that are based on the use of a mathematical or dynamic model of a process system. There is also clearly some potential in using knowledge-based models in all cases where analytical models are not available. Chapter 15 deals specifically with this approach and provides a thorough treatment and literature survey of expert systems techniques. Chapters 2 to 14 discuss the use of a wide range of analytical techniques based on functional redundancy for sensor or *instrument fault detection* (IFD), hypothesis generation and testing, component fault detection, predictive diagnosis for maintenance and fault detection threshold determination.

Chapter 1 provides an overall introduction to the field of fault diagnosis by presenting the reader an overview of the book as a whole and by making some important definitions. This Chapter also provides some discussion of the foremost criteria for assessing the performance of an FDI scheme given that fault monitoring systems must be tolerant to signal deviations caused by process parameter changes, disturbances, non-linearities etc. which are normal functions of the operation of most engineering processes requirements. The Chapter also gives an outline of two of the main themes of research contained in the book, namely *robustness* and *application* considerations. In order to assist the reader in his or her research in this subject Chapter 1 also provides a list of some significant contributions made in applying FDI methods either to real process plant or to realistic non-linear simulations. A brief introduction to the individual Chapters is also given in Chapter 1.

The material presented in the book is considered to be of value to all Scientists and Engineers concerned with the reliability and fault tolerance aspects of process systems. It will be of value, for example, to those working in the aerospace field, nuclear process plants, chemical process control plants, transport technology and for persons interested in reliable non-destructive testing methods.

The wide scope of the book means that it is also of value in educational establishments as a source of useful library reference information and case study examples. It is particularly valuable to have been able to provide one combined set of references at the end of the book for ease of access to many important theoretical and practical publications.

The editors are indebted to many people for suggestions and aid in this project. The editorial work of the book has been made possible through a NATO funded collaboration between the Universities of Duisburg (FRG) and York (UK) together with an established exchange link between the University of Washington, Seattle and the University of Duisburg. The project has also received support from the British Council. Special recognition is due to Frau Sabine Colnar and Susanne Hain at the University of Duisburg for her excellent effort in word processing. Grateful appreciation is also extended to Mr. Alan Gebbie of the University of York for the first class work in drawing and photography. Particular thanks are also given to the editor of the Prentice Hall Control Engineering Series of books, Professor Mike Grimble, for his assistance in making this book available.

R.J. Patton, P.M. Frank and R.N. Clark

LIST OF CONTRIBUTORS

C. Baskiotis
ADERSA
7, Boulevard du Marechal Juin
B.P. 52
91371 Verrieres-Le-Buisson Cedex
France
(Chapter 8)

Professor R.N. Clark
Department of Aeronautics and Astronautics FS-10
University of Washington
Seattle, Washington 98195
USA
(Chapters 1 and 2)

Mr. M.J. Corbin
Flight Systems, FS(F)1
Y20 Building
Royal Aerospace Establishment
Farnborough, Hants
UK
(Chapter 6)

Professor P.M. Frank
Universität -GH- Duisburg
Fachgebiet Meß- und Regelungstechnik
Bismarckstraße 81
4100 Duisburg 1
FRG
(Chapters 1 and 3)

Mr. P.A. Gorton
Litton Systems (Canada) Ltd.
Mailing Station 404-1-C
25 City Drive
Etobicoke, Ontario
Canada M9W 5A7
(Chapter 11)

Professor J.O. Gray
Department of Electronic and Electrical Engineering
University of Salford
Salford M5 4WT
UK
(Chapter 11)

Professor R. Isermann
Technische Hochschule Darmstadt
Institut für Regelsystemtechnik
Schloßgraben
6100 Darmstadt
FRG
(Chapter 7)

Professor J.G. Jones
Flight Systems, FS(F)1
Royal Aerospace Establishment
Farnborough, Hants
UK
(Chapter 6)

S.M. Kangethe
Department of Electronics
University of York
York YO1 5DD
UK
(Chapter 4)

Professor M. Kitamura
Department of Nuclear Engineering
Tohoku University
Aoba Sendai 980
Japan
(Chapter 9)

Professor K. Kumamaru
Department of Electrical Engineering
Kyushu University
Fukuoka
Japan
(Chapter 13)

Dr. R.C. Montgomery
Flight Systems Directorate
Advanced Controls Branch
NASA Langley Research Centre
Hampton, VA. 23665
USA
(Chapter 10)

Dr. R.J. Patton
Department of Electronics
University of York
Heslington
York YO1 5DD
UK
(Chapters 1 and 4)

Dr. A. Rault
ADERSA
7, Boulevard du Marechal Juin
B.P. 52
91371 Verrieres-Le-Buisson Cedex
France
(Chapter 8)

Professor S. Sagara
Department of Electrical Engineering
Kyushu University
Fukuoka
Japan
(Chapter 13)

Professor T. Söderström
Department of Electrical Engineering
Uppsala University
Uppsala
Sweden
(Chapter 13)

Professor S. Tanaka
Faculty of Engineering
Yamaguchi University
Ube
Japan
(Chapter 5)

Professor S. Tzafestas
The National Technical University
Division of Computer Science
Department of Electrical Engineering
15773 Zographou
Athen
Greece
(Chapter 15)

Professor B.K. Walker
College of Engineering
Department of Aerospace Engineering & Engineering Mechanics
Mail Location 70
University of Cincinnati
Cincinnati, OH 45221-0070
USA
(Chapter 14)

Professor K. Watanabe
College of Engineering
Shizuoka University
Johoku 3-5-1
Hamamatsu 432
Japan
(Chapter 12)

Professor J.P. Williams
Flight Systems Directorate
Advanced Controls Branch
NASA Langley Research Centre
Hampton, VA 23665
USA
(Chapter 10)

Dipl.-Ing. J. Wünnenberg
Universität -GH- Duisburg
Fachgebiet Meß- und Regelungstechnik
Bismarckstraße 81
4100 Duisburg
FRG
(Chapter 3)

Chapter One

INTRODUCTION

R.N. Clark, P.M. Frank, R.J. Patton

This Chapter provides an overall introduction to the field of fault diagnosis in dynamic systems. Some important definitions are provided and some introductory remarks are given to describe the content of each chapter.

The main aim of the book is to provide the reader with an overview of the state of the art of various fault detection or fault diagnosis methods. This has been achieved by presenting some recent research work by a number of well known international investigators. The material presented in the book includes many references to application case studies in order to provide the reader with an idea of how some methods are being considered for application to real process plant and aerospace systems. The book is therefore a study of the *theory and applications* of fault diagnosis.

The work of the authors of this book reflects the demands of the users of complex automatic systems for the reliability and security of those systems. The complex automatic systems so widely employed in modern commerce and industry can consist of hundreds of inter-dependent working parts which are individually subject to malfunction or failure. Total failure of these systems can present unacceptable economic loss or hazards to personnel. It is therefore necessary to provide the required operation of the entire system by (a) a plan of maintenance which will replace worn parts before they malfunction or fail and (b) a *scheme of monitoring* which detects a fault as it occurs, which identifies the malfunction of a faulty component, and which compensates for the fault of the component by substituting a configuration of redundant elements so that the system continues to operate satisfactorily. The principal concern of our authors is this monitoring function, i.e., the detection, prediction, identification, and correction of faults during on-line operation of a dynamic system.

A plant, to be automated can be considered to consist of three major types of subsystems, actuators, main structure (or process), and instrumentation or sensors. For example, in an aircraft flight control system the actuators are the servomechanisms which drive the control surfaces and the engines which provide the driving thrust. The actuators take their input signals from an autopilot controller. The main structure is the airframe with its cargo and appendages, along with the aerodynamic forces exerted on the control surfaces. The instrumentation consists of the several sensors or transducers attached to the airframe, which provide signals proportional to the vital motions of the airframe, such as airspeed, altitude, heading, accelerations, attitude and rates of change of attitude, control surface deflections, engine thrust, etc. This specific application has been discussed further in Chapters 4 and 6 and to a lesser extent in Chapters 3 and 12 through design examples.

Clearly, these sensor signals provide feedback information to the autopilot, but they are also used in the fault monitoring subsystem. A fault monitoring scheme is usually designed especially to detect and correct faults in only one of the three subsystems. Early proposed schemes were concerned primarily with detecting *sensor faults* which, once detected, could usually be corrected by electronic switching techniques not requiring the reconfiguration of mechanical parts. The compensation of faults in *actuators* is usually more difficult than the re-direction of electrical signals. The compensation of malfunction in the *main structure* is even less feasible, although recent research results indicate that some degree of feasibility can be expected here too. Some of the techniques developed in the early sensor fault detection systems have been employed in actuator fault detectors.

The chapters in this book report upon work done mainly in the realm of **Fault Detection and Fault Isolation**, denoted as FDI by many authors. The remaining task of the monitoring system - the reconfiguration of the system signals or component actions in order to permit continued operation of the system - is not of major concern in this book and has therefore not been addressed by most of these authors.

The fault diagnosis algorithms employed in FDI subsystems are normally implemented in a digital computer. These algorithms are essentially signal processing techniques combined with logical switching functions. The overall system reliability evidently depends upon the reliability both of the computer and of all the physical components of the system. The reliability of the computer is not addressed by most of our authors, the tacit assumptions being that computer faults are much less likely to occur than malfunction in the actuators, components or instruments since the computer can operate in a well-conditioned environment with a special layout for fail-safe operation, and that even more reliable computers can be expected in the future. This view is naturally held more widely amongst control system engineers than it is amongst computer engineers.

Fault-tolerance in dynamic systems is traditionally achieved through the use of *hardware redundancy*. Repeated hardware elements (actuators, measurement sensors, process components etc.) are usually distributed spatially around the system to provide protection against localised damage. Such schemes operate typically in a triplex or quadruplex redundancy configuration and redundant outputs (or measurements) are compared for consistency. For example, three (or more) sensors measuring the same variable are installed where one would be sufficient if it were completely reliable. The signals from these sensors are monitored by a logic circuit which ignores small differences in the signals due to electronic noise, manufacturing tolerances, and monitoring errors inherent in the instruments, but which declares that a sensor is faulty if its signal deviates too far from the average value of the others (assuming that the others remain within a small difference from one another). This approach to fault-tolerance can be simple and in some cases, reasonably straightforward to apply and is thus widely used. It is essential in the control of aircraft, space vehicles and in certain process plants which are safety-critical (nuclear power plants and plants handling dangerous chemicals).

The major problems encountered with hardware redundancy are the extra cost and software and, furthermore, the additional space required to accommodate the equipment. In aircraft, for example, the additional space could be used, for example, for more mission-oriented equipment. The pay-load required for hardware redundancy also limits the pay-load available for defensive equipment and, most particularly, for fuel. On the other hand, it has been recognised that since redundant sensors tend to have similar life expectancies, it is likely that when one of a set of sensors malfunctions the others will soon become faulty also.

To overcome these problems, at least in part and improve the overall system reliability and fault-tolerance in view of the availability of reliable and powerful computers, new approaches have been developed which eliminate some or all of the redundant hardware. The new developments have been prompted since the early 1970's by the high cost of the excess hardware and the weight and space penalties which the hardware redundancy imposes upon some systems.

These new approaches to Instrument Fault Detection (IFD) are based upon the idea that three (or more) *dissimilar* sensors measuring different variables, and therefore producing entirely different signals, can be used in a comparison scheme more sophisticated than simple majority-vote logic to detect a fault in one of the set. The rationale for this idea is that even though the sensors are dissimilar they are all driven by the same dynamic state of the system and are therefore *functionally* related. These newer schemes, initially called *inherent redundancy* or *functional redundancy* to distinguish them from *physical* or *hardware redundancy*, have also been termed *analytical redundancy* and *artificial redundancy*. All of the work reported in this book lies within the realm of this newer approach, although it is recognised that both approaches can be employed together to advantage in many cases.

The functionally-redundant FDI schemes are basically signal processing techniques employing state estimation, parameter estimation, adaptive filtering, variable threshold logic, statistical decision theory, and various combinatorial and logical operations, all of which can be performed in electric circuits or in high speed digital computers. Normally both the input signals to the actuators and the sensor signals, i.e., the input and the output signals of the monitored plant are available to the FDI subsystem. The FDI schemes are therefore designed under an assumption that either the dynamic nature of the system being monitored is known to a reasonable degree of precision or that it is possible to determine the values of certain physical parameters by on-line identification techniques applied to the input and output signals of the monitored plant. Using these techniques it is possible to generate signals within the FDI subsystem which serve the same purpose as the simple majority-vote signals used in hardware redundancy, that is, to generate signals for logical processing to detect a fault and to identify the faulty component.

This book provides a wide coverage of the various approaches to functionally-redundant FDI schemes. State estimation, parameter estimation and parametric modelling techniques are discussed in some detail. Reference is made on numerous occasions to state estimation methods as these are the most widely studied in the literature on this subject. There is, however, a significant contribution in the field of parameter estimation. In Chapter 7, Isermann provides a thorough study of the use of least-squares parameter estimation methods.

Least-squares parameter estimation provides a powerful way of detecting faults in dynamic systems by on-line monitoring of the estimates of physical system parameters, wherever these can be defined. The approach can be used for both component and sensor fault detection and isolation. Estimated parameters connected to specific subsystems of a plant can be used to monitor the condition of various components of the plant. On the other hand, estimates of instrument parameters can be used directly for detection of measurement system malfunctions. This is a particularly important method, when considering that process plants (chemical processes, nuclear reactors etc.) can have slow dynamic behaviour whilst parameter variations due to process faults often cause rapid parameter estimate changes.

In Chapter 9, Kitamura discusses the parameter estimation approaches to FDI further and defines the concepts of serial and parallel redundancy. This work shows how FDI-related information can be extracted through the direct use of a parametric model. The parametric model is used as an estimator of a process variable using other process variables as inputs. The applicability of FDI methods based on parameter modelling has been demonstrated through simulation and actual experiments at several nuclear power stations.

One Chapter in the book deals with the problem of predictive diagnosis. In Chapter 8 Rault and Baskiotis discuss the approach of diagnosis used for maintenance of Jet engines. The authors provide some useful data and an assessment of the method.

Kumamaru, Sagara and Söderström in Chapter 13, discuss the use of statistical signal processing for fault diagnosis by first defining a hierarchical two-level technique. In the first level a parametric monitoring scheme for on-line fault detection is introduced. On detecting the likelihood of a fault at the first level, the erroneous signal is then processed at the second level using a generalised likelihood ratio test (GLR). This approach constitutes a powerful diagnosis scheme, particularly when input-output data are to be used for fault monitoring.

As a simple example of an instrument fault detection (IFD) scheme, assume that there are m sensors, that one of them is known to be reliable, and that a state estimator driven by the reliable sensor signal (and the inputs to the monitored plant, of course) can be constructed. In this case it is possible to generate estimated values of the entire set of m sensor signals. These can then be compared with the corresponding actual sensor signals, and any differences will be due only to the small discrepancies described above as the errors in the state estimation process or noise in the sensors. If the state estimation is known to be reasonably accurate, simple threshold logic can then again be applied to the difference signals where the thresholds are unlike zero to prevent false alarms, whilst the difference signals remain sensitive to moderate, if not to small, faults. Many variations of this simple idea are possible, and several are discussed in this book.

The early literature in the area of fault detection by functional redundancy was based upon theoretical ideas and analysis. This literature precipitated further developments which are represented by the contents of this book. These developments show that progress has occurred in practical application of functionally redundant fault detection schemes. Much of the early work which was oriented toward practical application was done on simulators. Simulation is valuable in developing specific schemes of detection, but it cannot include all of the many unpredictable challenges which are inevitably present in practice. The application of new and developing fault detection schemes to actual operating systems is normally prohibited because of expense or safety. However, the operators of real systems are usually willing to provide the FDI experimenter access to the signals coming from the operating plant, either in the actual on-site, real time environment, or by recordings, provided the experimenter does not inject any control signals into the operating plant. This access offers an economical opportunity to test fault detection schemes in a more realistic setting than that of a simulated environment. Experiments using signals recorded from nuclear power reactors during transient operations, and signals recorded on-board jet transport aircraft during automatic landings are both examples of this type of testing. A sensor fault on a recorded signal can be emulated by adding an appropriate fault component to the signal and the behaviour of the fault detection device can then be observed. Such records can be used again and again to test and improve the fault detection scheme. One disadvantage of this testing technique is that an emulated fault on one of the recorded signals affects none of the other recorded signals, whereas a real fault in the operating plant might affect several of the measured quantities (through feedback action), even if that fault occurs in only one sensor.

There have been many successful laboratory experiments and field experiments reported in which fault detection schemes based on functional redundancy have been demonstrated to be feasible in practice, at least under restricted circumstances. Some operational *redundancy management* systems have been built and operated (see list at the end of this chapter). These usually employ a combination of hardware redundancy and functional (or analytical) redundancy. Real time fault detection schemes operating completely on functional redundancy are not yet common, as design techniques to make them robust are not yet fully developed. Some of the foremost criteria for assessing the performance of an FDI scheme are:

(a) Promptness of detection
(b) Sensitivity to incipient faults
(c) False alarm rate
(d) Missed fault detections
(e) Incorrect fault identification

A brief discussion of each of these will now be given. The basic function of a fault detection scheme is to register an alarm when an abnormal condition develops in the monitored system and to identify the abnormal component. To test the performance of an FDI scheme one can induce a fault, maintain it, and observe the reaction of the detection system. A binary response is normally desired, the detection system declaring either that a component has failed or that it has not. An intermediate declaration of *perhaps* is not useful to the person or system which must take corrective action when a fault exists. If the imposed fault induces no response from the detection system, this is a missed detection, which may be acceptable for inconsequential faults. An example of such a fault could well be a very small bias on the signal from a relatively unimportant sensor. It would, on the other hand, be quite unacceptable if the fault had a serious impact on the operation of the monitored system. Not all faults occur suddenly or persist when they do occur. Slowly developing or small faults (often known as *incipient* faults), for example, bias or drift in an instrument or intermittent faults must be assessed in a different way.

Assuming that a fault is detected successfully, the issue of promptness may be of vital importance. In some aerospace applications a fault which persists for a fraction of a second without detection and correction can destroy the mission of the operating system, if not the operating system itself. In other applications it may be more desirable to have reliable detection of minor faults at the sacrifice of speed in detection time or promptness.

In some systems it is desirable to detect small or slowly developing i.e. incipient faults. This is important in fault detection schemes intended to enhance maintenance operations in plants by early detection of worn equipment, in which case promptness of detection may be of secondary importance to sensitivity. In other systems sensitivity and promptness may both be required. This leads to more complex detection schemes, possibly requiring both hardware and analytical redundancy, for example by using fault-tolerant computer techniques.

False alarms are generally indicative of poor performance in a fault detection scheme. Even a small false alarm rate during normal operation of the monitored system is unacceptable because it quickly leads to a lack of confidence in the detection system. However, a detection system that has an acceptable false alarm rate during normal operation might register a false alarm when the monitored plant undergoes an unusual excursion, and this might be acceptable in some applications. In other applications small faults may be so serious that it is preferable to react to false alarms, replacing unfailed components with other unfailed components, than to suffer deteriorated performance from an undetected, though small, fault. The compromises in detection system design among false alarm rate, sensitivity to incipient faults, and promptness of detection are difficult to make because they require extensive knowledge of the working environment, and an explicit understanding of the vital performance criteria of the monitored system.

A number of the authors in this book have mentioned the problems associated with false alarms and methods used to reduce the false-alarm rate by design. For example, Clark in Chapter 2 has described a very effective approach to adapting the fault detection thresholds in a monitor scheme. The adaption takes account of large and rapid changes in the control or measured input signals of the plant which, together with the associated transient performance of the *fault detection* signal can lead to threshold exceedance and false-alarm. By adapting the thresholds the false alarm rate, due at least to deterministic signals, can be minimised. Walker describes a Markov modelling approach to threshold determination in Chapter 14. This can be a valuable technique when probabilistic information from sensors and process plant can be gathered together.

Other authors, for example Frank and Wünnenberg (Chapter 3), Patton and Kangethe (Chapter 4), Tanaka (Chapter 5) and Jones and Corbin (Chapter 6) focus their attention on the problem of *robust fault detection* by design. Whilst the approaches described in these four chapters differ slightly, the goal is essentially the same - that of maximising the sensitivity of the detector to actual sensor malfunctions, whilst discriminating between these faults and disturbance effects due to noise and uncertain dynamics. Chapters 3 and 4 are based on robust observer approaches whilst Chapter 6 by comparison makes use of redundancy available through narrow-band filtering of measurements. Chapter 5 adds a further dimension of *diagnosis* and *optimal sensor location*. The choice and use of the sensors forms an important part of the robust sensor fault detection problem. These techniques all aim to provide fault monitor schemes which will reduce the probability of false alarm in fault detection.

Another malfunction of the detection system, closely related to both the false alarm and the missed detection, is the incorrect identification of a failed component. In this case the detection system corectly registers that a fault has occurred but incorrectly identifies the component which has failed. The reconfiguration system then proceeds to compensate for the wrong fault, an action which could produce consequences as serious as a missed detection.

In assessing the performance of a given fault detection scheme, or in comparing two different schemes, it is necessary to understand the working environment of the operating system and to have measures of the five criteria listed above. It is further necessary to establish a repertoire of faults in the monitored system which are to be detected by the FDI scheme. This is a difficult task and is related to the problem of robustness. The robustness of an FDI scheme is of overwhelming significance in present day research dealing with functional (or analytical) redundancy.

A principal value of the work reported in this book is the insight the authors provide into significant areas of research. We outline only two of these areas here: **robustness** and **applications**.

The *robustness* of an FDI scheme is the degree to which its performance is unaffected by conditions in the operating system which turn out to be different from what they were assumed to be in the design of the FDI scheme. Therefore robustness must be measured with respect to the variables associated with those conditions. A major problem in the field of robustness in FDI schemes is caused by uncertainties in the physical parameters of the operating plant. In these schemes the redundant signals which are combined logically to identify faults are generated inside the FDI subsystem using for example, state estimation techniques usually founded upon linear system theory. These state estimators are essentially mathematical models of the operating plant. Therefore they depend critically upon the values of the many physical characteristics of the plant, such as properties of mass, moments of inertia, electric circuit parameters, aerodynamic or hydrodynamic forces and moments, heat transfer properties, etc. If these are all known with precision, the state estimates will be accurate and the FDI scheme may be remarkably sensitive to incipient faults and immune to false alarms. However, in most operating plants, even those that are modelled accurately as linear and time invariant (the simplest class of dynamic system), some physical parameter values are known only approximately. This is especialy true where fluid flows and heat transfers are involved.

Consequently, the state estimators must be designed using only nominal values for the uncertain parameters or using some accommodating mechanism to compensate for the uncertainty. The result is that the state estimates are always in error, the severity of which depends upon the manoeuvres of the monitored plant in ways which are not easily determined. The logic devices for processing the redundant signals (e.g. state estimates) may then produce false alarms, or if they are protected against this, they may fail to detect a fault. **This is the robustness problem with respect to parameter uncertainties.**

Robustness problems also occur with respect to other prominent features of the operating plant. We mention only three additional ones here. Robustness problems with respect to:

(a) unmodelled non-linearities or uncertain dynamics
(b) disturbances and noise
(c) fault types

All dynamic plants are nonlinear, whilst many behave almost linearly, provided they are not required to deviate from a narrow range around a nominal operating condition. An FDI scheme based on linear models might be quite satisfactory for these conditions. However, outside of this range the nonlinearities of the plant produce signals which are not modelled accurately by the FDI scheme, and these may then be interpreted as faults. **This is the robustness problem with respect to nonlinearities or uncertain dynamics.**

Dynamic plants are always subjected to inputs other than those intended by the system designer. These inputs, called *disturbances*, are usually random functions originating in the environment, such as fluctuations in the wind. Furthermore, the sensors usually have electronic noise superimposed on their signals. This noise is also random, but it originates from a different source and is normally uncorrelated with the disturbances. Most signal processing techniques used by designers to account for random fluctuations of this sort are based on the assumption that those fluctuations are stationary Gaussian processes having known parameters. If the actual disturbances and noise are non-stationary, non-Gaussian or even corrected in some way then the FDI scheme will perform below its expected level. **This is the robustness problem with respect to disturbances and noise.**

It can be shown that (a) and (b) are easily combined together in the situation where observers or state estimators are used for fault monitoring. This is particularly true in the case of the Instrument Fault Detection (IFD) problem. Chapters 3 and 4 show how the unmodelled or uncertain dynamics of a process plant can be contained in an uncertain disturbance distribution matrix. The robust fault detection problem becomes one of *disturbance-decoupling* by design. If an observer or state estimator is used modelling errors and dynamic uncertainty can be shown to act like a disturbance on a *linear* system. The fact that the observer is a linear system is important to the approach to disturbance de-coupling. The authors of Chapter 3 call this the Unknown Input Observer Scheme (UIOS). Chapter 4 uses the same concept, however, the authors approach the problem from a different line of reasoning and based on *eigenstructure assignment.*

A given component can malfunction in many ways. For example, a sensor can suffer a change of scale factor, a bias which may not be constant, a nonlinearity due to wear or friction, a stuck value at any level within its dynamic range, excess noise, or hysteresis. Some FDI schemes are designed to detect only specific types of faults, and these become more cumbersome as the number of fault types in the repertoire is increased. Clearly, if a malfunction should occur which is not in this repertoire or fault diagnosis *knowledge base* then the FDI scheme will not recognise it. **This is the robustness problem with respect to fault types.** It is better to have a scheme which detects any fault and identifies the faulty component even though the specific type of fault is not identified.

A fault monitor which is robust to fault types must include *hypothesis-generation* and *hypothesis-testing*. The hypothesis generation procedure is to build up a repertoire of known or hypothesised possible malfunctions or faults in system components or instruments (sensors).

Hypothesis-generation can be performed using either model-based data using, for example state or parameter estimation or through the use of expert system techniques. The model-based approach is part of the deep-reasoning knowledge based strategy, however model-based approaches can also use qualitative deep reasoning. The human operator, through his or her experience and knowledge of a specific system or process can generate hypotheses which are both quantitative and qualitative. The role of the human operator therefore cannot be completely superseded. It can be argued that the most powerful approaches to hypothesis generation should be based on both quantitative and qualitative reasoning. This is certainly a subject which is wide open for future developments. A detailed study of the knowledge-based approach to fault diagnosis, in the widest context, is given by Tzafestas in Chapter 15. Chapter 15 also provides a very comprehensive list of the published work in the field. A useful addition to this has been provided by Milne (1987).

The robustness problem with respect to fault types also includes the important question of how a given generated fault hypothesis should be tested. Once again, either quantitative or qualitative approaches can be employed. The reader is referred again to Chapter 15 and Milne (1987) for further details.

Quantitative approaches to hypothesis-testing have been described in Chapters 10 and 11. These authors show how Kalman filters based on multiple dynamic models can be used for hypothesis-testing by processing the residual or innovation signals. Both Chapters 10 and 11 contain application studies. Chapter 10, for example, deals with the problem of redundancy management for a class of light weight spacecraft having appreciable structural dynamics and with distributed sensors and actuators leading to a requirement to employ active control techniques. The hypothesis-testing in this particular study is achieved using sequential probability ratio testing (SPRT). The approach allows for the on-line detection and rapid re-configuration of a single Kalman filter based on the structural dynamics of, for example a flexible grid structure similar in form to a spacecraft flexible panel.

In Chapter 11 the authors focus attention on an approach to non-destructive testing for quality assurance during the manufacture of engineering materials. This work is based on multiple-model hypothesis-testing using a bank of Kalman filters. The real-time application of this form of non-destructive testing can be used very effectively to provide quality assurance information to the human operator through an appropriate form of visual display facility.

When one takes an FDI scheme whose feasibility has been demonstrated (including a laboratory setting) and attempts to implement that scheme into a practical operating device in a real system, numerous practical and unforeseen difficulties present themselves. To overcome these difficulties one must understand the detailed design of the FDI scheme as well as the nature of the practical problems. This usually requires the FDI designer to follow his work into the specific engineering field, either doing the implementation himself or working closely with those who do it. For this reason we also call *applications* a legitimate area of research.

There has been a significant number of application studies of FDI techniques including some actual applications to either process plant or laboratory experiments using real-time equipment. The following list provides some useful examples together with appropriate references which the reader may wish to pursue further.

FAULT DETECTION/DIAGNOSIS APPLICATION STUDIES

EXAMPLE	DETECTION PROCEDURE	REFERENCE	
Aircraft F-4	2 Kalman filters, consistency test of measurements and estimates	Meier, Ross, Glaser	1971
Satellite, Autopilot for the lateral axis dynamics of an aircraft	Detection filter design for deterministic systems and systems with stochastic disturbances	Jones	1973
F8-C	Linearised conditional mean estimator based on non-linear aircraft dynamics - several extended Kalman filters - decision logic with cost minimisation by hypothesis testing	Montgomery, Price	1974
Inertial navigation system	Definition of confidence areas for the deviation of the trajectories of the system and a state estimator. One Kalman filter has to be designed for every instrument	Kerr	1987

TF-30	Hard failures are detected via rate limit thesholds	Hrach et al.	1975
Hydrofoil boat	One observer is designed for every system output that estimates the full state of the system	Clark, Fosth,Walton	1975
Turbofan	Threshold check of weighted average estimates	Ellis	1976
Space shuttle	Extended Kalman filter, generalised likelihood test	Deyst, Deckert	1976
Inertial navigation system	Kalman filter and extended Kalman filter, statistical fault evaluation for a window of measurements	Maybeck	1976
Space shuttle	Kalman filter bank, filter design for hypothetical errors and for the nominal state	Montgomery, Caglayan	1976
Traffic on freeways	Kalman filter and generalised likelihood test for the detection of jumps in linear systems	Willsky, Jones	1976
Turbojet	Bayesian hypothesis testing	Wells, de Silva	1977
Aircraft A-7D	For each measurement 2 instruments and 2 Kalman filters. Detection of scale factor and bias-type errors by hypothesis testing. Bandpass-Kalman filters considered	Cunningham, Poyneer	1977
Jet engines	Conditioned Kalman filters, Bayes estimator for hypothesis tests	Wells, de Silva	1977

NASA F8-DFBW Aircraft	For each measurement 2 instruments. Detection of bias-type errors by sequential probability ratio test. Fault identification by Kalman filters	Deckert, Deyst Desai, Willsky	1977
Single spool	Maximum likelihood testing	De Hoff, Hall	1978
Navigation system	Kalman filter bank, where each instrument is driven by a subset of all sensors. Comparison of the estimates and calculation of the covariance matrices	Smestad Orpen	1978
Hydrofoil boat	DOS-scheme, reduced order observers, decision logic, robustness test	Clark	1978 (a)
Hydrofoil boat	Simplified DOS-scheme, one reduced order observer that is driven by one system output signal, robustness test	Clark	1978 (b)
Space shuttle	UD-factorised filter with word limited digital processing of the measurements. Minimisation of a Bayes-cost function	Montgomery, Tabak	1979
Third-order system with system and measurement noise	Fault diagnosis for systems in discrete time in the presence of parametric faults. Kalman filter for the normal operating condition, sequential probability ratio test, adaptive extended Kalman filter estimates the state and the parameters under parameter variations	Yoshimura, Watanabe, Konishi, Sodea	1979
Lockeed L-1011 Aircraft	One observer of full order for one flight condition in the closed-loop to drive the autopilot	Shapiro, Decarli	1979
HFB 320	Fault safety concept for a double	Onken,	1979

Aircraft	sensor configuration. One observer is designed for each sensor	Stuckenberg	
Internal navigation system	One Kalman filter for each instrument and the nominal system model. If the confidence areas of the estimated and the nominal states do no longer overlap a fault is concluded	Kerr	1980
Nuclear reactor	5 Kalman filters are designed for 5 instruments that are each driven by 3 instrument readings	Clark, Campbell	1982
Boiling water reactor	Extension of the DOS-scheme (Clark, 1978) for a system with measurement noise, Kalman filter bank	Kitamura et al. Kitamura	1979 1980
Single input/ single output system	2 observers with different sensitivity	Frank, Keller	1980
Hydrofoil boat	One Kalman filter is driven by one system output signal. Exponential smoothing of the output estimation error	Clark, Setzer	1980
F 100 non-linear simulation	Detection filter	Meserole	1981
Pendulum, hydrofoil boat, aircraft	Fixing of directions in the state space that are unsensitive against parameter variations. Extension of the DOS-scheme (Clark) according to (Frank/Keller) to a dual DOS-scheme	Chen	1981
Discrete single input, single output system	Kalman filter for the nominal state, extended Kalman filter for the state and parameter estimation. Hypothesis test of the filter innovations	Watanabe, Yoshimura, Sodea	1981

Chemical reactor with heat exchanger	Detection of instrument faults in the presence of parameter uncertainties, system- and measurement noise for a class of nonlinear systems with insensitive state estimators	Watanabe, Himmelblau	1982
Nuclear reactor	Robust observer scheme based on DOS (Clark)	Elmadbouly, Keller, Frank	1982
A-7 Aircraft	Considering several output signals to drive an observer, logic design	Clark, Hertel	1982
Nuclear reactor	One Kalman filter, calculation of the covariance innovation for a time window of estimated and sampled values. Parameter variations are considered	Belkoura	1983
Nuclear reactor	Difference of the estimation error of two observers that are driven by different measurements (TEOS). Parameter invariant observer scheme (PIOS)	Elmadbouly, Frank	1983
Boiling water reactor	One Kalman filter for each instrument. Consistency checks between outputs of the filters	Tylee	1983
Pressurised water reactor	Fault detection by decentralised state estimation	Janssen, Frank	1984
Pulsed extraction column	Kalman filtering and sequential probability testing	Bonivento, Tonielli	1984
F 100 linear aircraft simulation	Sequential probability ratio test	Weiss et.al.	1985
F 100 linear aircraft simulation	Hypothesis testing	Emami-Naeini et al.	1986

F 100 Non-linear aircraft simulation	Threshold checks	Brown et al.	1985
Nuclear reactor	DOS-scheme (Clark) applied to detect instrument and component faults	Niccoli et al.	1985
Inverted Pendulum	DOS-scheme applied to detect instrument faults	Wünnen-berg et al.	1986
Gas-Pipeline	Adaptive state estimation, fault detection by fault models and cross-correlation	Billmann	1985
Nonlinear aircraft model	Detection filter design by eigen-structure assignment	Patton et al.	1986 (a), (b), 1987

Chapter Two

STATE ESTIMATION SCHEMES FOR
INSTRUMENT FAULT DETECTION

Robert N. Clark

Several schemes for the prompt detection of incipient instrument
faults in dynamic systems are described. Based on functional
redundancy, these schemes require the use of state estimators in the
monitoring subsystem to generate redundant signals. These signals,
along with the raw instrument outputs, are processed logically to
register an alarm and identify the failed instrument when a fault
occurs. An analysis of the errors introduced by uncertainties in the
state estimator models reveals the basic robustness problem of these
fault detection schemes, which introduces an inevitable design
compromise between fault detection sensitivity and false alarm rate.
Some examples of detection logic, intended to improve this compromise, are described. Practical applications of functionally
redundant instrument fault detection schemes have been demonstrated
to be feasible in laboratory settings and by simulations. Extensive
field applications are yet to be realised but offer challenging
development problems.

2.1 INTRODUCTION

A diagram of the typical multivariable automatic control system in
which instrument fault detection (IFD) is important is drawn in
Figure 2-1. The instruments, or sensors, provide the feedback signals
which are essential for the control of the plant. A fault in one of
the instruments could cause the plant state variables to deviate
beyond acceptable limits unless the fault is detected promptly and
corrective reconfiguration of the system is accomplished in time to

avoid such deviations. It is the purpose of the IFD subsystem to detect any instrument fault which could seriously degrade the performance of the control system.

There are several approaches to the design of instrumentation systems to insure their reliability in spite of occasional sensor faults. Some of these require redundant sensors with the IFD subsystem acting simply as a monitor to detect a fault which causes one of the redundant signals to deviate from its unfailed counterparts. This is called *hardware redundancy* and in many applications it is an effective means for providing the reliability required. In other applications, where hardware costs are excessive or space is limited, approaches based on *functional redundancy* (also called *analytical redundancy*) offer protection against sensor faults without redundant sensors. These schemes use state estimation techniques, so that the physical form of the IFD subsystem shown in Figure 2-1 is a computer, or possibly a shared portion of the computer which serves as the feedback controller in Figure 2-1. Because these IFD schemes use state estimators (Luenberger Observers or Kalman Filters) the IFD subsystem must have available the complete set of instrument outputs, represented by the vector y_I, and either the actuator input signals, represented by the vector u, or the actuator outputs, represented by u_A, as indicated in Figure 2-1. It is worthwhile noting that although most functional redundancy schemes depend on the plant and instruments being essentially linear, so that conventional state estimator design principles can be utilised, it is not at all necessary that the feedback controller be linear, or that the actuators be linear when u_A is used as an input to the IFD scheme. It is not even necessary that the actuator-plant-instrument package be a part of a feedback system.

In fact, any linear plant-instrument part of a larger, nonlinear, system which can be isolated topologically as in Figure 2-1 may have its instruments monitored by a functionally redundant subsystem.

There is extensive literature on IFD schemes based on functional redundancy. References by Meier et al. (1971), Mehra et al. (1971), Maybeck (1976) and Clark et al. (1975) represent early works in this field. Recently Frank (1986), Watanabe (1985), Upadhyaya (1985) and Kerr (1987) have provided bibliographies of later work. In this chapter we can consider only a few examples which embody some of the features which are characteristic of many of the schemes proposed, and which illustrate some of the difficulties one encounters in applying these schemes in practice.

For the examples considered in this chapter we assume that prompt detection of incipient instrument faults is the object. We also wish to avoid false alarms and will have to sacrifice some promptness and some detection sensitivity to small faults to achieve an acceptable false alarm rate. We will assume that only a single instrument suffers a fault at any one time. Multiple faults, such as common mode faults, are not treated here although some successful experience with such faults has been reported, Clark (1978). We will not account for any faults in the IFD subsystem, such as computer malfunctions. We will consider the problem of detecting instrument faults and identifying the faulty instrument, but will not consider the reconfiguration which is initiated following the detection and identification.

Instruments of many different physical forms are used in dynamic systems and there are many possible ways for each type to become faulty and fail. An IFD scheme must be capable of detecting any possible fault and should not be restricted in its design to detect only a limited repertoire of fault modes. It is usually not important for the IFD subsystem to identify the type of fault which has occurred, or the magnitude of the fault, so long as the faulty instrument is identified promptly.

In the following section entitled **DETERMINISTIC IFD**, we consider the system shown in Figure 2-1, in which there are no disturbance inputs to the plant and there is no instrument noise. In this case the state estimators used in the IFD subsystem are simply Luenberger Observers, Luenberger (1971). In Section 2.3 entitled **IFD WITH PLANT DISTURBANCES** we consider the more realistic case in which the plant has disturbance inputs in addition to the control inputs u_A shown in Figure 2-1, and also has normal noise on the instrument signals. In this case the state estimators in the IFD subsystem are Kalman filters, Meditch (1969).

2.2 DETERMINISTIC IFD

In this Section we assume that the Dynamic Plant, Instruments, and Faults shown in Figure 2-1 can be adequately modelled, at least for moderate fluctuations of the actuator outputs u_A from their equilibrium values (taken to be zero here for convenience), in the following manner. The plant is of dynamic order n and its state vector $x(t)$ plays a central role in the model (see Brogan (1974)). $x(t)$ is represented by the n x 1 matrix:

$$x(t) = \begin{bmatrix} x_1(t) \\ x_2(t) \\ \vdots \\ x_n(t) \end{bmatrix} \tag{2-2.1}$$

where $x_1(t)$, $x_2(t)$,..., $x_n(t)$ are the state variables. There are r actuators, the outputs of which are the elements of the r x 1 matrix $u_A(t)$:

$$u_A(t) = \begin{bmatrix} u_1(t) \\ u_2(t) \\ \vdots \\ u_r(t) \end{bmatrix} \tag{2-2.2}$$

These r variables are the inputs to the plant, and the dynamic equations of motion relating these r control inputs to the n state variables are represented in the standard *vector-matrix* differential equations:

$$\dot{x}(t) = Ax(t) + Bu_A(t) \tag{2-2.3}$$

Here A is an n x n matrix:

$$A = A_o + \Delta A \tag{2-2.4}$$

where A_o is a constant matrix, the elements of which are derived from the physical properties of the plant as it operates in its assumed *nominal* mode. ΔA is the uncertainty in A which is induced by uncertainties in the plant model. In case the physical parameters of the plant are known to an unusual degree of precision ΔA will be small, in some sense, as compared to A_o and may be relatively unimportant in the design of the IFD subsystem. But normally this is not the case. ΔA may be time variable if the plant parameters shift during the operating interval in which equation (2-2.3) is a valid model of the plant dynamics. We assume that the plant model remains linear, although uncertain, during that operating interval.

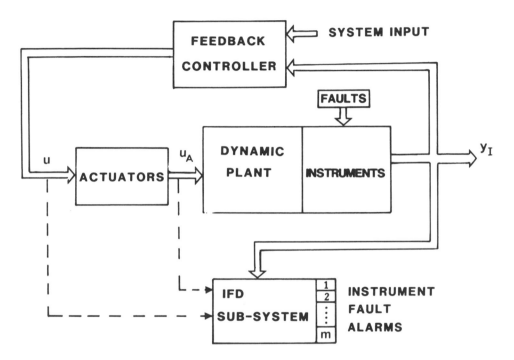

Figure 2-1: Control system using IFD

B is the n x r matrix:

$$B = B_o + \Delta B \qquad\qquad (2-2.5)$$

where B_o is a constant matrix derived from the nominal operating mode of the plant, as in the case of A_o. ΔB is the uncertainty in B due to the uncertainties in the physical parameters of the plant, as in the case of ΔA.

There are m instruments, or sensors, mounted on the plant to provide the signals required for feedback control and other purposes.

We assume that none of these m instruments are redundant, although nothing in the sequel actually prohibits such redundancy. We

require m to be 3 or greater. The m signals from these instruments
are represented as the elements of the plant output vector y_I:

$$y_I(t) = \begin{bmatrix} y_{1I}(t) \\ y_{2I}(t) \\ \vdots \\ y_{mI}(t) \end{bmatrix} \tag{2-2.6}$$

and this vector is dependent on the state vector and a fault vector
y_F:

$$y_I(t) = C\, x(t) + y_F \tag{2-2.7}$$

Here the influence of the instrument faults is modelled as an
additive component in $y_I(t)$. The fault vector is:

$$y_F = \begin{bmatrix} y_{1F} \\ y_{2F} \\ \vdots \\ y_{mF} \end{bmatrix} \tag{2-2.8}$$

where y_{iF} models the fault in the ith instrument. y_{iF} is zero except
when a fault exists on the ith instrument. When such a fault occurs
y_{iF} becomes a non-zero function which can be a simple term dependent
only on t (if the fault is simply a bias independent of $x(t)$) or it
can be a more intricate function if the fault is a nonlinear scale
factor deviation, a stuck value, or some other fault which gives a
false signal dependent on $x(t)$. We assume that only one component of
y_F will be non-zero at any time.

C is also subject to uncertainty for the same reasons that A and
B are:

$$C = C_0 + \Delta C \tag{2-2.9}$$

We note that some instrument faults could be modelled by ΔC, but we
prefer to reserve ΔC to represent plant uncertainties and include all
possible instrument faults in y_F. We also note that in some systems
there is a term $Du_A(t)$ in $y_I(t)$ which has been omitted from Equation
(2-2.7) here for sake of simplicity. This omission does not affect
the principles of IFD design discussed here, Clark (1985).

In the sequel we may see the shortened forms x, x_1, x_2,..., x_n, y_I,
y_{1I}, etc., for $x(t)$, $x_1(t)$, etc.

The IFD subsystem shown in Figure 2-1 usually consists of two major components. The first of these, the state estimator (or state estimators) is designed to produce a set of signals that are somehow redundant. These are estimated state variables, estimated instrument signals, or differences between estimated instrument signals and the actual instrument signals (which are available as $y_I(t)$ to the IFD subsystem as inputs).

The second component is a logic device which processes the redundant signals generated by the state estimators to detect when a non-zero element of y_F occurs, and to identify which of the m elements is non-zero. This detection and identification constitute the fault alarm. We discuss several configurations of IFD schemes in the sequel which are based on this general estimation-detection scheme.

For state estimation we use a Luenberger Observer (or several such observers) which for simplicity here is taken to be a 'full order' observer. With u_A and y_I as inputs to the observer the plant state may be estimated by the following algorithm:

$$\dot{\hat{x}} = [A_O - LC_O]\hat{x} + Ly_I + B_O u_A \qquad (2\text{-}2.10)$$

Here $\hat{x}(t)$ is an estimate of the plant state $x(t)$. The error in this estimate is $\epsilon(t)$:

$$\epsilon(t) = x(t) - \hat{x}(t) \qquad (2\text{-}2.11)$$

which is not directly measurable within the IFD subsystem, of course, because $x(t)$ is not available as an input to the IFD subsystem. $\epsilon(t)$ would be identically zero during the entire interval of operation of the system (in the absence of instrument faults) if the uncertainties represented by the ΔA, ΔB, and ΔC were not present and if the initial value of $\hat{x}$ could, somehow, be made to match that of x at the beginning of the operation interval. In most realistic cases, however, $\epsilon(t)$ will be non-zero at all times, its behavior being determined by combining equations (2-2.10) and (2-2.11) with Equations (2-2.3) and (2-2.7) to yield:

$$\dot{\epsilon} = [A_O - LC_O]\epsilon + [\Delta A - L\Delta C]x + \Delta Bu_A - Ly_F \qquad (2\text{-}2.12)$$

Equation (2-2.12) makes clear that it is the combined effects of the

instrument faults, the plant uncertainties, and the initial error
that cause $\epsilon(t)$, the solution to equation (2-2.12), to be non-zero.

The (n x m) matrix L must be selected by the designer to optimise
the performance of the IFD sub-system, in some sense. Prompt
detection and identification of the onset of a non-zero y_{iF} is, of
course, a prime performance measure, but the avoidance of false
alarms, induced by ΔA, ΔB, and ΔC, is equally important. The design
of the logic device as well as that of the state estimator must be
considered in this problem. It is apparent from Equation (2-2.12)
that the nature of the fluctuations of x(t) and of $u_A(t)$, which in
turn depend upon the system inputs (Figure 2-1), will also have a
bearing on this design. y_F is dependent on the type of fault which
occurs, as noted above. Further, ΔA, ΔB, and ΔC are all linked by the
underlying physical uncertainty which produce them, and cannot
therefore be considered separately except, possibly, in some special
cases. Because of these complicated interrelationships, which are
different for each particular system considered, little progress has
been made in the search for a scheme for designing both the state
estimator and the logic components as an integrated IFD system which
will be generally useful. On the other hand, several approaches to
such designs, for special cases, have yielded promising results, and
these might serve as guideposts along the path toward more general
results.

It is clear that the selection of the observer matrix L depends
on several nebulous factors pertaining to the IFD problem. We
therefore do not expect to apply the same criteria in its selection
that are used in observers designed for purposes other than IFD. One
reason for this is that those other criteria may be based on less
severe operating conditions than we assume in IFD work, e.g. ΔA, ΔB,
and ΔC are assumed to be negligibly small. Another reason is that in
IFD work, the eigenvalues of the $[A_o - LC_o]$ matrix need to be *prompt*
compared to the duration of the operating interval over which IFD
monitoring is required (hours, say) and not necessarily in comparison
with the dominant time constants of the plant (minutes or seconds,
say). It should also be noted, as a general property of observer
design, that the pair (A_o, C_o) need not, in all cases, be observable
in order for the observer to successfully estimate x (neglecting the
errors induced by ΔA, etc.). It is only necessary that L can be

chosen such that the eigenvalues of $[A_o-LC_o]$ are satisfactory. The reason for this is that the observer is essentially a model of the plant and its state vector is able to mimic the plant state because both have the same input $u_A(t)$. The $y_I(t)$ input to the observer contributes to the state estimation only in ensuring that the $[A_o-LC_o]$ matrix is satisfactory (which could occur in some cases without any $y_I(t)$ input at all).

As a first approach to the design of an observer based IFD scheme we temporarily ignore the uncertainties of the plant. Further, we define the estimated instrument output vector to be $\hat{y}(t)$:

$$\hat{y}(t) = C_o \hat{x}(t) \tag{2-2.13}$$

Now Equation (2-2.10) becomes:

$$\dot{\hat{x}} = A_o \hat{x} + B_o u_A + L(y_I - \hat{y}) \tag{2-2.14}$$

This shows that in the ideal case ($\Delta A=0$, etc.), except for a brief initial interval during which $y_I \neq \hat{y}$, the observer is a perfect model of the plant, as noted above, and the estimated vector $\hat{x}$ will be upset only if an instrument fault appears, manifested of course, as an undesired component y_{iF} in y_I. This fault occurring on a single instrument will induce errors in all n elements of $\hat{x}$ and so also in all m elements of $\hat{y}$. We seek a method of processing the elements of $\hat{y}$ logically so as to resolve the faulty element of y_I and thereby identify the instrument fault. This can be accomplished if we use only one of the instrument signals, say y_{iI}, to drive the observer. In this case the observer equation becomes

$$\dot{\hat{x}} = [A_o - L_i C_{oi}]\hat{x} + B_o u_A + L_i y_{iI} \tag{2-2.15}$$

where:

$$y_{iI} = C_{oi} x + y_{iF} \tag{2-2.16}$$

and C_{oi} is the (1 x n) matrix which is the ith row of C_o. L_i is the (n x 1) design matrix for the observer. Figure 2-2 is a block diagram of this estimator-logic scheme. The detection logic forms $\Delta(t)$, an (m - 1) x 1 matrix of decision functions $\Delta_i(t)$:

$$\Delta(t) = \begin{bmatrix} \Delta_1(t) \\ \Delta_2(t) \\ \vdots \\ \Delta_{i-1}(t) \\ \Delta_{i+1}(t) \\ \vdots \\ \Delta_m(t) \end{bmatrix} = \left\{ |y_{jI} - \hat{y}_j| \right\} \qquad \begin{array}{l} j=1,2,\ldots,m \\[6pt] j \neq i \end{array} \qquad (2\text{-}2.17)$$

If there is no instrument fault all (m - 1) of the decision functions $\Delta_j(t)$ will be small provided ΔA, ΔB, and ΔC are small. Thresholds can be placed on these decision functions to prevent false alarms due to the fluctuations in the $\Delta_j(t)$ caused by the plant uncertainties. If a fault develops on instrument k, where $k \neq i$, then only $\Delta_k(t)$ will be affected. If the fault is large enough $\Delta_k(t)$ will exceed its threshold and register the alarm. The remaining $\Delta_j(t)$, $j \neq k$, will remain unaffected by the fault on instrument k because only y_{kI} is affected by the fault and none of the $\hat{y}_j$ are affected because the observer is driven by a normal instrument signal y_{iI}. If the fault should occur on instrument i then all of the $\hat{y}_j$ signals will be in error and therefore all of the $\Delta_k(t)$ decision functions will be affected. In this case the logic is programmed to register an alarm on instrument i. This scheme is obviously feasible if there are no uncertainties in the plant. Its feasibility has also been demonstrated, by simulation, in one particular example where modest uncertainties ($\pm 10\%$) are present in some of the physical parameters of the plant (Clark (1978)). Here it was found that small constant thresholds on the decision functions prevented false alarms while permitting successful detection of small instrument faults, provided the plant manoeuvres were restricted. In later work these restrictions were removed by making the thresholds variable and dependent on the manoeuvres of the plant (Yu (1985)). This later work is described below under **ROBUSTNESS QUESTIONS**.

A wide variety of fault detection logic is made available if more than one observer is used in the IFD subsystem. An example of this is shown in Figure 2-3. Here there are m observers, one for each of the instruments to be monitored. Each observer is specially designed to operate on a single instrument signal, so this configuration has been called the *dedicated observer scheme*, or DOS (Clark (1975),(1978)). Assuming that an L_i can be found for each of the observers for which

$[A_o-L_iC_{oi}]$ has satisfactory eigenvalues, each observer generates n estimated state variables for use by the detection logic. Note that this has n times as many signals than is required minimally, so this scheme is replete with excess redundancy.

Two logic schemes using the basic DOS configuration are described here. One of these uses only one of the state variables from each of the m estimated state vectors. For the sake of illustration assume there are only three instruments (m=3) and we are using estimated state variable number 7 from each of the three observers as the most suitable for the IFD purpose. Define three functions

$$\Psi_1 = |\hat{x}_{71}-\hat{x}_{72}|$$
$$\Psi_2 = |\hat{x}_{71}-\hat{x}_{73}| \qquad\qquad (2\text{-}2.18)$$
$$\Psi_3 = |\hat{x}_{72}-\hat{x}_{73}|$$

Here $\hat{x}_{ij}$ is the estimated value of the ith state variable from the jth observer. Next, form three 'decision functions', one for each of the three instruments which are to be monitored:

$$\eta_1 = \Psi_1\Psi_2$$
$$\eta_2 = \Psi_1\Psi_3 \qquad\qquad (2\text{-}2.19)$$
$$\eta_3 = \Psi_2\Psi_3$$

If there are small uncertainties in the A, B, or C matrices there will be small errors in each of the estimated state variables and Ψ_1, Ψ_2 and Ψ_3 will fluctuate slightly from zero even when there is no instrument fault. Similarly, the decision functions η_1, η_2 and η_3 will also fluctuate from their zero values, but only slightly if the Ψ's are small. Now assume a fault develops on instrument number 1. Ψ_1 and Ψ_2 will be increased because they each depend upon state estimates from an observer which is driven by a faulty signal. Ψ_3 remains small because it depends only on estimates from observers which are operating normally. Consequently η_1 becomes very large compared to η_2 and η_3 and this trips the alarm on instrument number 1.

This logic is easily extended for the case m > 3. In this case there are more differences Ψ_i possible so that each of the m decision functions η_i can have more terms. This enhances the immunity of the IFD schemes to false alarms caused by small uncertainties in the plant parameters because the η_i are nonlinear functions of the Ψ_i.

The small fluctuations in each Ψ_i due to the ΔA, ΔB, and ΔC in normally operating systems are reflected in the η_i functions as the products of small quantities, which make the fluctuations in the η_i very small and these can be effectively masked against false alarms using thresholds, as described above. In like fashion, when a fault occurs on instrument 1 all of the Ψ_i factors in η_1 become large and thus η_1 becomes very large, making the detection of an incipient fault possible despite the small ΔA, ΔB, and ΔC.

The second logic scheme uses a comparison between the actual instrument signals and their estimated counterparts, as in the one observer system. Let:

$$\hat{y}_I^i = \begin{bmatrix} \hat{y}_{1i} \\ \hat{y}_{2i} \\ \vdots \\ \hat{y}_{mi} \end{bmatrix} = C_o \hat{x}_i \qquad (2\text{-}2.20)$$

be the estimated instrument vector from the ith observer. Form the vector:

$$\Delta y^i = \begin{bmatrix} \Delta y_1^i \\ \Delta y_2^i \\ \Delta y_3^i \end{bmatrix} = \begin{bmatrix} y_{1I} - \hat{y}_{1i} \\ y_{2I} - \hat{y}_{2i} \\ y_{3I} - \hat{y}_{3i} \end{bmatrix} \qquad (2\text{-}2.21)$$

Here m is again taken to be 3 for illustration purposes. Note that all three actual instrument outputs are used in Δy^i but that estimated instrument outputs only from observer i appear in Δy^i. Now the decision function for the ith instrument fault alarm is formed as:

$$\Delta_i(t) = |\Delta y_j^i \cdot \Delta y_k^i| \qquad\qquad j,k \neq i \qquad (2\text{-}2.22)$$

The term Δy_i^i does not appear in $\Delta_i(t)$ because it is usually zero. A fault in instrument 1, for example, will have no influence on $\Delta_2(t)$ and $\Delta_3(t)$ because those decision functions are unaffected by the estimated instrument signals from observer 1. However, $\Delta_1(t)$ becomes large because both Δy_2^1 and Δy_3^1 become large due to the faulty estimates $\hat{y}_{21}$ and $\hat{y}_{31}$. Thus the fault on instrument 1 is identified. The benefit of the nonlinear decision functions $\Delta_i(t)$ is realised here, as it was for the η_i, for reducing the chance of false alarms, as the number of instruments increases.

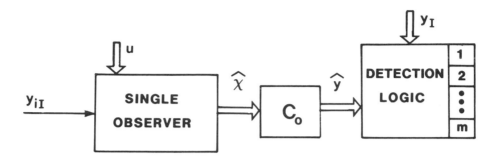

Figure 2-2: A simplified DOS

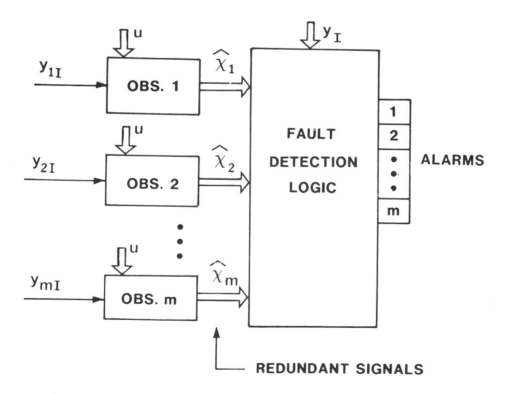

Figure 2-3: The dedicated observer scheme (DOS)

The basic DOS illustrated in Figures 2-2 and 2-3 offers many possibilities for generalisation. Frank and his coworkers have described a configuration, called the *Generalised Observer Scheme* (GOS) in which each of the m observers is driven by a set of inputs, different for each observer, and chosen to enhance the reliability of the IFD subsystem, Frank (1986). This scheme has the advantage that the observer becomes simpler to implement as the number of inputs increases, and more design parameters become available in the observer L matrix to influence the performance of the detection logic.

2.3 IFD WITH PLANT DISTURBANCES

If the dynamic plant has disturbance inputs in addition to the actuator inputs shown in Figure 2-1 the state estimation in the IFD subsystem will be in error, even if there are no uncertainties in the plant model. This is because the disturbance inputs are not available to the IFD subsystem, only the actuator inputs are available. In many cases the plant disturbances are random fluctuations and at the best are only known by their statistical parameters. When they are known in this fashion it is possible to design a state estimator in which the errors will still exist but will be minimised. The Kalman Filter is the most common such estimator, and there are many examples of its use in IFD systems (Mehra and Peschon (1971), Frank (1986), Watanabe, Tylee (1983), Clark (1976), Clark and Campbell (1982), Clark (1980), Willcox and Patton (1985)). An example of the use of fixed gain Kalman Filters in an IFD system using a variation of the GOS is the system shown in Figure 2-4. The Plant in this case is the pressuriser in a small scale pressurised water nuclear reactor (Clark and Campbell (1982)).

A simplified model of the pressuriser permits the state vector to be represented as:

$$x = \begin{bmatrix} q \\ p \\ t_p \end{bmatrix} \qquad\qquad (2\text{-}3.1)$$

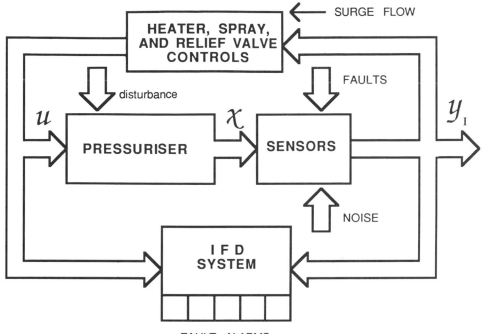

SURGE FLOW

HEATER, SPRAY,
AND RELIEF VALVE
CONTROLS

disturbance

FAULTS

u

x

y_I

PRESSURISER

SENSORS

NOISE

I F D
SYSTEM

FAULT ALARMS

Figure 2-4: PWR pressuriser and IFD system

where q is the quality of the (assumed) saturated mixture in the pressuriser, p is the pressure, and t_p the average temperature of the mixture. The input vector is:

$$u = \begin{bmatrix} w_{sur} \\ w_{sp} \\ w_{rv} \\ Q_h \end{bmatrix} \tag{2-3.2}$$

where w_{sur} is flow in the surge line connecting the pressuriser to the main coolant system *hot leg*, w_{sp} is the spray flow (from the *cold leg*), w_{rv} is the relief valve flow and Q_h is the power flow into the heaters. The dynamic model relating x to x and u in this case is nonlinear due to the complex evaporation and condensation processes which occur in response to fluctuations in the input variables. Nevertheless, it is possible to establish a linear model of the pressuriser which represents its dynamic behaviour if the state variables fluctuate only slightly from a known nominal operating condition. That condition is taken to be the one corresponding to the reactor operating at full power.

There are five sensors to be monitored by the IFD system in this case. These are identified by the output vector y_I:

$$y_I = \begin{bmatrix} l_1 \\ l_2 \\ l_3 \\ p_I \\ t_I \end{bmatrix} \tag{2-3.3}$$

where l_1, l_2, and l_3 are the signals from three redundant water level instruments, p_I is the pressure signal and t_I the signal corresponding to t_p. y_I contains the natural noise component of the sensors. An inspection of data from the operating plant permitted these noise components to be described by a matrix R, whose diagonal elements are the covariances of the noise components of sensor signals, and with zero off-diagonal elements. Similarly, a covariance matrix Q, representing the disturbances shown in Figure 2-4 is also determined from observations of the plant data.

The IFD system in this case can be a hybrid between hardware redundancy and functional redundancy. The three redundant level sensors can be monitored with a simple majority vote logic detector. This detector not only identifies faults in the level sensors but also provides a valid level signal. The pressure and temperature sensors are monitored by two Kalman Filters. Each of these filters must use three input signals, l_I, p_I, and t_I where l_I is the validated level signal provided by the monitor of the three level instruments. However these two filters must be sensitised to discriminate between faults in the pressure and temperature sensors. This is accomplished by detuning each of these filters. The filter corresponding to the pressure sensor is rendered relatively insensitive to the pressure signal p_I by artificially increasing the element in the R matrix corresponding to the pressure signal. This makes the filter much more dependent on the l_I and t_I signals for its estimation of the state vector. Similarly, the filter corresponding to the temperature signal is desensitised to that signal in the same fashion. The decision function for the pressure sensor is:

$$D_p = (p_I - \hat{p}_I)^2 \tag{2-3.4}$$

where $\hat{p}_I$ is the estimated pressure signal from the filter designed, as described above, for the pressure sensor. A fault in the pressure sensor causes p_I to deviate from its proper value but affects $\hat{p}_I$ to a much lesser extent because $\hat{p}_I$ depends primarily on the level and temperature sensors inputs, the filter having been desensitised to the pressure input. In like manner the decision function for the temperature sensor is derived from its filters as:

$$D_t = (t_I - \hat{t}_I)^2 \tag{2-3.5}$$

Simulation tests in which the pressuriser and its controls were represented by a nonlinear model and the detection filters were designed from a linear approximation to this model were performed to determine the potential feasibility of this scheme. The pressuriser was subjected to a standard manoeuvre during these trials. The manoeuvre was induced by a pulse of surge flow, large enough to activate both the heaters and spray, but not the relief valve. Eighteen tests were conducted in which various types of instrument faults were simulated (Clark and Campbell (1982)). A typical result is shown in Figure 2-5. Here a sudden bias fault of 2 psi occurs on the pressure sensor at t=60 sec. The decision function D_p would have a threshold of 2 or 3 to prevent a false alarm by the normal fluctuations caused by estimation errors due to instrument noise and the difference between the nonlinear plant and the linear Kalman filter which are apparent prior to the time the fault occurs. We also note that the D_t decision function is unaffected by the fault on the temperature sensor, and that its threshold against false alarms would also be 2 or 3 during this transient.

2.4 ROBUSTNESS QUESTIONS

Fault detection schemes utilising functional or analytical redundancy, such as the DOS scheme and its variations, work well in those situations in which the modelling uncertainties in A, B, and C are small and where the disturbance and sensor noise are not excessive and are accurately modelled as Gaussian processes. These ideal conditions are seldom met in practical applications, so it is necessary to address the probelem of robustness of these schemes in non-ideal settings. While experience has shown that some of these schemes remain effective in some situations which are clearly outside

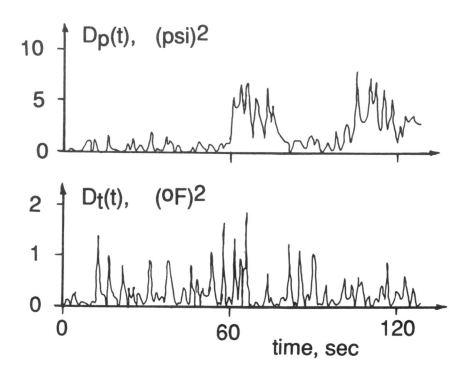

Figure 2-5: Decision Functions, bias fault of 2 psi occurring
pressure sensor at t = 60. sec

the envelope for which they were designed, there is no reliable guide
from this experience which permits one to assess, quantitatively, the
robustness of a given fault detection scheme with respect to the
variables of the situation.

The principal variables in a situation which influence the
performance of an IFD scheme are:

(a) *The uncertainties in the model of the plant*. In a linear plant these are
   represented by ΔA, ΔB, and ΔC and by unknown characteristics of
   the disturbance inputs and sensor noise. In a nonlinear plant the
   uncertainties are not so easily represented.
(b) *The system inputs*. The plant may be subjected to input commands
   other than simple step functions, harmonic oscillations, or well
   known stochastic processes. Consequently, the plant may manoeuvre
   outside the limits for which the IFD system has been designed.

(c) *Instrument Fault Type*. Instruments fail in many different ways. Some IFD schemes are designed to detect only a limited repertoire of fault types and may not be effective against unexpected, but realistic, faults.

(d) *Implementation*. The IFD scheme is usually realised as a computer program which must operate on-line to be effective. Some compromise with the theoretical design is normally necessary to satisfy the practical constraints imposed by the available computer.

The performance of an IFD scheme is measured by its sensitivity to incipient faults, its propensity to issue false alarms, and whether it permits a failure to go undetected. It is qualitatively clear that a balance between detection sensitivity and false alarms must exist in any IFD scheme. What is lacking is a useful, quantitative method for describing such a balance and hence providing design guidance for achieving a specified measure of performance. Kerr (1987) has provided some pertinent criticism on some of these questions.

A specific example illustrates an attempt to improve the robustness of an IFD scheme with respect to changing system inputs. In this case there is an uncertainty in the linear model of the plant so that thresholds are used on the decision functions. If constant thresholds are used and set low enough to be triggered by small faults during certain prescribed manoeuvres of the plant, false alarms will result when the plant is driven to manoeuvres outside the prescribed envelope, because the fluctuations in the decision functions during no-fault operation become larger as the plant manoeuvres become large. To improve the robustness in this case it is possible to use thresholds which are varied according to the control activity of the plant. One such scheme has been used effectively on the simplified DOS (Yu (1985)). This scheme is depicted in Figure 2-6.

In this scheme the same decision functions defined in equation (2-2.22) are used with each being masked by a variable threshold. To illustrate the design of the adaptive threshold we consider an example system in which there are two plant input variables and four instrument outputs (Clark (1978), Yu (1985)). With a 30 percent change in an important physical parameter of the plant, without a corresponding change in the design of the observer, a system input, $\theta(t)$, consisting of a pulse of duration 5.5 seconds is applied. The

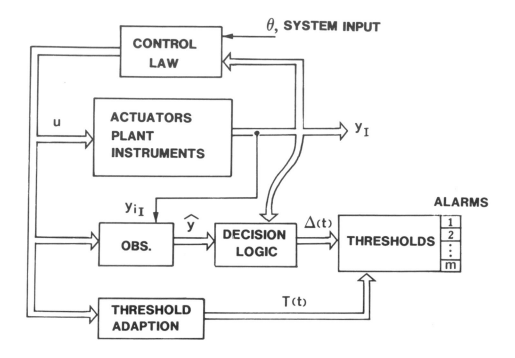

Figure 2-6: Simplified DOS using adaptive threshold detection logic

resulting control inputs $u_1(t)$ and $u_2(t)$ are shown in Figure 2-7.
Because there is an uncertainty in the model the three decision
functions $\Delta_1(t)$, $\Delta_3(t)$, and $\Delta_4(t)$ fluctuate from zero, as shown in
Figure 2-7 even though there is no instrument fault. Here the
observer is driven by instrument number 2 so there is no $\Delta_2(t)$. We
note from Figure 2-7 that the modal content of the three $\Delta(t)$
functions resembles that of $u_1(t)$ and $u_2(t)$. This suggests that the
thresholds might have the form:

$$T = T_0 + T_1|u_1(t)| + T_2|u_2(t)| \qquad (2-4.1)$$

where T_0, T_1, and T_2 are constants, to be selected as appropriate for
each of the three decision functions. In the case illustrated here,
for example, an iterative procedure is used to select the set (T_0,
T_1, and T_2) which would prevent false alarms for uncertainties in the
physical parameters from +30 % to -30 % of their nominal values. It
is possible to find such sets of constants for two of the three
decision functions. For the third it is necessary to use a threshold

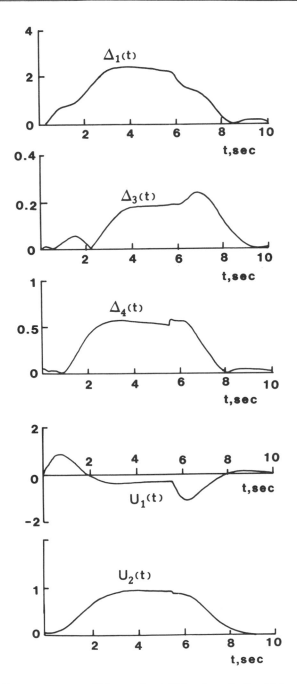

Figure 2-7: Responses u(t) and Δ(t) to 5.5 sec. input pulse, no instrument faults, 30 percent parameter uncertainty

of the form:

$$T = T_0 + T_1|u_{1F}(t)| + T_2|u_{2F}(t)| \qquad\qquad (2-4.2)$$

where $u_{1F}(t)$ and $u_{2F}(t)$ are the output signals from two first order filters having time constants τ_1 and τ_2 and having inputs $u_1(t)$ and $u_2(t)$.

With eleven design parameters established for the adaptive threshold algorithm, using the criterion that no false alarms should occur when the system input is a pulse, it is then necessary to test the performance of the adaptive thresholds under more realistic system inputs. A random signal having a realistic power spectral density is applied at $\theta(t)$. A short section of the resultant $\Delta_i(t)$ decision functions, along with the adaptive threshold, is displayed in Figure 2-8. Here T^i is the threshold. Clearly no false alarms are induced by the fluctuations in the decision functions caused by the uncertainty.

Finally, it is necessary to test the scheme to see if it can detect instrument faults during this random manoeuvre. The fault is simulated and the difference between $\Delta_i(t)$ and $T^i(t)$ (i=1, 3, 4) is monitored.

$$D_i(t) = 0 \quad \text{if } T^i(t) > \Delta_i(t)$$
$$D_i(t) = [\Delta_i(t) - T^i(t)] \quad \text{if } \Delta_i(t) \geqslant T^i(t) \qquad (2-4.3)$$

Bias faults, dead-zone faults, and scale factor faults are all used in the tests. The result of a typical test is shown in Figure 2-9. Here the physical parameter is 30 % less than its nominal value and a bias fault appears on instrument number 1 at t=10 sec. The fault magnitude is 2 % of the dynamic range of the instrument. We note the prompt, and prominent, indication of the fault by $D_1(t)$ and the lack of any such response in $D_3(t)$ or $D_4(t)$. Thus we see that this particular fault is successfully detected by this algorithm which has also proved its immunity to false alarms, at least under the circumstances presented here. A series of such tests, too long to report here, showed results for faults on instrument number 3 and instrument number 4 similar to those in Figure 2-9. For faults on instrument number 2 all three $D_i(t)$ functions became non-zero, which indicates a fault on instrument 2 (Yu (1985)).

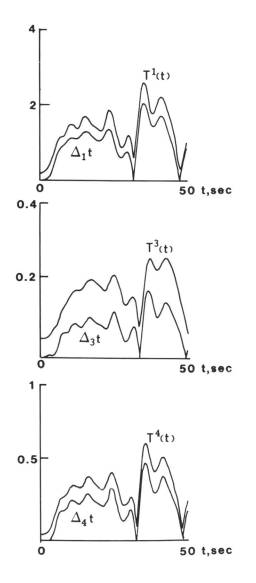

Figure 2-8: $\Delta(t)$ and $T(t)$ responses to random input, no instrument faults, 30 percent parameter uncertainty

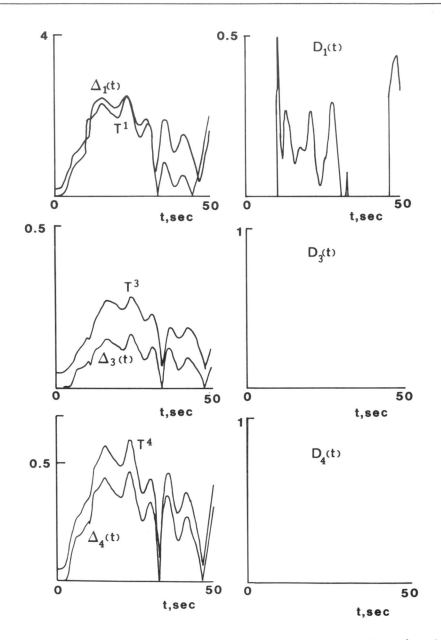

Figure 2-9: $\Delta(t)$, $T(t)$, and $D(t)$ responses to random input, 30 percent parameter uncertainty, small bias fault occurring on sensor 1 at t = 10 sec

2.5 PRACTICAL APPLICATIONS

In the early development work on IFD schemes based on functional redundancy very few, if any, practical applications were made. Simulators were used as the dynamic plant-instrument combination shown in Figure 2-1 and the IFD scheme was realised as a digital computer program, usually without regard to requirements that would have to be met in practice such as real-time computation. Some work has been accomplished using recorded input and output signals from operating plants, with simulated instrument faults added to the recorded output signals to test the IFD system (Clark et al. (1976)). This approach adds a significant practical aspect to development studies, especially in testing the IFD system for its false alarm immunity. Some realism is sacrificed when instrument faults are applied since these faults cannot affect the recorded input or output signals, as a real fault would.

Applications of IFD schemes to real plants operating in real time are difficult because of the robustness questions cited above. For this reason practical application is itself a legitimate area for research, and an important one. Some applications to real systems which are approximately linear with small uncertainties and fairly well known disturbances have been reported (Wünnenberg et. al. (1985), Brockhaus (1986), Stuckenberg (1985), Horak and Allison (1987), De Laat and Merrill (1984)). In these cases remarkably small instrument faults have been detected promptly, even in those cases where precautions against false alarms have been incorporated into the detection logic. More practical applications are being attempted as schemes developed in the laboratory become mature.

Chapter Three

ROBUST FAULT DIAGNOSIS USING UNKNOWN INPUT OBSERVER SCHEMES

Paul M. Frank, Jürgen Wünnenberg

This chapter describes a unified approach to the design of robust observer schemes for sensor, actuator and component fault detection and isolation (FDI) in dynamic systems. It focuses on the problem of residual generation with the goal of providing effective discrimination between different faults in the presence of unknown inputs such as system disturbances, modelling uncertainties, process parameter variations and measurement noise. The approach is based on the theory of unknown input observers providing complete fault decoupling and disturbance invariance under certain conditions independent of the modes of the faults and disturbances. For the case that the underlying conditions are not given we present a method to find an optimal compromise that minimises the ratio of a norm of the sensitivity with respect to unknown inputs to the norm of the sensitivity with respect to faults in linear systems. A brief outline of the extension to nonlinear systems is given.

3.1 INTRODUCTION

The general procedure of fault detection, isolation and accommodation (FDIA) in dynamic systems with the aid of analytical redundancy consists of the following three steps:

(a) *Generation* of so-called *residuals*, i.e. of functions that carry information about the faults.
(b) *Decision* on the occurrence of a fault and *isolation* of the faulty element, i.e. localisation of the fault.
(c) *Accommodation* of the faulty process to normal operation, i.e. correction of the effect of a fault.

This chapter focuses attention on the first step, i.e., the process of residual generation using state estimation techniques with emphasis on the robustness with respect to unknown inputs.

The crucial point in any model-based FDI scheme is the influence of unmodelled disturbances as, for example, parameter uncertainties, changes in the system parameters, and system and measurement noise. All these influences can be summarised as *unknown inputs* acting on the system. The effect of unknown inputs obscures the performance of fault detection and isolation and acts as a source of false alarms. Therefore, in order to minimise the false alarm rate, one should design the observer such that it becomes robust with respect to the unknown inputs. In recent years intensified attempts have been made to achieve better applicability of observer-based FDI schemes by providing increased robustness with respect to unknown inputs; for example, see Frank and Keller (1980), Watanabe and Himmelblau (1982), Belkoura (1983), Chow and Willsky (1984), Weiss et al. (1985), Lou et al.(1986), Frank (1987), Wünnenberg and Frank (1987, 1988), Patton et al. (1987, 1988, see Chapter 4), Clark (see Chapter 2).

The most challenging achievement in the robustness of fault detection and isolation schemes is to reach complete decoupling and hence invariance between different fault effects or between the effects of faults and of unknown inputs independent of the fault modes. There are several different ways to approach this goal:

(a) the detection filter approach (Beard 1971, and Jones 1973)
(b) the parity space approach (Chow and Willsky 1984, Lou et al. 1986)
(c) the eigenstructure assignment approach (White and Speyer 1986, Patton et al. 1987, 1988)
(d) the unknown input observer approach (Watanabe et al. 1982, Wünnenberg and Frank 1987, Viswanadham and Srichander 1987, Fang and Ge 1988).

In this chapter we focus our attention on the *unknown input observer* approach. It will be shown that this approach can be considered a unified one which includes the results obtained by the approaches (a)-(c) to such an end that they can, in part, be interpreted as special cases.

At this point it should be noticed, however, that certain mathematical conditions have to be fulfilled in order to reach the property of *full decoupling*. For the case that these conditions cannot be fulfilled in a given situation we propose an alternative design procedure that achieves at least an *optimal* compromise between sensitivity to faults to be detected and robustness to unknown inputs.

It is emphasised that one of the great advantages of the unknown input observer approach is that the properties of decoupling and invariance are not dependent on the *mode*, i.e., the time evolutions and sizes of the faults and the unknown inputs. This property is logically lost in the alternative (optimal) approach, in which perfect decoupling is, in general, not achievable.

3.2 FAULT DETECTION BASED ON STATE ESTIMATION

3.2.1 Basic principle of observer-based fault detection

The basic principle of fault detection and isolation (FDI) using state estimation is illustrated in Figure 3-1. In general, faults can occur either in the *actuators* or in the *functional units* (components) inside the system or in the *sensors*.

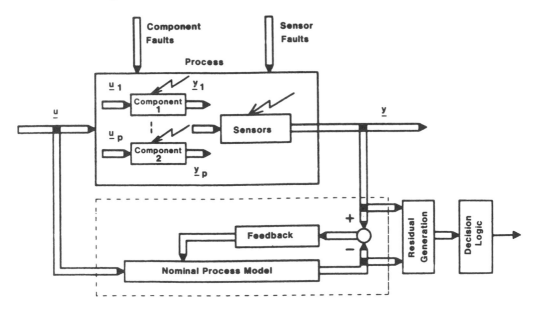

Figure 3-1: Basic principle of FDI using state estimation

It is useful to distinguish between these sources of faults, because, as will be seen later, their isolation requires different observer configurations, i.e., different detection schemes.

It is therefore common to make a distinction between actuator fault detection (AFD), component fault detection (CFD) and instrument fault detection (IFD).

The fault *isolation* task may require different observer configurations depending on the location of the faults to be discovered, however, the basic principle of observer-based *residual generation* is always the same: A set of measured variables, y_i, of the actual system is compared with the corresponding signals of the nominal model, $\hat{y}_i$. The differences $e_i = y_i - \hat{y}_i$ are fed back to improve tracking of the model. This configuration constitutes an observer or a Kalman filter (depending on the way of calculating the feedback gain), where the vector $\underline{e}$ of the e_i represents the observation error or innovation, respectively. For the purpose of fault diagnosis, $\underline{e}$ or a function of $\underline{e}$ can be used as a residual, $\underline{r}$, that carries information about the fault: If no fault occurs, the observer will track the process so that $\underline{r}$ only depends on the unknown inputs, i.e. the unmodelled disturbances such as system and measurement noise, parameter variations or modelling uncertainties. If, however, a fault in the system occurs, the observer models the system with less accuracy and the magnitude of the components of $\underline{r}$ will be increased. Hence, the fault can be detected by checking the increments of $\underline{r}$ caused by the fault (or faults). This is the basic idea of FDI with the aid of state estimation. The state observers or Kalman filters that are used for this task, must properly be designed to provide an optimal discrimination between the effects of faults and unknown inputs. Such observers or Kalman filters that are particularly designed for the purpose of fault detection are called *fault detection* (or *fault diagnosis*) *observers* (FDO).

It is well known from observer theory that for state estimation one can use linear or nonlinear, full or reduced-order state observers (in the deterministic case) or Kalman filters (when noise is considered). Note, that in either case a mathematical model of the process is involved. The standard observer configuration for the case of a full-order observer is shown in Figure 3-2.

One may ask why the feedback in the observer is necessary in the

FDI application and why a simple parallel model is not sufficient. The feedback in the observer is important for the following reasons:

(a) to compensate for the effects of different initial conditions
(b) to stabilise the observer which is particularly important in the case of an unstable system
(c) to provide freedom for the design of the observer, for example, to decouple the desired effects of faults, from the effects of unknown inputs, or the effects of different faults from each other.

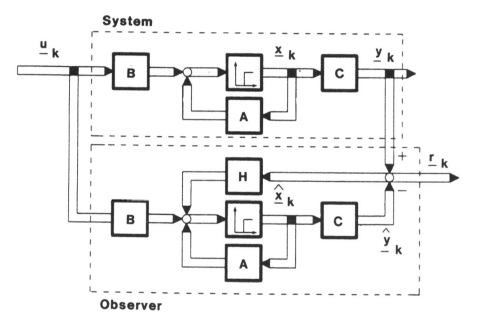

Figure 3-2: Residual generation with full-order observer

One should notice, however, that these benefits have to be paid by a coupling of a fault that appears in a single output to all other outputs. This may complicate the taste of isolation of a fault.

The key problem of the observer-based fault detection and isolation procedure is the generation and evaluation of a *set* of residuals, that permits not only the detection but also a unique distinction between different faults in the face of unknown input. In general, this goal can be achieved by a *bank* of observers or an observer *scheme*, where each observer is made sensitive to a different fault or group of faults whilst being insensitive to unmodelled disturbances, noise, modelling uncertainties and process parameter variations. This last category of system excitation is known collectively as *unknown inputs* (UI).

This chapter concentrates on observer-based fault detection. Starting from a generalising framework, new results for a robust observer-based fault detection are derived. In addition, various tools are given to support the design procedure and to reduce the computational effort arising when the algorithms are implemented for practical application.

The first essential step in the development of an observer-based FDI scheme is a realistic representation of the physical process under consideration, which includes system dynamics, faults and all kinds of possible unknown inputs. The resulting state equations are then transformed into a canonical form, the so-called *Kronecker Canonical Form* (KCF) which specifically serves as a basis for the mathematical derivation of the FDI procedure described here. The design of Unknown Input Observers (UIO) using the KCF is outlined and it is shown, how the UIOs can be used for the robust detection of faults. A major result of the derivation is a necessary and sufficient condition for the existence of an Unknown Input Fault Detection Observer (UIFDO). The design procedure for the filter algorithm for a computerized on-line system supervision is then developed. The advantage of using the KCF approach is that the software for the computer-aided proof of existence as well as design procedure is already available. As a most important result we also consider the fact that this approach leads to a unified FDI theory that contains the results of the other approaches to observer-based fault detection. The cross-connections with these approaches are shown mathematically.

If the mathematical conditions for the existence of a UIFDO cannot be fulfilled, we present a design procedure for a UIFDO that is optimal with respect to a properly defined performance index. In addition to the derivation of the general method we describe the complete procedure of numerical evaluation, so that the design can also be computerised. In the case that the minimum of the *performance index* can become zero the optimal UIFDO is identical with the de-coupled UIFDO.

The advantage of using the KCF lies in its *block-diagonal*

structure. Hence once the basis of the approach is defined, the proofs are easy to carry out and straightforward. We intentionally avoid the use of the geometric approach, which is, as the authors believe, not yet widely enough spread among engineers so as to be readily applied.

3.2.2 System specification

To tackle the problem from the most general point of view we start with a system description that includes all kinds of disturbances that can occur in practise. Among these are the faults to be detected, and the unknown inputs with respect to which the observers should be invariant.

In order to obtain an algorithm which is best suited for a computer implementation, the system under consideration is described by the following discrete-time state equations:

$$\underline{x}_{k+1} = \underline{A}\,\underline{x}_k + \underline{B}\,\underline{u}_k + \underline{E}\,\underline{d}_k + \underline{K}\,\underline{f}_k \quad, \quad \underline{x}_{k=0} = \underline{x}_0 \qquad (3\text{-}2.1)$$

$$\underline{y}_k = \underline{C}\,\underline{x}_k \qquad (3\text{-}2.2)$$

$\underline{x}_k$ represents the state vector at the k-th sampling instant, $\underline{A}$ the nxn dimensional system matrix, $\underline{u}_k$ the known input signal of the system and $\underline{B}$ the nxr dimensional known input distribution matrix. $\underline{E}$ represents the nxs dimensional distribution matrix of the unknown input signal, $\underline{d}_k$. $\underline{K}$ denotes the nxl-dimensional distribution matrix of the so-called event vector, $\underline{f}_k$. The only information commonly available about the faults is the location of their possible appearance, whilst no assumptions can be made about their mode, i.e., their time evolution and size. $\underline{C}$ is the mxn dimensional measurement matrix of the system and $\underline{y}_k$ the measurement vector. It is assumed that the matrices $\underline{E}$ and $\underline{K}$ are perfectly known. However, no such assumption is made for the modes of the time evolutions of $\underline{d}_k$ and $\underline{f}_k$.

Note that the system equations (3-2.1) and (3-2.2) also include a unique representation of all possible parameter variations and faults appearing in the system. To show this the matrices are partitioned as follows:

$$\underline{A} = \begin{bmatrix} \underline{A}_O & \underline{0} & \underline{0} \\ \underline{0} & \underline{A}_E & \underline{0} \\ \underline{0} & \underline{0} & \underline{A}_E^* \end{bmatrix} \qquad (3\text{-}2.3)$$

$$\underline{E} = \begin{bmatrix} \Delta\underline{A}, & \Delta\underline{B}, & \overline{\underline{E}}, & \overline{\underline{K}}, & \underline{K}_E \end{bmatrix} \qquad (3\text{-}2.4)$$

$$\underline{d}_k = \begin{bmatrix} \underline{x}_k^T, & \underline{u}_k^T, & \overline{\underline{d}}_k^T, & \overline{\underline{f}}_k^T, & \underline{f}_{Ek}^T \end{bmatrix}^T \qquad (3\text{-}2.5)$$

$$\underline{K} = \begin{bmatrix} \underline{K}^*, & \underline{K}_E^* \end{bmatrix} \qquad (3\text{-}2.6)$$

$$\underline{f}_K = \begin{bmatrix} \underline{f}_k^{*T}, & \underline{f}_{Ek}^{*T} \end{bmatrix}^T \qquad (3\text{-}2.7)$$

$$\underline{C} = \begin{bmatrix} \underline{C}_O, & \underline{C}_E, & \underline{C}_E^* \end{bmatrix} \qquad (3\text{-}2.8)$$

In the input distribution matrix $\underline{E}$ the submatrix $\Delta\underline{A}$ represents parameter variations in the dynamics of the system, $\Delta\underline{B}$ parameter variations in the actuators, and $\overline{\underline{E}}$ represents the distribution of disturbances, $\overline{\underline{d}}_k$, in the system. $\overline{\underline{K}}$ is the distribution matrix of all faults in the system that need not be detected and should therefore not affect the residual error. In addition to this, there are uncritical *instrument* deviations that we also do not intend to detect because they cause merely an uncertainty in the system measurements; they affect only the output equation and have no influence on the system dynamics other than through an external feedback. Because no knowledge can be assumed about the time histories of these signals, it is reasonable to model them by a dynamic system, represented by the input distribution matrix $\underline{K}_E$, the dynamics matrix $\underline{A}_E$ and the measurement matrix $\underline{C}_E$, which is driven by a disturbance signal $\underline{f}_{Ek}$.

The faults that we want to detect are denoted by the event vector $\underline{f}_K$ which is distributed in the system via the distribution matrix $\underline{K}$. The submatrix $\underline{K}^*$ together with the event vector $\underline{f}_k^*$ characterise the so-called *component faults* as well as the *actuator faults*, i.e., faults acting directly onto the system dynamics. On the other hand, *instrument faults* (sensor faults) are characterised by the extended system matrix $\underline{A}_E^*$, the distribution matrix $\underline{K}_E^*$, the event vector $\underline{f}_{Ek}^*$ and the measurement matrix $\underline{C}_E^*$.

It can easily be seen that instrument faults can as well as actuator and component faults be represented as inputs to the dynamic system. The instrument faults are modelled as outputs of an augmented state-space description such that they finally act on the measurement equation. By this procedure the system description is simplified and this, in turn, simplifies to proofs necessary to show the existence of the UIFDOs.

3.2.3 Residual generation via state estimation

The overall procedure of fault detection by state estimation consists of the following two steps:

(a) Generation of residuals by state estimation
(b) Evaluation of the residuals by a decision logic.

As mentioned above, the idea of residual generation via state estimation is to reconstruct the outputs of the process with the aid of observers or Kalman filters and to use the *estimation error* or *innovation*, respectively, or some function of them as a *residual*. The general relationship between the residual and the faults $\underline{f}_k$ as well as the unknown inputs $\underline{d}_k$ can be determined as follows.

In the case of a system with the state equations (3-2.1) and (3-2.2), the state $\hat{\underline{x}}_k$ and output $\hat{\underline{y}}_k$ of a full-order observer obey the equations:

$$\hat{\underline{x}}_{k+1} = (\underline{A} - \underline{H}\,\underline{C})\,\hat{\underline{x}}_k + \underline{B}\,\underline{u}_k + \underline{H}\,\underline{y}_k \quad , \quad \hat{\underline{x}}_{k=0} = \hat{\underline{x}}_0 \qquad (3\text{-}2.9)$$

$$\hat{\underline{y}}_k = \underline{C}\,\hat{\underline{x}}_k \qquad (3\text{-}2.10)$$

where $\underline{H}$ denotes the observer feedback gain matrix. With equations (3-2.1), (3-2.2), (3-2.9) and (3-2.10) the relations for the state estimation error, $\underline{\varepsilon}_k = \underline{x}_k - \hat{\underline{x}}_k$, and the output estimation error, $\underline{e}_k = \underline{y}_k - \hat{\underline{y}}_k$, become:

$$\underline{\varepsilon}_{k+1} = (\underline{A} - \underline{H}\,\underline{C})\,\underline{\varepsilon}_k + \underline{E}\,\underline{d}_k + \underline{K}\,\underline{f}_k \qquad (3\text{-}2.11)$$

$$\underline{e}_k = \underline{C}\,\underline{\varepsilon}_k \qquad (3\text{-}2.12)$$

When $\underline{e}$ is taken as the residual $\underline{r}$, one can see from equations (3-2.11) and (3-2.12) that $\underline{r}$ is a function of both $\underline{f}$ and $\underline{d}$. In a similar way one can find the residuals for reduced-order observers, nonlinear observers or Kalman filters due to the well established state estimation theory (Frank 1987).

3.2.4 Problem formulation

In this chapter we focus attention on the use of unknown input observers for the generation of the residual vector $\underline{r}$. The objective of the unknown input observer design is to determine the feedback gain matrix, $\underline{H}$, such that, for the method described, the effects of the individual faults, f_{ik}, on the residual are decoupled from each other and from the effects of the unknown inputs, d_{ik}. For the solution of this problem a repeated design procedure is derived consisting of the following steps.

The *first* step provides the design of a single observer, that generates a residual which is insensitive to the first possible fault f_{1k} only. In the *second* step the same design procedure is repeated for the second fault f_{2k}. This procedure is continued until one reaches an observer scheme that allows the detection and isolation of all possible faults. The fundamental theory of the robust residual generation with the aid of unknown input observers is derived in sections 3.3 to 3.6. The resulting structures of observer schemes consisting of banks of observers with the desired properties are presented and discussed in Section 3.7.

For the case that the conditions of the above design are not given, an alternative method is derived that provides an optimal compromise between the sensitivity of the residual with respect to the faults on one hand, and to the unknown inputs on the other hand.

3.3 ROBUST STATE OBSERVATION IN THE FACE OF UNKNOWN INPUTS

Because of the importance of state observers for the supervision of dynamic systems it appears reasonable to give a brief introduction to the observability and the design of observers for systems operating under unknown inputs. In order to find the solution to the design problem of observers that provide complete decoupling between the effects of faults, $\underline{f}$, and unknown inputs, $\underline{d}$, we make use of a transformation of the state equations (3-2.1) and (3-2.2) of the system into block-diagonal Kronecker Canonical Form (KCF). A side effect of this transformation is that it gives generally a deep insight and understanding of the behaviour of linear systems operating under unknown inputs.

In addition, software is available for the transformation of a state-space description to a triangular form that is similar to the KCF and contains all information necessary for the design (van Dooren, Dewilde (1983), van Dooren (1979)). It is therefore possible to computerise the design procedure.

3.3.1 Transformation of the state equations to Kronecker Canonical Form (KCF)

To obtain the KCF, we define the matrices:

$$T_1 = \left[\ M, \ C^R \ \right] \qquad\qquad (3\text{-}3.1)$$

$$T_2 = \begin{bmatrix} N \\ E^L \end{bmatrix} \qquad\qquad (3\text{-}3.2)$$

where M represents a basis for the kernel of C, C^R a right inverse of C, N a left annihilator of E, and E^L a left inverse of E. Expressed in equations, this means:

$$C \ T_1 = [\ 0, \ I_m \ ] \qquad\qquad (3\text{-}3.3)$$

$$T_2 \ E = \begin{bmatrix} 0 \\ I_s \end{bmatrix} \qquad\qquad (3\text{-}3.4)$$

where I_m is a m-dimensional and I_s a s-dimensional unity matrix. Multiplying equation (3-2.1) throughout by the matrix T_2 from the left and partitioning the state vector accordingly into:

$$x_k = \left[\ M \quad C^R \ \right] \begin{bmatrix} x_k^* \\ y_k \end{bmatrix} \qquad\qquad (3\text{-}3.5)$$

we then obtain a partitioning between the measurable part, y_k, and the nonmeasurable part $\underline{x}_k^*$. It then follows that:

$$\underline{NM}\ \underline{x}_{k+1}^* - \underline{NAM}\ \underline{x}_k^* = \underline{NB}\ \underline{u}_k - \underline{NC}^R\ \underline{y}_{k+1} + \underline{NAC}^R\ \underline{y}_k + \underline{NK}\ \underline{f}_k \qquad (3\text{-}3.6)$$

and:

$$\underline{d}_k = -\underline{E}^L[\underline{AM}\ \underline{x}_k^* - \underline{M}\ \underline{x}_{k+1}^* + \underline{B}\ \underline{u}_k + \underline{K}\underline{f}_k - \underline{C}^R\underline{y}_{k+1} + \underline{AC}^R\ \underline{y}_k] \qquad (3\text{-}3.7)$$

Note that equation (3-3.6) no longer contains the unknown input, $\underline{d}_k$. The left hand side contains the *unmeasured* and therefore unknown part of the state vector $\underline{x}_k$. The right hand side of the equation contains known matrices and measurable variables as well as the event vector $\underline{f}_k$.

The equation (3-3.6) can be written in operator form (with z the shifting operator) yielding:

$$\underline{S}_{(z)}\ \underline{x}_k^* = \underline{u}_k^* \qquad (3\text{-}3.8)$$

where:

and:

$$\underline{S}_{(z)} = z\ \underline{N}\ \underline{M} - \underline{N}\ \underline{A}\ \underline{M} \qquad (3\text{-}3.9)$$

$$\underline{u}_k^* = \underline{N}\ \underline{B}\ \underline{u}_k - \underline{N}\ \underline{C}^R\ \underline{y}_{k+1} + \underline{N}\ \underline{A}\ \underline{C}^R\ \underline{y}_k + \underline{N}\ \underline{K}\ \underline{f}_k \qquad (3\text{-}3.10)$$

It is well known (e.g. Gantmacher (1959)) that the so-called *matrix pencil* $\underline{S}_{(z)}$ can be transformed to Kronecker canonical form by a suitable choice of the matrices $\underline{N}$ and $\underline{M}$.

$\underline{S}_{(z)}$ is then of KCF, namely:

$$\underline{S}_{(z)} = \text{block-diag}\ [\underline{S}_{\mu(z)}, \underline{S}_{\infty(z)}, \underline{S}_{\Lambda(z)}, \underline{S}_{\Lambda(z)}^{\perp}, \underline{S}_{\epsilon(z)}] \qquad (3\text{-}3.11)$$

where the blocks are block-diagonal by themselves, namely

$$\underline{S}_{\mu(z)} = \text{block diag}\ [\underline{S}_{\mu_i(z)}] \qquad (3\text{-}3.12)$$

$$\underline{S}_{\epsilon(z)} = \text{block diag}\ [\underline{S}_{\epsilon_i(z)}] \qquad (3\text{-}3.13)$$

with the kx(k+1)-dimensional matrix (k = ϵ_i, μ_i)

$$
\underline{S}_k = \begin{bmatrix}
z & -1 & & & & \\
 & z & -1 & & \Large 0 & \\
 & & z & -1 & & \\
 & & & \ddots & \ddots & \\
 & \Large 0 & & & \ddots & \ddots \\
 & & & & & z & -1
\end{bmatrix}
\tag{3-3.14}
$$

The integers μ_i, i = 1,...,p, are called Kronecker row indices and the integers ϵ_i, i = 1, ...,q, Kronecker column indices and are uniquely determined by the transformation to KCF.

The matrix $\underline{S}_{\infty(z)}$ is given by:

$$
\underline{S}_{\infty(z)} = \text{block diag } [\underline{I} - z \, \underline{J}_{i(z)}^{\infty}]
\tag{3-3.15}
$$

with $\underline{J}_{i(z)}^{\infty}$ (i = 1,...,ξ) a $n_i^{\infty} \times n_i^{\infty}$ dimensional Jordan matrix with all eigenvalues equal to zero. The n_i^{∞} are again determined by the transformation to KCF.

The matrices $\underline{S}_{\Lambda(z)}$ and $\underline{S}_{\Lambda(z)}^{\perp}$ are defined as:

$$
\underline{S}_{\Lambda(z)} = \text{block diag } [\underline{I} \, z - \underline{J}_{\Lambda i(z)}]
\tag{3-3.16}
$$

$$
\underline{S}_{\Lambda(z)}^{\perp} = \text{block diag } [\underline{I} \, z - \underline{J}_{\Lambda i(z)}^{\perp}]
\tag{3-3.17}
$$

with $J_{\Lambda i(z)}$ (i = 1,...,ζ) $n_i^{\Lambda} \times n_i^{\Lambda}$ dimensional Jordan matrices with stable eigenvalues (all eigenvalues inside the unit circle). $J_{\Lambda i(z)}^{\perp}$ (i = 1,...,η) represent $\bar{n}_i^{\Lambda} \times \bar{n}_i^{\Lambda}$ dimensional Jordan matrices with unstable eigenvalues. The eigenvalues of these matrices are called *finite elementary divisors* and the inverse of the eigenvalues of $\underline{J}_i^{\infty}$ are the infinite elementary divisors of the matrix pencil $\underline{S}_{(z)}$.

For further developments it is also useful to partition $\underline{M}$ and $\underline{N}$ according the block-diagonal form of the matrix pencil:

$$
\underline{M} = [\, \underline{M}_{\mu}, \, \underline{M}_{\infty}, \, \underline{M}_{\Lambda}, \, \underline{M}_{\Lambda}^{\perp}, \, \underline{M}_{\epsilon} \,]
\tag{3-3.18}
$$

$$
\underline{N} = [\, \underline{N}_{\mu}^{T}, \, \underline{N}_{\infty}^{T}, \, \underline{N}_{\Lambda}^{T}, \, \underline{N}_{\Lambda}^{\perp T}, \, \underline{N}_{\epsilon}^{T} \,]^{T}
\tag{3-3.19}
$$

The submatrices are further partitioned into:

$$
\underline{M}_{\mu} = [\, \underline{M}_{\mu 1}, \, \dots\dots\dots, \, \underline{M}_{\mu p} \,]
\tag{3-3.20}
$$

and:

$$\underline{N}_\mu = [\ \underline{N}_{\mu 1}^T, \ \dots\dots\dots, \ \underline{N}_{\mu p}^T \ ]^T \tag{3-3.21}$$

Equation (3-3.6) can be seen as a system of difference equations that has to be solved for the $\underline{x}_k^*$. Because of the KCF, one can determine the solutions almost directly from the equation.

The properties important for this approach are stated with the help of Lemma 1 (below). The unknown part of the state vector is therefore partitioned according to the block structure of the pencil and denoted with corresponding indices yielding:

$$\underline{x}_k^{*T} = \left[\underline{x}_{\mu k}^T, \ \underline{x}_{\infty k}^T, \ \underline{x}_{\Lambda k}^T, \ \underline{x}_{\Lambda k}^{\perp T}, \ \underline{x}_{\epsilon k}^T \right] \tag{3-3.22}$$

Lemma 1 (Wilkinson (1978))

The elements of $\underline{x}_{\mu k}$ of the state vector $\underline{x}_k$ are completely determined by the input signal $\underline{u}_k^*$. This holds for each point on the time axis and for all initial conditions $\underline{x}_0$.

The proof of Lemma 1 is straight forward, if one considers equation 3-3.6 as a system of difference equations. The equations corresponding to the i-th Kronecker row index can be written in the form:

$$x_{\mu i,1,k+1} = u_{\mu i,1,k}^*$$

$$x_{\mu i,2,k+1} - x_{\mu i,1,k} = u_{\mu i,2,k}^*$$

$$x_{\mu i,3,k+1} - x_{\mu i,2,k} = u_{\mu i,3,k}^*$$

$$\vdots$$

$$x_{\mu i,\mu i,k+1} - x_{\mu i,\mu i-1,k} = u_{\mu i,\mu i,k}^*$$

$$- x_{\mu i,\mu i,k} = u_{\mu i,\mu i+1,k}^* \tag{3-3.23}$$

$x_{\mu i,j,k}$ denotes the j-th element ($j = 1, \dots, \mu_i$) of the part $\underline{x}_{\mu i,k}$ corresponding to the μ_i row index of the state vector $\underline{x}_k^*$. $u_{\mu i,j,k}^*$ denotes the j-th element $j = (1, \dots, \mu_i+j)$ of the vector $\underline{u}_k^*$, corresponding to the μ_i row indices. From equation (3-3.23) it becomes clear that all $x_{\mu i,j,k}$ are completely determined by the

$u^*_{\mu i,j,k}$. In addition, the $u^*_{\mu i,j,k}$ must fulfil the compatibility condition:

$$u^*_{\mu i,1,k} + u^*_{\mu i,2,k+1} +\ldots+ u^*_{\mu i,\mu i,k+\mu i-1} + u^*_{\mu i,\mu i+1,k+\mu i} = 0 \qquad (3\text{-}3.24)$$

Equations (3-2.23) and (3-2.24) are independent from the initial condition in $\underline{x}_k$, which completes the proof.

For the design of a state estimator for FDI a second lemma is needed. This is stated here for a better understanding of the following results.

Lemma 2 (Wilkinson (1978), van Dooren, Dewilde (1983))

The solution $\underline{h}_i(z)$ of *minimal order* to the problem:

$$\underline{h}_i(z)\ \underline{S}(z) = \underline{0} \qquad (3\text{-}3.25)$$

is given by:

$$\underline{h}_i(z) = (\ \bar{\underline{0}}_{\mu i},\ 1,\ z,\ z^2,\ \ldots\ldots,\ z^{\mu i},\ \underline{0}_{\mu i}\ ) \qquad (3\text{-}3.26)$$

with $\bar{\underline{0}}_{\mu i}$ a zero row vector of dimension $\sum\limits_{j=1}^{i-1} \mu_j+1$, and $\underline{0}_{\mu i}$ a zero row vector of dimension $n - s- \sum\limits_{j=1}^{i} \mu_j+1$ with $i = 1,\ldots,p$. This shows that p linearly dependent solutions exist to equation (3-3.25). In addition, one can find a solution to equation (3-3.25) for fixed values of z where z is chosen as a finite elementary divisor of the matrix pencil $\underline{S}(z)$.

The reader may note, that equation (3-3.25) may be re-written as:

$$\underline{h}_i(z)\ \underline{S}(z) = \underline{h}_i(z)\underline{u}^*(z) = 0 \qquad (3\text{-}3.27)$$

which is equivalent to the compatibility condition mentioned in the proof of Lemma 1 and expressed by equation (3-3.24). This fact will be used in Section 3.4.4.

It would be beyond the scope of this presentation to give all the illustrative proofs. We therefore restrict the scope here in order to provide the reader with a simple idea of the design procedure. For more details we refer to the literature on this subject (see, e.g. Engell, Konik (1986)).

3.3.2 Observability of systems under unknown inputs

The investigation of the *observability* of systems under unknown inputs is best carried out by using the KCF representation of the system. The term *observability under unknown inputs* means observability in the presence of an arbitrary disturbance, $\underline{d}_k$. For the sake of simplicity we assume that the system is free of faults, i.e. $\underline{f}_k = \underline{0}$ throughout this section. Equation (3-3.6) represents an implicit system of difference equations the solutions of which reveal the observability of the system. The results can be summarised by the following theorem:

Theorem 1: (Konik (1986), Wünnenberg and Frank (1987))

The rows $\underline{M}_\mu$ of the matrix $\underline{M}$ (equation 3-3.18) form a basis for the *reconstructible* part of the unknown states $\underline{x}_k^*$.

The rows $\underline{M}_\Lambda$ form a base for the *detectable* part of the unknown states $\underline{x}_k^*$.

The rows $\underline{M}_\infty$ form a basis for the part of the unknown state vector $\underline{x}_k^*$ that is reconstructible with anticipation.

Reconstructability and detectability are defined e.g. in Kwakernaak, Sivan (1972).

Anticipation in this context implies, that the state is reconstructible with a delay of a finite number of time steps.

3.3.3 Unknown input observer (UIO)

In this section we discuss the state observers for systems with unknown inputs. This is done with the aid of the transformation to KCF. The consideration is limited to the case that the state of the system is either reconstructed or detected. *Reconstruction* of the state vector implies, that the eigenvalues of the corresponding observer are freely assignable. *Detection* of the states implies, that the eigenvalues of the resulting observer are not freely assignable. If unknown inputs are acting on the system, the state vector can, under certain conditions, only be observed with an anticipation. This problem will not be treated further in this chapter, because the anticipation is not critical in an open-loop application of the observer as is the case for the purpose of fault detection. Note however, that this is not possible in the case of the application to a closed-loop feedback control. Even though the anticipation allows a state reconstruction it does not contribute to the residual generation directly and therefore to the fault detection problem, as defined in Section 3.4. If anticipated state reconstruction is admitted, the corresponding structure of the observer can readily be derived from equation (3-3.6). The solution to this problem is given e.g. in Wünnenberg, Frank (1987).

The *unknown input observer* is defined as follows.

Definition 1 (Engell, Konik (1986))

A dynamic system:

$$\underline{z}_{k+1} = \underline{F}\,\underline{z}_k + \underline{J}\,\underline{u}_k + \underline{G}\,\underline{y}_k \ , \quad \underline{z}_{k=0} = \underline{z}_o \qquad (3\text{-}3.28)$$

$$\underline{r}_k = \underline{L}_1\,\underline{z}_k + \underline{L}_2\,\underline{y}_k \qquad (3\text{-}3.29)$$

with $\underline{z}_k$ being the t-dimensional state vector of the system and $\underline{r}_k$ being the v-dimensional output signal, is called an *unknown input observer* (UIO) of the system given by equations (3-2.1) and (3-2.2) if it estimates a linear combination of the state vector according to $\underline{r}_k = \underline{R}\,\underline{x}_k$. This is only possible, if the following conditions are satisfied for all $\underline{u}_k$ and $\underline{d}_k$.

(a) for all $\underline{x}_o$, $\underline{z}_o$

$$\lim_{k\to\infty}\ (\underline{r}_k - \underline{R}\,\underline{x}_k) = 0 \qquad (3\text{-}3.30)$$

(b) for a (t x n)-dimensional matrix $\underline{T}$

$$\underline{z}_o = \underline{T}\,\underline{x}_o \Rightarrow \underline{z}_k = \underline{T}\,\underline{x}_k \quad \text{for all k} \qquad (3\text{-}3.31)$$

The above conditions ensure that the observer estimates the system state exactly even in the presence of unknown inputs. This holds for any instant of time if the initial conditions of the system are known. If the initial conditions are unknown the estimate *converges* to the true state. It should be kept in mind that this result was derived under the assumption that no fault occurs.

The observer description according to equations (3-3.28) and (3-3.29) is the most general formulation to the observation problem. If $\underline{T}$ is chosen to be the n-dimensional identity matrix, the full order Luenberger observer is derived. It also allows the design of reduced order observers, as well as the design of observers for augmented system states, the so-called disturbance observers.

The question now arises: Under what conditions does such an observer exist? The answer is given in Theorem 2:

Theorem 2 (Engell, Konik (1986))

An unknown input observer of the form described by the equations (3-3.28), (3-3.29) for the system (3-2.1), (3-2.2) in the presence of unknown inputs $\underline{d}_k$ and the absence of faults ($\underline{f}_k = \underline{0}$) exists *iff*:

$$(\text{Ker } \underline{T}) \cap \text{Ker } \underline{C} \subset \text{Ker } \underline{R} \qquad (3\text{-}3.32)$$

where the matrix $\underline{T}$ is given by:

$$\underline{T} = \underline{T}^*\underline{N} \qquad (3\text{-}3.33)$$

and the row vectors $\overline{\underline{t}_j}$ of $\underline{T}^*$ satisfy the equations:

$$\overline{\underline{t}_j} (\lambda_{fi} \underline{N}\,\underline{M} - \underline{N}\,\underline{A}\,\underline{M}) = \underline{0}^T \qquad \text{for} \qquad j = \sum_{k=1}^{i} n_{fi} \qquad (3\text{-}3.34)$$

$$\overline{\underline{t}_{j-1}} (\lambda_{fi} \underline{N}\,\underline{M} - \underline{N}\,\underline{A}\,\underline{M}) = -\overline{\underline{t}_j}\,\underline{N}\,\underline{M} \quad \text{for} \quad j = \sum_{k=1}^{i} n_{fi} \cdots \sum_{k=1}^{i-1} n_{f(i+1)},$$

$$i = 1,\ldots,s_f \qquad (3\text{-}3.35)$$

with $\sum_{i=1}^{s_f} n_{fi} = q$ for s_f complex numbers λ_{fi} with an absolute value smaller than one. The numbers λ_{fi} are the eigenvalues of the matrix $\underline{F}$ given in Jordan canonical form with Jordan blocks of dimension n_{fi}.

The resulting estimation error of the unknown input observer is governed by the difference equation:

$$\underline{e}_{k+1} \overset{\Delta}{=} \underline{z}_{k+1} - \underline{T}\,\underline{x}_{k+1}$$

$$= \underline{F}\,\underline{z}_k + \underline{J}\,\underline{u}_k + \underline{G}\,\underline{y}_k - \underline{T}\underline{A}\,\underline{x}_k - \underline{T}\underline{B}\,\underline{u}_k - \underline{T}\underline{E}\,\underline{d}_k - \underline{T}\underline{K}\,\underline{f}_k \qquad (3\text{-}3.36)$$

From this equation we can draw the following conclusion: If the relations:

$$\underline{T} \, \underline{A} - \underline{F} \, \underline{T} = \underline{G} \, \underline{C} \qquad\qquad (3\text{-}3.37)$$

$$\underline{J} = \underline{T} \, \underline{B} \qquad\qquad (3\text{-}3.38)$$

$$\underline{T} \, \underline{E} = \underline{0} \qquad\qquad (3\text{-}3.39)$$

$$\underline{L}_1 \underline{T} + \underline{L}_2 \underline{C} = \underline{R} \qquad\qquad (3\text{-}3.40)$$

hold, then $\underline{e}_k$ depends only on $\underline{f}_k$. Note that equation (3-3.33) is obtained directly from equation (3-3.39). Assuming $\underline{T}$ as to be given by equation (3-3.33) and as $\underline{F}$ is in Jordan canonical form we arrive directly at equations (3-3.34) and (3-3.35). If equation (3-3.37) is multiplied from the right with the regular matrix $\underline{T}_1$, as defined in equation (3-3.1), $\underline{G}$ is given by:

$$\underline{G} = \underline{T} \, \underline{A} \, \underline{C}^R - \underline{F} \, \underline{T} \, \underline{C}^R \qquad\qquad (3\text{-}3.41)$$

From Theorem 2 and the Lemma 2 it becomes apparent that the matrix $[\underline{N}_\mu{}^T, \ \underline{N}_\Lambda{}^T]^T$ forms a basis for all possible rows of $\underline{T}$. For the special case of nonrepeated eigenvalues in $\underline{F}$ and $\underline{N}_\Lambda = \underline{0}$, $\underline{T}^*$ becomes a block-diagonal matrix, where each block consists of van der Monde matrices. It is important to note that $\underline{T}^*$ is of full rank.

This concludes the design of an UIO, where the KCF serves as a tool for the derivation of the complete solution to the problem of state observation for systems, acting under unknown inputs. The resulting observer is in Jordan canonical form and all design freedom is used. It therefore represents the most general solution possible.

3.4 RESIDUAL GENERATION IN THE FACE OF UNKNOWN INPUTS

The general theory of unknown input observers is described in section 3.3. This section shows how these observers can be applied to generate residuals, that are equal to zero (or in some sense small) as long as no fault in the system appears, but are different from zero as soon as a fault appears. The resulting observers are then termed *unknown input fault detection observers* (UIFDO).

In the first part of this section we describe the observer-based approach. In the second part it is shown that limiting the observer poles to equal zero, which has no influence on the information contained in the residual, leads to the so-called *dead beat observer* which is mathematically easy to handle and provides solutions to the design of optimal fault detection filters, even if the conditions for an exact decoupling as described in the first part are not given.

3.4.1 Unknown input fault detection observers (UIFDO)

By analogy with *definition 1* one can define the structure and the corresponding residual of an unknown input observer for fault detection as follows:

Definition 2
Consider the following discrete dynamic system:

$$z_{k+1} = F\, z_k + J\, u_k + G\, y_k\ ,\qquad z_{k=0} = z_0 \tag{3-4.1}$$

$$r_k\ = L_1\, z_k + L_2\, y_k \tag{3-4.2}$$

with z_k being the t-dimensional state of the system and r_k being the v-dimensional output signal called the residual. This system is called an unknown input fault detection observer (UIFDO) for the system (3-2.1) and (3-2.2) *iff*:

(a) for all x_0, z_0

$$\lim_{k\to\infty} r_k = 0 \qquad \text{if } f_k = 0 \tag{3-4.3}$$

(b) for $z_0 = T\, x_0$ and any arbitrary $f_k \neq 0$

$$r_k \neq 0 \tag{3-4.4}$$

(c) for a (t x n)-dimensional matrix T

$$z_0 = T\, x_0 \Longrightarrow z_k = T\, x_k \quad \text{for all k and } f_k = 0 \tag{3-4.5}$$

The conditions must hold for any signal u_k and d_k.

For a full understanding of the relevance of the above *definition 2*, one should notice the following. Condition (a) ensures that the residual of the filter is equal to $\underline{0}$, if no fault is present in the system, even if the input signal of the system is not zero and if an unknown input of an arbitrary mode - i.e., arbitrary size and time function - is present. Condition (b) guarantees that the residual is controllable by the event vector $\underline{f}_k$. It also shows that there might be a fault that cannot be detected if the time history of $\underline{f}_k$ compensates the effect of the unmatched initial conditions. Condition (c) states that the filter is a model of the system in the presence of disturbances $\underline{d}_k$ and known inputs $\underline{u}_k$ as long as no fault is present. However, if the fault appears this acts like a disturbance acting on this model.

From this definition of the UIFDO it becomes clear that the UIFDO can be interpreted as a *Luenberger observer* for a system acting under unknown inputs, $\underline{d}_k$, where the residual is the measurable part of the estimation error affected by the fault. The mathematical formulation of the problem is again most general and offers various degrees of freedom for the UIFDO design.

3.4.2 Conditions for the existence of an UIFDO

A neccessary and sufficient condition for the existence of the UIFDO is given by the following theorem:

Theorem 3
A UIFDO of the form described by the equations (3-4.1) exists, *iff*:

$$\text{Rank } \underline{N}_\mu \underline{K} = \text{Rank } \underline{K} \qquad (3\text{-}4.6)$$

with $\underline{N}_\mu$ as defined by equation (3-3.19) and $\underline{K}$ as defined by equation (3-2.1).

Proof of Theorem 3
From equation (3-3.36) it follows that a necessary condition for the fact that the estimation error is affected by the fault, is:

$$\text{Rank } \underline{T} \underline{K} = \text{Rank } \underline{K} \qquad (3\text{-}4.7)$$

The requirement that no linear combination of the state vector is estimated by the UIFDO can be expressed in terms of the UIFDO output equation (3-4.2) in conjunction with equation (3-4.5) yielding:

$$L_1 \, T + L_2 \, C = 0 \tag{3-4.8}$$

or:

$$\begin{bmatrix} L_1 & L_2 \end{bmatrix} \begin{bmatrix} T \\ C \end{bmatrix} = 0 \tag{3-4.9}$$

Now it has to be shown under what conditions equation (3-4.9) can be solved in addition to the equations (3-3.37) to (3-3.39) and what rows of the matrix T contribute to a non trivial solution. For this purpose we distinguish between two cases:

(a) T is of full row rank.

Without loss of generality it can be assumed that only one block of Kronecker row indices exists. Furthermore, it is assumed that the eigenvalues of the matrix F as well as of the finite elementary divisors of the matrix pencil (3-3.6) are all distinct. Then the matrix T (equation (3-4.5)) can be written in accordance with equations (3-3.34) and (3-3.33) as:

$$T = T^* \begin{bmatrix} N_\mu^T, & N_\Lambda^T \end{bmatrix}^T \tag{3-4.10}$$

with:

$$T^* = \begin{bmatrix} T^{**}, & I_\Lambda \end{bmatrix} \tag{3-4.11}$$

and:

$$T^{**} = \begin{bmatrix} 1 & \lambda_{f1} & \cdots & \lambda_{f1}^{\mu i} \\ \vdots & & & \\ 1 & \lambda_{f\mu+1} & \cdots & \lambda_{f\mu+1}^{\mu i} \end{bmatrix} \tag{3-4.12}$$

and I_Λ an identity matrix of the same dimension n^Λ as the matrix $S_\Lambda(z)$ given in equation (3-3.16). Hence one has only to show, what rows of the matrix $[N_\mu^T, N_\Lambda^T]^T$ enter into the solution of equation (3-4.9). Therefore, the Matrix $[N_\mu^T, N_\Lambda^T]^T$ is multiplied from the right with matrix M, which is a basis for the Kernel of C. Because of the canonical form of the pencil, this product can be re-written as:

$$\begin{bmatrix} \underline{N}_\mu \\ \underline{N}_\Lambda \end{bmatrix} \underline{M} = \begin{bmatrix} \begin{bmatrix} \underline{I}_{\mu i} \\ \hline \underline{0}_{\mu i} \end{bmatrix} & \underline{0} & \underline{0} \\ \hline \underline{0} & \underline{I}_{n\Lambda} & \underline{0} \end{bmatrix} \qquad (3\text{-}4.13)$$

according to equations (3-3.11), (3-3.14), (3-3.15) and (3-3.16). $\underline{I}_{n\Lambda}$ and $\underline{I}_{\mu i}$ are identity matrices of dimension n^Λ and μ_i and $\underline{0}_{\mu i}$ a zero row vector of dimension μ_i.

The vector $\underline{0}_{\mu i}$ shows that only the last row of the matrix $\underline{N}_\mu$ depends linearily on the rows of $\underline{C}$, and that the rows of $\underline{N}_\Lambda$ do not contribute to the solution of equation (3-4.9).

(b) $\underline{T}$ is not of full row rank.

$\underline{T}$ can only be of less than full row rank if the order of the observer is larger than $\mu_{i+1} + \beta_\Lambda$. Obviously, this can always be achieved by adding extra rows to the matrix $\underline{T}^{**}$.

Concerning the existence of the solution of equation (3-4.9) it can be seen that for a regular $\underline{T}^{**}$, equation (3-4.9) can be solved in a straight-forward manner through the following choice:

$$\underline{L}_2 \underline{C} = \underbrace{\begin{bmatrix} 0 & \dots & 0 & 1 \end{bmatrix}}_{\mu_i + 1} \underline{N}_{\mu i} \qquad (3\text{-}4.14)$$

Using $\underline{C}^R$, the right inverse of $\underline{C}$, it follows that:

$$\underline{L}_2 = \begin{bmatrix} 0 & \dots & 1 \end{bmatrix} \underline{N}_{\mu i} \, \underline{C}^R \qquad (3\text{-}4.15)$$

Similarly:

$$\underline{L}_1 \underline{T} = -\begin{bmatrix} 0 & \dots & 1 \end{bmatrix} \underline{N}_{\mu i} \qquad (3\text{-}4.16)$$

Hence:

$$\underline{L}_1 = -\begin{bmatrix} 0 & \dots & 1 \end{bmatrix} \underline{T}^{**^{-1}} \qquad (3\text{-}4.17)$$

If $\underline{T}$ is of higher dimension, some freedom is left in solving equation (3-4.9).

The results reveal that the UIO is identical with a Luenberger observer. However, to be applicable as a UIFDO the fault has to have an influence on the observable subspace of the state space. If $\underline{T}^{**}$ is regular, the residual is the undisturbed, measurable part of the estimation error. If $\underline{T}^{**}$ is larger than necessary, the resulting observer is either the so-called disturbance observer as described by O'Reilly (1978) or a full order Luenberger observer. It was earlier shown by Wünnenberg and Frank (1986, 1987) that the disturbance observer as well as the full order observer can be employed successfully for residual generation in the face of model uncertainties.

A possible re-interpretation of the results obtained shows the connection to the eigenstructure approach for fault detection as used by Patton et al. (1987). From equation (3-4.4) it is easily seen, that the matrix $\underline{T}$ acts like a modal analyser for the assignable eigenvectors of the resulting UIFDO. Because $\underline{F}$ is in JCF, the matrix $\underline{T}$ contains all eigenvectors assigned to the observer, either with multiplicity one, if $\underline{T}$ is of full rank, or with higher multiplicity, if $\underline{T}$ is not of full row rank. Because of the requirement expressed by equation (3-3.29), the rows of $\underline{T}$ lie in the left null space of the unknown input distribution matrix $\underline{E}$. In addition the eigenvectors contained in $\underline{T}$ are affected by the faults and the corresponding estimation error is measurable, which are the requirements for UIFDOs.

For a full row rank matrix $\underline{T}$, the UIFDO is of minimal order, so that in this case the effort to realise the filter on-line is minimal. The design therefore is a systematic procedure for the design of UIFDOs, using all degrees of freedom available.

To give the reader an idea, whether a UIFDO exists, the following guidelines should be regarded. If the number of independent measurement is *larger* than the number of disturbances (i.e. m > s), the KCF will contain at least one block of row indices. A residual generation is, in that case, possible and the question remaining is, whether the residual is affected by the faults. This can be proved using Theorem 3. Even if the number of disturbances is larger than or equal to the number of measurements a residual generation is possible, if row indices exist.

Another important fact concerning the existence of a UIFDO is, that even if Theorem 3 is not fulfilled a residual generator might exist, as long as the rank of the matrix $N_\mu K$ is larger than zero. The restriction in that case is, that not each fault f_k might be detected, because some of the directions expressed by the matrix K lie in the null space of T.

3.4.3 Pole placement

The key for the discussion of the properties and for the design of the UIFDO is the choice of the poles and the knowledge of the relationship between the fault f as the input and the residual r as the output of the filter. The free parameters for the designer are the poles of the UIFDO which are identical with the eigenvalues of the matrix F of the observer. Since the eigenvalues can be chosen arbitrarily it is important to know how their placement affects the shape of the residual. In this section we derive the z-transfer function matrix of the UIFDO and discuss the problem of the proper choice of the poles.

Because of the block-diagonal structure of the system KCF, the derivation can be restricted to the case of a single block of *Kronecker row indices* in the KCF. The generalisation to the multiple block case is straightforward.

From the estimation error equation (3-3.36) and the equations corresponding to the filter output (3-4.9), (3-4.15) and (3-4.17) it follows that the relationship between the scalar residual r_k and the fault vector f_k can be written as:

$$e_{k+1} = F\, e_k + T\, K\, f_k \tag{3-4.18}$$

$$r_k = -[0\ 0\ 0\ \ldots\ 0\ 1]\, T^{**^{-1}}\, e_k \tag{3-4.19}$$

With the aid of the z-transformation we obtain:

$$r_{(z)} = -[0\ \ldots\ 0\ 1] T^{**^{-1}} (zI - F)^{-1}\, T^{**}\, N_\mu\, K\, f(z) \tag{3-4.20}$$

or:

$$r_{(z)} = -[0\ \ldots\ 0\ 1]\, (zI - T^{**^{-1}}\, F\, T^{**})^{-1}\, N_\mu\, K\, f(z) \tag{3-4.21}$$

Equation (3-4.21) shows how the z-transform r(z) of the residual depends on the z-transform of the fault, $f(z)$. On realising that F is in Jordan canonical form and the van der Monde matrix T^{**} transforms F into the observability canonical form (OCF) it is evident that the whole system is now in OCF. One should further notice that only $\underline{F}$ and the transformation matrix $\underline{T}^{**}$ depend on the poles, whilst the new input distribution matrix $\underline{N}_{\mu}\underline{K}$ is independent of the pole-placement.

This can be interpreted such that the zeros of the numerator polynomial of each single transfer function given by:

$$G_i(z) = \frac{r(z)}{f_i(z)}$$

(3-4.22)

are completely determined by the choice of $\underline{E}$, where f_i denotes the i-th component of $\underline{f}(z)$ (i = 1...s). Even though it has not been shown explicitly, the residual of the UIFDO can be considered in the light of this, except for a delay of $\mu_i + 1$ steps, as the residual of a dead-beat UIFDO fed through a lag system with amplification 1 and the time constants defined by the observer poles. A dead-beat UIFDO is an observer where all poles of the observer are placed at $\lambda_{fi} = 0$ (i = 1 ... $\mu_i + 1$). Thus the dead-beat UIFDO can be understood as a special case in the sense that it is a finite impulse response filter, having the broadest bandwidth possible.

This interpretation shows, that a scalar residual r of any UIFDO can be written in the form:

$$r(z) = \left[\frac{1-\lambda_{f1}}{z-\lambda_{f1}} \frac{1-\lambda_{f2}}{z-\lambda_{f2}} \cdots \frac{1-\lambda_{f\mu i+1}}{z-\lambda_{f\mu i+1}} \right] \underline{G}^*(z) \, \underline{f}(z)$$

(3-4.23)

where:

$$\underline{G}_i^*(z) = \frac{r(z)}{f_i(z)}$$

(3-4.24)

represents the i-th transfer function of the transfer function vector $G^*(z)$ (i = 1, ..., s) corresponding to a dead-beat UIFDO. Note, that r(z) of equation (3-4.23) contains all information about the fault, no matter where the poles λ_{fi} are being placed. A possible re-interpretation is, that the choice of the observer poles is independent of the decoupling properties of the UIFDO, which is a well known result when decoupling controllers are designed.

3.4.4 Cross connection with the parity-space approach

In this section, the fundamental relationship between the unknown input observer approach and the parity-space approach, well-known from the literature, (Chow and Willsky (1984), Lou et al. (1986)) is discussed. It will be shown that the parity-space approach is similar to the design of a dead-beat UIFDO, so that the latter could also be termed "dead-beat observer" approach.

Following the work of Chow and Willsky (1984) and Lou et al. (1986), the subspace of $(s'+1) \times m$-dimensional vectors given by:

$$P : \left\{ \underline{v} \mid \underline{v}_0^T \begin{bmatrix} \underline{C} \\ \underline{C}\,\underline{A} \\ \vdots \\ \underline{C}\,\underline{A}^{s'} \end{bmatrix} = \underline{0} \right\} \qquad (3\text{-}4.25)$$

is called *the parity-space of order s'*. At any time, k, every vector, $\underline{v}_0^T$ in equation (3-4.25) can be used for a parity check:

$$r_k = \underline{v}_0^T \left[\begin{bmatrix} \underline{y}\ k\text{-}s' \\ \underline{y}\ k\text{-}s'\text{+}1 \\ \vdots \\ \underline{y}_k \end{bmatrix} - \begin{bmatrix} \underline{0} & & & \underline{0} \\ \underline{C}\,\underline{B} & & \underline{0} & \\ \vdots & & & \\ \underline{C}\,\underline{A}^{s'\text{-}1}\,\underline{B} & \dots & \underline{C}\,\underline{B} & \underline{0} \end{bmatrix} \begin{bmatrix} \underline{u}\ k\text{-}s' \\ \underline{u}\ k\text{-}s'\text{+}1 \\ \vdots \\ \underline{u}\ k \end{bmatrix} \right] \qquad (3\text{-}4.26)$$

Note that equation (3-4.26) describes a *finite impulse response filter* that generates a residual which is independent of the input signal $\underline{u}_k$, however, it may be affected by both faults and disturbances. Lou (1982) has shown, that the parity-space can be viewed as the left null-space of the matrix:

$$\begin{bmatrix} \underline{C} \\ z\,\underline{I} - \underline{A} \end{bmatrix} \qquad (3\text{-}4.27)$$

according to:

$$\begin{bmatrix} \underline{p}(z) & , & -\underline{q}(z) \end{bmatrix} \begin{bmatrix} \underline{C} \\ z\,\underline{I} - \underline{A} \end{bmatrix} = \underline{0} \qquad (3\text{-}4.28)$$

or:

$$\underline{p}(z)\underline{C} - \underline{q}(z)\,(z\,\underline{I} - \underline{A}) = \underline{0} \qquad (3\text{-}4.29)$$

Multiplication of equation (3-4.29) from the right with the regular matrix $\underline{T}_1$ as given by equation (3-3.1) leads to:

and:
$$\underline{q}(z) \ (\ z \ \underline{M} - \underline{A} \ \underline{M} \ ) = \underline{0} \qquad\qquad (3.4\text{-}30)$$

$$\underline{p}(z) - \underline{q}(z) \ (\ z \ \underline{C}^R - \underline{A} \ \underline{C}^R \ ) = \underline{0} \qquad\qquad (3\text{-}4.31)$$

The problem of finding a parity relation reduces to finding a solution to equation (3-4.30). Equation (3-4.31) then becomes redundant. It is easy to see that equation (3-4.30) expresses the same as equation (3-3.25) for the special case $\underline{E} = \underline{0}$. The proof of the equivalence between the parity relation approach is now a direct derivation of the results of Lemma 1 and Lemma 2. Equation (3-3.23) describes a dead-beat observer and the compatibility condition describes its estimation error, that is affected by the faults. From Lemma 2 it is apparent that the compatibility condition is a basis of the parity space according to equation (3-4.30). The two approaches are therefore equivalent. This result was already addressed by Massoumnia (1986a) for the special case of an undisturbed single output system and is herby generalised.

3.4.5 Design procedure for UIFDO

This section serves as a guideline to design unknown input observers (UIOs) for FDI in dynamic systems. No attention is placed upon the numerical calculations that are necessary to obtain the KCF. The required software to solve this problem is available (see, for example van Dooren, Dewilde, (1983) and van Dooren (1979)).

Step 1:
Definition of a linear system according to equation (3.2.1) and (3.2.2). The matrices $\underline{A}$, $\underline{B}$, $\underline{E}$, $\underline{K}$ and $\underline{C}$ have to be determined from the given situation. It is important to note, that the design result does not depend on the choice of an appropiate coordinate system, neither does it depend on the scaling of the matrices $\underline{E}$ and $\underline{K}$ or the modes of $\underline{d}$ and $\underline{f}$. The important information is only contained in the *direction* of the base vectors of the matrices $\underline{E}$ and $\underline{K}$.

Step 2:
Transformation of the system equation into KCF by an appropriate choice of the matrices $\underline{N}$ and $\underline{M}$. This can be done with the software mentioned above. The next step is the UIFDO existence proof according

to Theorem 3 (equation (3-4.6)). The matrix $\underline{T}$ according to equation (3-4.5) is then completely determined by equations (3-3.34) and (3-3.35) or equation (3-4.10). The KCF constitutes the basic structure of the observer including the decoupling. Only the poles of the UIFDO are left to be chosen which leads to the coefficients of the matrix $\underline{F}$.

Step 3:
Solution of equation (3-3.41) as well as of equation (3-3.38) determines the UIFDO input matrices $\underline{G}$ and $\underline{J}$. Solution of equation (3-4.15) and (3-4.17) determines the UIFDO output equations. This completes the design of the UIFDO that is then directly implementable on the computer. The next step could then be the implementation of an appropriate residual evaluation procedure (threshold logic etc.).

3.4.6 Example

To illustrate the design procedure let us consider the following example. Let the state space description of the process be given by

$$\underline{x}_{k+1} = \begin{bmatrix} 1 & 0 & 0 \\ 0 & 2 & 0 \\ 0 & 0 & 3 \end{bmatrix} \underline{x}_k + \begin{bmatrix} 1 \\ 1 \\ 0 \end{bmatrix} \underline{d}_k + \begin{bmatrix} 0 \\ 1 \\ 1 \end{bmatrix} \underline{f}_k \qquad (3-4.32)$$

$$\underline{y}_k = \begin{bmatrix} 1 & 1 & 0 \\ 0 & 1 & 1 \end{bmatrix} \underline{x}_k \qquad (3-4.33)$$

If $\underline{N}$ and $\underline{M}$ are chosen according to equation (3-3.1) and (3-3.2) as

$$\underline{N} = \frac{1}{3} \begin{bmatrix} 3 & -3 & -3 \\ -1 & 1 & 2 \end{bmatrix} \qquad (3-4.34)$$

$$\underline{M} = \begin{bmatrix} 1 & -1 & 1 \end{bmatrix}^T \qquad (3-4.35)$$

The matrix pencil in Kronecker Canonical Form (equation (3-3.11)) becomes

$$\underline{S}(z) = z \begin{bmatrix} 1 \\ 0 \end{bmatrix} - \begin{bmatrix} 0 \\ 1 \end{bmatrix} \qquad (3-4.36)$$

The appropriate choice of the matrices $\underline{N}$ and $\underline{M}$ is actually the key problem of the whole design procedure. For larger systems, the support by a digital computer is necessary to determine these matrices.

From equation (3-4.26) it can be seen that the matrix pencil consists of a single Kronecker row index, $\mu = 1$. Therefore, the submatrix $\underline{N}_\mu$ (equation (3-3.19)) is determined by:

$$\underline{N}_\mu = \underline{N} \tag{3-4.37}$$

The proof of existence of the UIFDO is performed with the help of Theorem 3. This yields:

$$\text{Rank } \underline{N}_\mu \ \underline{K} = \text{Rank} \begin{bmatrix} -2 \\ 1 \end{bmatrix} = \text{Rank } \underline{K} = \text{Rank} \begin{bmatrix} 0 \\ 1 \\ 1 \end{bmatrix} \tag{3-4.38}$$

This means that the observer exists.

The observer is now obtained with the aid of the corresponding equations. Choosing $\lambda_{f1} = 0.5$ and $\lambda_{f2} = 0.25$ for the observer poles, $\underline{F}$ becomes

$$\underline{F} = \begin{bmatrix} 0.5 & 0.0 \\ 0.0 & 0.25 \end{bmatrix} \tag{3-4.39}$$

$\underline{T}$ is then determined by equation (3-4.10):

$$\underline{T} = \underline{T}^{**}\underline{N}_\mu = \begin{bmatrix} 1 & 0.5 \\ 1 & 0.25 \end{bmatrix} \underline{N}_\mu = \frac{1}{12} \begin{bmatrix} 10 & -10 & -8 \\ 11 & -11 & -10 \end{bmatrix} \tag{3-4.40}$$

If the right inverse $\underline{C}^R$ to the measurement matrix $\underline{C}$ is chosen to be:

$$\underline{C}^R = \begin{bmatrix} 1 & 0 \\ 0 & 0 \\ 0 & 1 \end{bmatrix} \tag{3-4.41}$$

$\underline{G}$ can be determined by equation (3-3.41). We obtain

$$\underline{G} = \underline{T} \ \underline{A} \ \underline{C}^R - \underline{F} \ \underline{T} \ \underline{C}^R = \frac{1}{48} \begin{bmatrix} 20 & -80 \\ 33 & -110 \end{bmatrix} \tag{3-4.42}$$

With this the matrix $\underline{J}$ can be calculated from equation (3-3.38) which leads to

$$\underline{J} = \underline{T}\,\underline{B} = \underline{0} \tag{3-4.43}$$

The remaining problem is the determination of the observer output matrices. From equation (3-4.17) we obtain

$$\underline{L}_1 = -\begin{bmatrix} 0 & 1 \end{bmatrix}\,\underline{T}^{**-1} = \begin{bmatrix} -4 & 4 \end{bmatrix} \tag{3-4.44}$$

and from equation (3-4.15):

$$\underline{L}_2 = \begin{bmatrix} 0 & 1 \end{bmatrix}\,\underline{N}_\mu\,\underline{C}^R = \begin{bmatrix} -\dfrac{1}{3} & , & \dfrac{2}{3} \end{bmatrix} \tag{3-4.45}$$

This completes the design procedure. The filter equation according to the equations (3-4.1) and (3-4.2) are completely determined and can be implemented on the computer for on-line system supervision.

3.5 OPTIMALLY ROBUST RESIDUAL GENERATION

So far, in this chapter we have discussed the problem of complete disturbance decoupling. This implies that the residual can be made totally unaffected by the unknown input signal, $\underline{d}_k$, whilst being affected by the event vector, $\underline{f}_k$. In many practical situations, complete decoupling may not be achievable, because there may be too many disturbances present in the system. The best compromise that can be achieved in such a case is to make the residual "optimal" in the sense of minimising a performance index which relates the effects of disturbances to the effects of faults. A proper performance index can be defined as follows:

Definition 3
Consider the FDO of the *dead-beat* type as characterised by equation (3-4.26). This FDO is optimal in the sense of a minimisation of the ratio of the Euclidian norm of the effects of disturbances to the Euclidian norm of the effects of faults. Actually, the value of the minimum is equal to zero.

In mathematical notation this can be expressed by the

minimisation of the following performance index:

$$P = \frac{\| \, \underline{v}_o^T \, \underline{H}_2 \, \|_2}{\| \, \underline{v}_o^T \, \underline{H}_3 \, \|_2} \qquad (3\text{-}5.1)$$

with:

$$\underline{H}_2 = \begin{bmatrix} \underline{0} & & & & & \\ \underline{C}\,\underline{E} & & \underline{0} & \cdot & \cdot & & \underline{0} \\ \underline{C}\,\underline{A}\,\underline{E} & & \underline{C}\,\underline{E} & & \underline{0} & & \\ \vdots & & & & & & \\ \underline{C}\,\underline{A}^{s'-1}\,\underline{E} & & \cdot & \cdot & \cdot & \cdot \; \underline{C}\,\underline{E} & \underline{0} \end{bmatrix} \qquad (3\text{-}5.2)$$

$$\underline{H}_3 = \begin{bmatrix} \underline{0} & & & & & \\ \underline{C}\,\underline{K} & & \underline{0} & \cdot & \cdot & & \underline{0} \\ \underline{C}\,\underline{A}\,\underline{K} & & \underline{C}\,\underline{K} & & \underline{0} & & \\ \vdots & & & & & & \\ \underline{C}\,\underline{A}^{s'-1}\,\underline{K} & & \cdot & \cdot & \cdot & \cdot \; \underline{C}\,\underline{K} & \underline{0} \end{bmatrix} \qquad (3\text{-}5.3)$$

and:

$$\underline{v}_o^T = \underline{w}^T \, \underline{V} \qquad (3\text{-}5.4)$$

where $\underline{V}$ is a basis for all solutions $\underline{v}_o^T$ to the equation:

$$\underline{v}_o^T \begin{bmatrix} \underline{C} \\ \underline{C}\,\underline{A} \\ \vdots \\ \underline{C}\,\underline{A}^{s'} \end{bmatrix} = \underline{0} \qquad (3\text{-}5.5)$$

and $\underline{w}^T$ is a selector for the "optimal" $\underline{v}_o^T$.

To interpret the performance index definition (3-5.1) we re-write the input-output relationship of the system (equations (3-2.1) and (3-2.2)) in the following form:

$$\begin{bmatrix} \underline{y}_{k-s'} \\ \underline{y}_{k-s'-1} \\ \vdots \\ \underline{y}_k \end{bmatrix} = \begin{bmatrix} \underline{C} \\ \underline{C}\,\underline{A} \\ \vdots \\ \underline{C}\,\underline{A}^{s'} \end{bmatrix} \underline{x}_{k-s'} + \underline{H}_1 \begin{bmatrix} \underline{u}_{k-s'} \\ \vdots \\ \underline{u}_k \end{bmatrix} + \underline{H}_2 \begin{bmatrix} \underline{d}_{k-s'} \\ \vdots \\ \underline{d}_k \end{bmatrix} + \underline{H}_3 \begin{bmatrix} \underline{f}_{k-s'} \\ \vdots \\ \underline{f}_k \end{bmatrix} \qquad (3\text{-}5.6)$$

with:

$$
\underline{H}_1 = \begin{bmatrix}
\underline{0} & & & & & \mathbf{0} \\
\underline{C}\,\underline{B} & \underline{0} & \cdot & \cdot & \cdot & \overline{} \\
\underline{C}\,\underline{A}\,\underline{B} & \underline{C}\,\underline{B} & & \underline{0} & & \\
\vdots & & & & & \\
\underline{C}\,\underline{A}^{s'-1}\,\underline{B} & & \cdot & \cdot & \cdot & .\underline{C}\,\underline{B} & \underline{0}
\end{bmatrix}
\tag{3-5.7}
$$

and:

$$
\underline{H}_2 = \begin{bmatrix}
\underline{0} & & & & & \mathbf{0} \\
\underline{C}\,\underline{E} & \underline{0} & \cdot & \cdot & \cdot & \overline{} \\
\underline{C}\,\underline{A}\,\underline{E} & \underline{C}\,\underline{E} & & \dot{0} & & \\
\vdots & & & & & \\
\underline{C}\underline{A}^{s'-1}\,\underline{E} & & \cdot & \cdot & \cdot & .\underline{C}\,\underline{E} & \underline{0}
\end{bmatrix}
\tag{3-5.8}
$$

$$
\underline{H}_3 = \begin{bmatrix}
\underline{0} & & & & & \mathbf{0} \\
\underline{C}\,\underline{K} & \underline{0} & \cdot & \cdot & \cdot & \overline{} \\
\underline{C}\,\underline{A}\,\underline{K} & \underline{C}\,\underline{K} & & \underline{0} & & \\
\vdots & & & & & \\
\underline{C}\,\underline{A}^{s'-1}\,\underline{E} & & \cdot & \cdot & \cdot & .\underline{C}\,\underline{E} & \underline{0}
\end{bmatrix}
\tag{3-5.9}
$$

Note that this is a complete description of the system, which can simply be found by an iterative solution of the state equations (3-2.1), (3-2.2).

In these terms the resulting residual in the presence of disturbances and faults takes the form:

$$
r_K = -\underline{w}^T\,\underline{V}\,\underline{H}_2 \begin{bmatrix} \underline{d}_{k-s'} \\ \vdots \\ \underline{d}_k \end{bmatrix} -\underline{w}^T\,\underline{V}\,\underline{H}_3 \begin{bmatrix} \underline{f}_{k-s'} \\ \vdots \\ \underline{f}_k \end{bmatrix}
\tag{3-5.10}
$$

This leads to the following conclusion: Independently of the initial conditions, $\underline{x}_{k-s'}$, the residual generator checks permanently with the aid of available inputs, $\underline{u}_k$, and outputs, $\underline{y}_k$, whether the input-output relationship of the system holds.

The performance index P as defined by equation (3-5.1) may now be used for the case where no complete decoupling is

possible. Then the first term in equation (3-5.10) does not disappear
and the minimum of the performance index P is not zero. The task is
then to find the minimum.

Because no assumptions are made about the mode (size or time
evolution) of the disturbance vector, $\underline{d}_k$, and the fault event vector,
$\underline{f}_k$, one must choose the vector $\underline{v}_0^T$ so that its direction becomes as
close as possible parallel to that of the matrix $\underline{H}_3$ and orthogonal to
that of the matrix $\underline{H}_2$. Viewing the performance index in terms of
sensitivities, the task is to find the vector $\underline{w}^T$ which selects the
vector $\underline{v}_0^T$ out of all possible vectors contained in the matrix $\underline{V}$ that
is most sensitive with respect to a particular fault and least
sensitive with respect to unknown disturbances.

To show, how the numerical procedure to achieve the minimum of
the performance index can be carried out, we re-formulate the problem
as follows: Find $\underline{w}^T$ such that:

$$P = \frac{\underline{w}^T \underline{V}\ \underline{H}_2\ \underline{H}_2^T\ \underline{V}^T\ \underline{w}}{\underline{w}^T \underline{V}\ \underline{H}_3\ \underline{H}_3^T\ \underline{V}^T\ \underline{w}} \qquad (3-5.11)$$

becomes minimal.

Taking the derivative with respect to $\underline{w}^T$ and re-arranging the
equations leads to:

$$\underline{w}^T\ (\ \underline{V}\ \underline{H}_2\ \underline{H}_2^T\ \underline{V}^T - P\ \underline{V}\ \underline{H}_3\ \underline{H}_3^T\ \underline{V}^T\ ) = \underline{0} \qquad (3-5.12)$$

This is a *necessary* condition for the existence of an extremum.
Consequently, the search for the optimum selector $\underline{w}^T$ reduces to the
solution of equation (3-5.12). This is a *generalised* eigenvector -
eigenvalue problem, where the minimal eigenvalue P is the value of
the performance index and the corresponding eigenvector is the
optimal selector. Actually, equation (3-5.12) again represents a
matrix pencil which may be solved with the aid of the KCF described
in section 3.3. It turns out that this result is similar to the
result obtained by Lou et al. (1986) but with a significant reduction
of the complexity and expenditure of calculation and a residual that
is equal to zero if no disturbance and no fault occurs.

Let us now recall that by the derivation of section 3.4.4 it was
shown that the dead-beat UIFDO can be specified with the help of
equation (3-4.26). From this, it becomes clear that the optimal FDO

is a UIFDO if the performance index can be made equal to zero. This happens if the residual becomes totally unaffected by the unknown input, whilst being sensitive to faults. In this case, both approaches are equivalent and one can find out, how the order s' of the parity-space has to be chosen: From the matrix pencil in KCF (equation 3-3.11) it is seen that the largest possible row index μ_i is $\mu_{imax} = n - m$. Equation (3-3.11) also shows that one then needs n-m+2 measurements to generate the residual. This result can also be derived directly from the order of the UIFDO which is determined by the rank of matrix $\underline{T}^{**}$ according to equation (3-4.12). From this observation it can be concluded that the maximal order of the parity-space for the generation of a residual that is completely decoupled from the unknown input is:

$$s'_{max} = n - m + 1 \qquad\qquad (3-5.13)$$

Increasing the order s' beyond this bound does not provide additional freedom for the generation of a completely decoupled residual. The equivalent result for systems without unknown inputs concerning the order of the parity space was obtained by Mironowski (1979). If only an optimal solution with a performance index $P \neq 0$ (equation 3-5.1) is possible, the order s' may have an influence on P. This is caused by the least-squares characteristic of the computation of P. The designer therefore has to try by a repeated design, to see whether the performance of the filter improves by an increase of s'. The price for the improvement is additional computational effort, which may be critical when the filter algorithm (equation 3-4.26) is calculated on-line.

3.5.1 Design procedure

Step 1:
Choice of a system description (equations 3-2.1, 3-2.2). For the design of the optimal FDO the matrices $\underline{E}$ and $\underline{K}$ act like weighting matrices, so that their numerical values influence the design result. An appropriate scaling is therefore necessary for a meaningful solution. Multiplying the i-th column of the matrix $\underline{E}$ with a constant that is larger than 1 causes the fault detection filter to be less sensitive with respect to the corresponding unknown input,

$\underline{d}_{ik}$. An example to illustrate the behaviour of the solution is given in Section 3.5.2.1.

Step 2:
Now the basis for the parity-space of order s' according to equation (3-4.25) has to be determined. This can be performed by the search for all solutions of a homogeneous matrix equation which can be easily solved by a computer. The choice of the order s' of the parity-space is again to the freedom of the designer, where the minimal value is given by equation (3-5.13).

Step 3:
Calculation of the matrices $\underline{H}_2$ and $\underline{H}_3$ which determine, together with matrix $\underline{V}$, the eigenvalue-eigenvector problem according to equation (3-5.12). The minimal eigenvalue determines the value of P and the corresponding eigenvector the vector $\underline{v}_o^T$.
   Equation (3-4.26) describes the final relationship for which an algorithm has to be implemented on the computer.

3.5.2. Two examples

This section demonstrates the application of the above derived method in terms of two examples. The first is a numerical example, where the system does not necessarily have a physical interpretation. It is just chosen to demonstrate the design procedure. The second example applies to a model of the longitudinal motion of an aircraft. In both cases the perfect decoupling from the unknown input is not possible and it is therefore shown how the optimal compromise according to Section 3.5.1 can be achieved.

3.5.2.1 Numerical example
This example gives an idea of the principal behaviour of the residual generation. Consider the state-space model of the system:

$$\underline{x}_{k+1} = \begin{bmatrix} a & 0 \\ 0 & b \end{bmatrix} \underline{x}_k + \begin{bmatrix} 1 & 0 \\ 0 & c \end{bmatrix} \underline{d}_k + \begin{bmatrix} 1 \\ 1 \end{bmatrix} \underline{f}_k \qquad (3\text{-}5.14)$$

$$\underline{y}_k = \begin{bmatrix} 1 & 0 \\ 0 & 1 \end{bmatrix} \underline{x}_k \qquad (3\text{-}5.15)$$

Notice that the system is given in modal form, where both states are measurable. Each state is disturbed by a linearly independent disturbance. A scalar fault signal acts on both states in the same way. A basis for the parity-space is found by solving equation (3-4.25) and is given by:

$$\underline{v} = \begin{bmatrix} a & 0 & -1 & 0 \\ 0 & b & 0 & -1 \end{bmatrix} \qquad (3-5.16)$$

This means, that the residual generator checks both state equations whether the dynamic equation of the undisturbed and unfaulty system is valid.

The eigenvalue-eigenvector analysis according to equation (3-5.12) has to be performed on the following matrix pencil:

$$\begin{bmatrix} 1 & 0 \\ 0 & c^2 \end{bmatrix} - p \begin{bmatrix} 1 & 1 \\ 1 & 1 \end{bmatrix} \qquad (3-5.17)$$

the eigenvalues of which are:

$$p_1 \to \infty \quad \text{and} \quad p_2 = \frac{c^2}{c^2 + 1} \qquad (3-5.18)$$

The eigenvector corresponding to p_2 is thus:

$$\underline{w}_2^T = [\, c^2,\ 1 \,] \qquad (3-5.19)$$

This shows, that the *optimal* residual generator checks both state equations, putting a weight onto the result, according to a quadratic function. If $c \to 0$, an undisturbed residual is generated and the residual generator relies on the second state equation only, whilst, if $c \to \infty$, the filter relies only on the first state equation.

To illustrate the cross-connection of the decoupling UIFDO, we state the result that is obtained with the procedure outlined in Section 3.4.5 for $c = 0$. For an arbitrary observer pole λ_f, the UIFDO for the system (3-5.14) and (3-5.15) is described in the state space by the equations:

$$z_{k+1} = \lambda_f\, z_k + \begin{bmatrix} 0, & b - \lambda_f \end{bmatrix} \underline{y}_k$$

$$r_k = (1 - \lambda_f)\, z_k + \begin{bmatrix} 0, & \lambda_f - 1 \end{bmatrix} \underline{y}_k$$

Re-writing this equation as a transfer function in the z-domain yields:

$$r(z) = \frac{1 - \lambda_f}{z - \lambda_f} \begin{bmatrix} 0, & b - z \end{bmatrix} \underline{y}(z)$$

For $\lambda_f = 0$ this transfer function is a description for the finite impulse response filter, associated with the parity check (equation (3-4.26)) of the optimal UIFDO for this system if $c = 0$. It therefore serves as an example illuminating the equivalence between the optimal UIFDO and the decoupling UIFDO if the decoupling UIFDO exists.

3.5.2.2 A physical example

As a second example consider the nonlinear physical system of the longitudinal motion of a small unmanned aircraft. The mathematical model is completely known and a simulation program for the digital computer exists (see Aslin and Patton (1983)). To develop a simple and fast fault detection observer, the longitudinal motion was modelled by two state equations, representing the short-period subsystem of the aircraft. This model was derived by a linearisation of the nonlinear aerodynamic system equations associated with a model reduction. The resulting nominal continous-time state-space equations are in normalised notation given by (Patton et al., 1987):

$$\dot{\underline{x}}_o = \underline{A}_o \, \underline{x}_o + \underline{B}_o \, u \qquad\qquad (3-5.20)$$

$$\underline{y}_o = \underline{C}_o \, \underline{x}_o \qquad\qquad (3-5.21)$$

with:

$$\underline{A}_o = \begin{bmatrix} -2.34 & 32.98 \\ -0.453 & -0.0499 \end{bmatrix} \qquad\qquad (3-5.22)$$

$$\underline{B}_o = \begin{bmatrix} 5.317 \\ -13.5789 \end{bmatrix} \qquad\qquad (3-5.23)$$

$$\underline{C}_o = \begin{bmatrix} 1 & 0 \\ 0 & 1 \end{bmatrix} \qquad\qquad (3-5.24)$$

where the subscript o denotes nominal values. The first component x_1

of the state vector $\underline{x}$ represents the vertical velocity of the aircraft and x_2 the pitch rate of the aircraft.

In the actual system there are modelling uncertainties and parameter changes in the $\underline{A}$, $\underline{B}$ matrices around the nominal values $\underline{A}_0$, $\underline{B}_0$. They can be modelled in the equation for the estimation error $\underline{\varepsilon}$ by a term $\underline{E}\ \underline{d}$. Hence the resulting state equation of the estimation error reads:

$$\dot{\underline{\varepsilon}} = (\underline{A}_0 - \underline{H}\ \underline{C}_0)\ \underline{\varepsilon} + [\Delta\underline{A} \quad \Delta\underline{B}]\ [\underline{x}_0{}^T\ u]^T + \underline{K}\ \underline{f}$$

$$\dot{\underline{\varepsilon}} = (\underline{A}_0 - \underline{H}\ \underline{C}_0)\ \underline{\varepsilon} + \underline{E}\ \underline{d} + \underline{K}\ \underline{f} \qquad (3\text{-}5.25)$$

For this specific example it is reasonable to assume (see Chapter 4) that $\Delta\underline{A} = \underline{0}$ and that the disturbance distribution matrix $\underline{E}$ can be chosen to be equal to the input distribution vector $\underline{B}_0$, i.e., the uncertainties act like an input signal u. The system input is the elevator position. The manoeuvre, that is performed is a change in the absolute height of the airplane. Figure 3-3 shows the response of a fault detection filter to a change of the height of 2.5 m within 2 sec at t = 0 and the effect of a fault in the instrument that measures the vertical velocity.

The simulated fault is an offset of 1 m sec^{-1} in the vertical velocity measurement, appearing at t = 6 sec and disappearing at t = 9 sec. From figure 3-3 one can observe the effect of the nonlinearities of the system. These cause a significant excitation of the residual at t = 0. After t = 2 sec the effect of the nonlinearities excited by the manoeuvre has decreased to minor amplitude. This means that the observer is not robust with respect to the modelling errors associated with the neglected nonlinearities. This is clear, because the direction, into which the filter is robust can only be the one described by $\underline{E}$. All other model uncertainties introduced by using the reduced order linear model will then affect the residual. Nevertheless, the filter also shows a significant response to the instrument fault at t = 6 sec, which could be detected by a simple threshold logic.

Both examples (as well as the analysis of Section 3.5) show that the achievable robustness depends on the given situation and only limited robustness against unknown inputs and limited sensitivity against faults can be obtained. The performance of the linear fault

detection filter can often be improved by the use of a more
sophisticated model. If the system is nonlinear and does not operate
at certain operating points, one should take into account the
nonlinearities (Frank 1987). This will be the subject of the
following section.

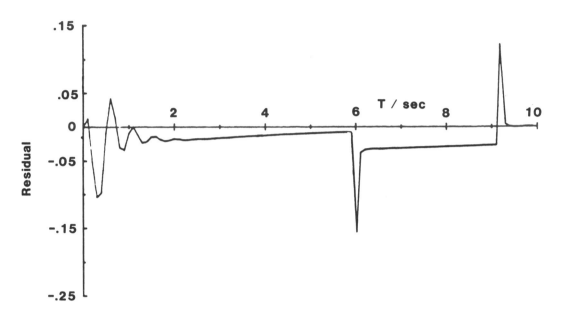

Figure 3-3: Linear residual for aircraft example

3.6 EXTENSION TO NONLINEAR RESIDUAL GENERATION FOR NONLINEAR SYSTEMS

The results presented in the previous section show that the residual
of a dead-beat observer is generated by re-calculating the state
equation of the system. As a result, the unknown disturbances are
eliminated from the corresponding equations. So far we restricted the
examination to linear systems. The same procedure is now briefly
outlined for nonlinear systems in terms of two examples. A general
systematic treatment of the nonlinear case is far beyond the scope of
this chapter and will therefore not be included.

3.6.1 First Example: Tank system

Consider a single tank system as shown in Figure 3-4. The input is the incoming mass flow q_1, the cross section of the outlet is Q. System output is the fluid level h in the tank

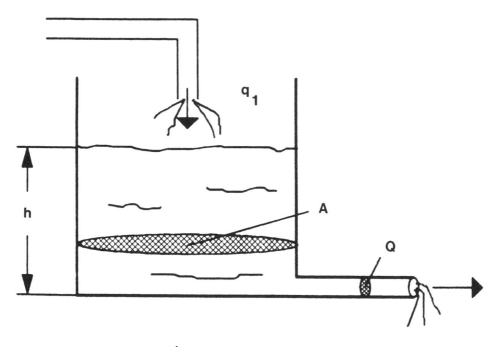

Figure 3-4: Tank system

An approximate discrete description of the dynamics of this system is a nonlinear state equation of the form:

$$x_{k+1} = f(x_k, u_k) \qquad (3-6.1)$$

$$y_k = h(x_k) \qquad (3-6.2)$$

In the present case the equations become:

$$x_{k+1} = x_k + \frac{\Delta T}{A} (q_1 - Q \sqrt{\rho 2ghx_k}) \qquad (3-6.3)$$

$$y_k = x_k \qquad (3-6.4)$$

with the parameters:

$x_k \triangleq$ system state variable
$\Delta T \triangleq$ sampling time
$A \triangleq$ cross section of the tank
$q_1 \triangleq$ incoming mass flow (known input)
$Q \triangleq$ cross section of outlet pipe
$g \triangleq$ gravity constant
$\rho \triangleq$ density of fluid
$y_k \triangleq$ tank level.

The residual resulting from the design procedure for the above state and output equation is given by:

$$r_k = \frac{y_k - y_{k-1}}{\Delta T} - \frac{1}{A} \left[q_1 - Q \sqrt{\rho 2g\, h\, y_{k-1}} \right] \qquad (3\text{-}6.5)$$

This shows that the residual generator for this simple system is a discrete differentiator acting on the system output signal from which the inward and outward mass flow is substracted.

3.6.2 Second Example: Nonlinear aircraft model

In principle, the same design strategy can be applied to the completely nonlinear aircraft example, treated before in Section 3.5.2.2. The two state equations belonging to the vertical velocity and the pitch rate of the airplane are rearranged such that the input signal aileron position is eliminated from the equation. The remaining dynamic equation of second order then represents the residual generator. Figure 3-5 shows the simulation result that is obtained with the same manoeuvre and the same instrument fault as explained in Section 3.5.2.1 except considering the nonlinearities. (Notice that the residual is drawn with inverted sign.) Comparing Figure 3-3 and Figure 3-5 one realises that the effect of the residual due to the nonlinearities has become significantly smaller, so that the part of the signal which actually dominates is that caused by the fault. In other words, the fault detection abilities are therefore significantly improved, and the fault can be detected by a simple threshold logic with a very low rate of false alarms.

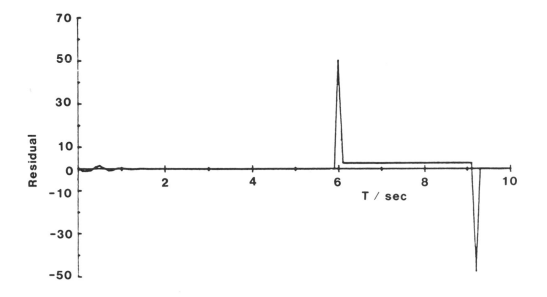

Figure 3-5: Residual for nonlinear aircraft model

3.7 UNKNOWN INPUT OBSERVER SCHEMES (UIOS)

In Sections 3.3 to 3.6 we discussed the unknown input observer as a basis for robust residual generation under the assumption that faults represented by the fault vector, $\underline{f}_k$, and disturbances respresented by the unknown input vector, $\underline{d}_k$, are present in the system. When faults occur and are to be detected in different sectors of the system, e.g., in different sensors, or actuators or other components, one needs, in general, *a bank of observers* rather than a single observer in order to localise the faults uniquely. A bank of observers is generally called an *observer scheme*. This section deals with the design of unknown input observer schemes (UIOS) providing structured sets of residuals for fault isolation.

3.7.1 General Structure of UIOS for AFD, IFD and CFD

In general, the number of observers in an observer scheme can be arbitrary. To become specific and for the sake of simplicity of presentation let us assume that m different faults f_i (i=1,...,m) can occur in the system where m is the number of measurements available. The general structure of the unknown input observer scheme - whether used for instrument or actuator or component fault detection - is shown in Figure 3-6.

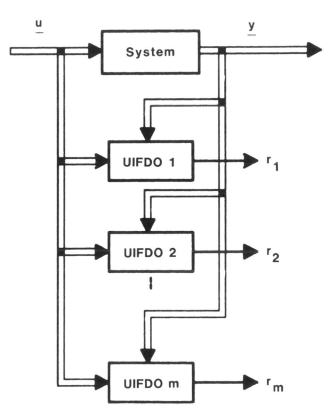

Figure 3-6: General structure of the Unknown Input Observer Scheme

The set of m UIFDOs is driven by the input vector, $\underline{u}$, and the available measurement vector, $\underline{y}$. Concerning the design of the UIOs one has to partition the faults into subsets of $\underline{f}$, specified by vectors $\underline{f}_i$. Each UIFDO of the observer scheme is now assigned to be sensitive to a different fault or set of faults and invariant to the rest of faults. The remaining design freedom can then be used to provide invariance to unknown inputs. This is done, in turn, to such an extent that properly structured sets of residuals are obtained which enable a unique decision of the time of occurrence and location of the faults. The final design of the observers depends on the number of faults that are to be detected at the same time. The three extremes are:

(a) faults have to be detected, but not isolated
(b) only a single fault is to be detected and isolated at a time
(c) all faults are to be detected and isolated even if they occur simultaneously

Case (a):

The FDI-Scheme reduces to a single UIFDO that generates a residual that is sensitive to all faults whilst being robust to m-1 unknown inputs. The full design freedom can be used to generate a residual that is robust to unknown inputs.

Case (b):

One can achieve the most robust FDI scheme that allows fault isolation because the maximum design freedom is left for the generation of robustness to unknown inputs. Here, the i-th observer (i=1,2,...,m) is designed such that it becomes invariant to the i-th fault, f_i, and of m-2 unknown inputs. In other words, f_i is interpreted as an unknown input and the remaining design freedom can be used for generating invariance with respect to m-2 other (genuine) unknown inputs, e.g., for parameter variations. Repeating this design m times provides a UIOS according to Figure 3-6. Here the first residual, r_1, depends on all but the first fault, the second residual, r_2, on all but the second fault and so on. The resulting relations are:

$$r_1 = q_1(\underline{f}_2, \underline{f}_3, \ldots, \underline{f}_m)$$

$$r_2 = q_2(\underline{f}_1, \underline{f}_3, \ldots, \underline{f}_m)$$
$$\vdots$$
$$r_i = q_i(\underline{f}_1, \ldots, \underline{f}_{i-1}, \underline{f}_{i+1}, \ldots, \underline{f}_m) \tag{3-7.1}$$
$$\vdots$$
$$r_m = q_m(\underline{f}_1, \ldots, \underline{f}_{m-1})$$

The decision function for the logical evaluation of the residuals could then be as follows:

(a) if r_2 and r_3 and ... and r_m are unequal to zero, (and $r_1 = \underline{0}$), it is concluded that the first fault, $\underline{f}_1$, occurs,

(b) if r_1 and r_3 and ... and r_m are unequal to zero, (and $r_2 = \underline{0}$) it is concluded that the second fault, $\underline{f}_2$, occurs,

and so forth. It is seen that only a single fault at a time can be detected.

Case (c):

To be able to detect and isolate all faults occurring simultaneously, one has to interpret all but the i-th fault in the i-th observer (i=1,2,...,m) as unknown inputs. Therefore the rank of the matrix $\underline{E}$ is increased by m-1, which is the largest possible rank of $\underline{E}$ for which complete invariance can be achieved in many practically arising cases. In this case, however, the observer scheme cannot be made robust against any unknown input since no design freedom is left. The residuals depend on the faults due to the following relations:

$$
\begin{aligned}
r_1 &= q_1' \,(\underline{f}_1) \\
r_2 &= q_2' \,(\underline{f}_2) \\
&\vdots \\
r_i &\stackrel{!}{=} q_i' \,(\underline{f}_i) \\
&\vdots \\
r_m &\stackrel{!}{=} q_m' \,(\underline{f}_m)
\end{aligned}
\qquad (3\text{-}7.2)
$$

This permits a unique detection and isolation of m faults even if they occur simultaneously. The price to pay is the loss of robustness with respect to unknown inputs acting on the system. Note that this UIOS is equivalent to the Fault Detection Filter first presented by Beard (1971) and Jones (1973).

3.7.2 UIOS for component fault detection

The application of the UIFDOS to component fault detection in large, complex systems can be improved by providing more structural insight in the process. If no assumption on the fault mode can be made, a logical approach is to decompose the process and apply a scheme of local observers to the system under consideration (Frank 1987).

The problem of local state observation lies in the couplings among the components. If the couplings are sufficiently weak or measurable, the local observer scheme can be configured such that a malfunction in any of the components affects only the residual of the corresponding local observer. It is thus possible to identify the faulty component uniquely, and the design freedom can be used to improve the robustness to unknown inputs. The resulting local observer scheme corresponds to the so-called Available-State-Coupled Observer Scheme (ASCOS) proposed by Janssen and Frank (1984).

If the subsystems interact considerably with each other and the coupling signals are not measurable, one may interpret these couplings as unknown inputs acting on the local sub-system. The design freedom for the local observer can then be used to decouple from these coupling signals. They are therefore considered as unknown inputs, and by this the capability of the detection and isolation of the faulty component can be improved. The resulting observer scheme corresponds to the so-called Estimated-State-Coupled Observer Scheme (ESCOS), also presented by Frank and Janssen (1986).

In large complex systems one has to build a hierarchical structure to supervise the whole system. This hierarchical structure consists of different UIFDOs, operating on different system levels.

The most basic structure is that of the local observer according to case (a) as described in the previous section, indicating that a fault happened in the corresponding subsystem. The next step is a bank of filters according to case (b), indicating, that a single fault occurred in this subsystem. The observer bank, designed according to case (c) is able to detect simultaneously appearing faults. Note that by structuring the observers for one single subsystem in this manner, one reaches the highest degree of robustness possible.

The next level in the hierarchical structure has, principally, the same structure while interpreting a set of subsystems as a new subsystem with the corresponding input and output signals.

This may be repeated until the highest level possible is reached, which is when one designs the UIFDO for the overall system. The resulting structure corresponds to the Hierarchical Observer Scheme (HOS), described by Frank (1987).

3.7.3 UIOS for actuator fault detection (AFD)

The detection of actuator faults can be viewed as a special case of the component fault detection and can therefore be achieved by a UIOS that is equivalent to the structure shown in Figure 3-6. Yet it can be simplified because the location of the fault is equivalent to that of an input signal. For simplification it is assumed that m inputs act onto the system and a single actuator fault may possibly appear in any of the input channels. The structure then obtained is as shown in Figure 3-7.

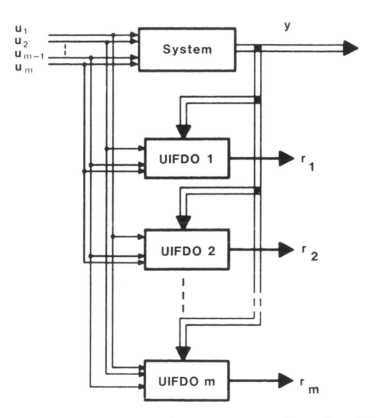

Figure 3-7: Generalised Observer Scheme for AFD

One can see that any UIFDO is driven by all but one inputs and all outputs of the system. The i-th residual is sensitive to all but the i-th actuator fault and the detection of a single fault at a time is possible. The i-th input that acts on the system must be treated like an unknown input and the i-th UIFDO must be robust to this missing signal.

   If m actuator faults are to be detected at a time, the resulting structure of the UIOS is as shown in Figure 3-8.

   The i-th UIFDO is driven by the i-th input only and the residual is therefore affected by the i-th actuator fault. The other m-1 input signals must be interpreted as unknown inputs and the residual is therefore robust to these missing signals.

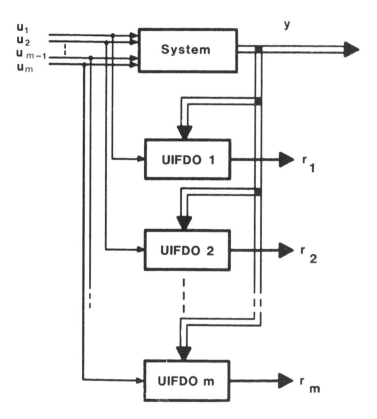

Figure 3-8: Dedicated Observer Scheme for AFD

3.7.4 UIOS for instrument fault detection

The application of UIOS to the special case of instrument fault detection allows more transparency in the FDI structures, similar to AFDI, because the faults are associated with direct changes in the output signals. Instrument faults are, therefore, easier to detect than component faults.

Suppose a single fault at a time in one of the m sensors of a system is to be detected with the aid of UIOS providing maximal robustness with respect to as many unknown inputs as possible. Then one arrives at the so-called *generalised observer scheme* (GOS) for IFD (Frank 1987). This is illustrated in Figure 3-9.

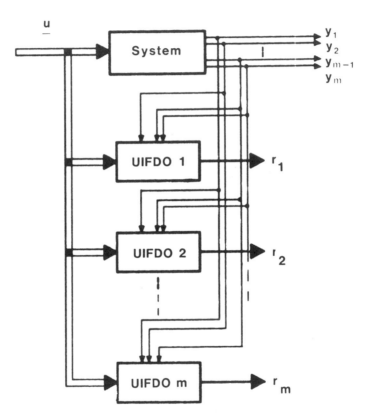

Figure 3-9: Generalised Observer Scheme (GOS) for IFD

The i-th observer (i=1,2,...,m) is driven by all but the i-th measured variable (i.e., $y_1, \ldots, y_{i-1}, y_{i+1}, \ldots, y_m$) and made invariant to m-2 unknown inputs. y_i is not used in the i-th observer because y_i is assumed to be corrupted by the fault and therefore does not carry information about the system.

When, on the other hand, the system can be considered undisturbed, and, hence, no robustness has to be provided, but simultaneous faults in all m instruments are to be detected according to equation (3-7.2), the design of UIOS leads to the Dedicated

Observer Scheme (DOS) for IFD proposed by Clark (1979), as shown in
Figure 3-10.

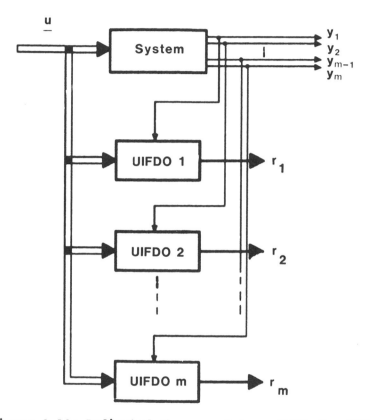

Figure 3-10: Dedicated Observer Scheme (DOS) for IFD

Here, the i-th fault only affects the i-th residual, r_i, which can
easily be evaluated in a decision logic.

 If the system is observable from any of the outputs, each UIFDO
is a full order Luenberger observer estimating the full state and the
i-th residual is the output estimation error $y_i - \hat{y}_i$.

3.8 CONCLUSION

This Chapter has discussed the technique of model-based fault
diagnosis based on the unknown input observer approach. The unknown

input observer approach was adopted to achieve the maximally reachable robustness in the face of unknown inputs which comprise system and measurement noise as well as modelling uncertainty and unmodelled process parameter variations. This leads to one of the most powerful model-based fault detection and isolation approaches for disturbed and uncertain dynamic systems. Another great advantage is that no assumptions have to be made on the modes (i.e. time evolution and size) of the faults to be detected.

Experimental results have shown that the resulting observer schemes have reached a rather high degree of maturity. There are, in particular, a number of encouraging results in the application to mechanical systems as, for example, aircraft or advanced transport systems. It should be noted, however, that in cases where only poor or unprecise models are available, as, for example, in some chemical processes the model-based FDI approach is still problematical. In such cases the support by knowledge-based methods may be helpful. In conclusion, the question of application of any observer based FDI schemes is primarily a question of the quality of the available model of the system. In addition to this, it should always be realized that the quality of fault identification decisively depends on the number of available measurements. This has been evidenced in this Chapter.

The Chapter has developed the general theory for the unified approach to the design of robust observers for fault diagnosis. The next Chapter serves to re-inforce these ideas by showing how an UIFDO can be designed directly from an eigenstructure assignment consideration. On careful examination it can be seen that several of the ideas expressed in both Chapters 3 and 4 lead to similar results.

Chapter Four

ROBUST FAULT DIAGNOSIS USING
EIGENSTRUCTURE ASSIGNMENT OF OBSERVERS

Ronald J. Patton, Samuel M. Kangethe

For a particular fault detection/diagnosis application the modal properties of the fault monitor can be designed according to the known process dynamics and the bandwidth of the sensor channels under consideration. The required frequency domain characteristics of an observer used for fault monitoring can be designed using *eigenstructure assignment* either in continuous or sampled form. The eigenvalues and eigenvectors of the closed-loop observer matrix are also chosen to provide disturbance de-coupling. In the continuous-time case an asymptotic form of robust fault monitor can be designed whilst, in the discrete-time case the dead-beat observer can be designed directly through assigning p zero-valued eigenvalues corresponding to p sensor channels. The dead-beat observer design has time-optimal transient performance and is made sensitive to specific fault channels. A bank of similarly designed robust observers using the Generalised Observer Scheme (GOS) provides a very effective *fault detection and isolation* (FDI) system. The observer robustness properties are demonstrated through application to a realistic non-linear simulation of a light aircraft with full aerodynamic coupling, turbulence and non-linear sensor fault characteristics.

4.1 INTRODUCTION

The main problem associated with many fault detection and isolation (FDI) schemes lies in the certainty required in modelling the physical system, in the design and testing stages. Most techniques make certain assumptions about the physical system. These assumptions may not be accurate, and may degrade the performance of the monitor scheme from the theoretically predicted performance. Clearly, any fault monitor scheme must have a very high reliability but, in particular must be robust against disturbances and parameter deviations which could cause uncertain errors and false-alarm rates to occur.

These limitations can cause linear model-based designs to be inadequate for many real applications. Sensitivity to input-induced parameter variations cause uncertain errors between redundant state estimation vectors and, in an FDI scheme these errors would, in turn, cause false-signalling of a fault. The solution proposed by some investigators is to widen the threshold band for the fault detection signal, thereby reducing the false-alarm rate. This increase in threshold level lengthens the time taken to detect a fault. The increased detection time may prove unacceptable in many fault-critical systems, such as, for example, those required for many aerospace applications.

In open-loop unstable aircraft, for example the vertical short take-off and landing (VSTOL) Harrier, a pitch rate gyro malfunction may require fault detection within a 200 msec time frame. This is particularly true when the aircraft is operating in a hovering flight mode using vectored thrust. A malfunction in the pitch rate gyro at low altitude can cause rapid instability and loss of control. A fault of this kind is most likely to be outside the human pilot's handling capability due to the very short time scale involved. It is thus important to be able to detect and locate the faulty instrument and reconfigure the feedback system within 200 msec. This is quite a challenge!

For such an application some hardware redundancy is essential for reconfiguration of the flight controls on detecting a fault. However, higher reliability can be achieved through the use of a lower hardware redundancy index, for example using duplex instead of triplex or quadruplex systems with each system lane being monitored independently through an analytical redundancy scheme. Each hardware lane has a dissimilar instrument set and these can be used very effectively in an instrument fault detection (IFD) scheme using a bank of observers.

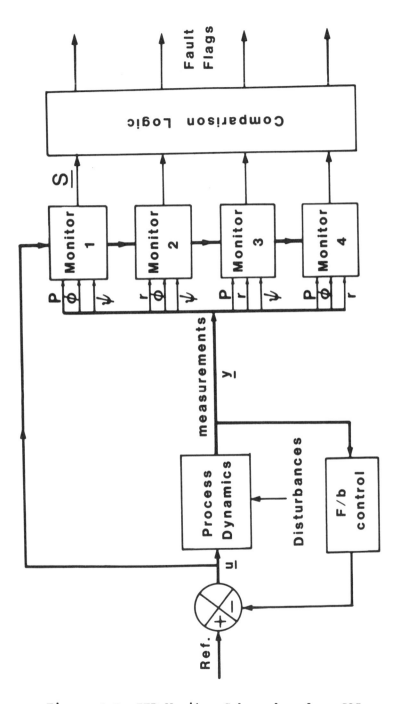

Figure 4-1: IFD Monitor Scheme based on GOS
(Generalised observer scheme) with four observers

Figure 4-1 shows a typical FDI scheme which is appropriate for Instrument Fault Detection (IFD) using the Generalised Observer Scheme (GOS). The measurements have been labeled with the variables typically measured for the lateral motion of an aircraft. The actual example is given in Section 4.4.3, however the principle of using a bank of observers for fault diagnosis is well known and has already been discussed in Chapters 2 and 3. The diagram shows that dissimilar measurements are used as inputs to each observer.

The diagram shows only one channel of hardware and over-lapping sets of measurements are used for the observer inputs. In the example shown four measurements are used, however, each of four observers takes three measurements (in general, it is necessary to use p + 1 observers in the GOS scheme where p is the number of measurements applied to each observer). In this way, for example, a malfunction occurring in the roll-rate gyro (p) will produce an erroneous signal in monitors 1, 3 and 4. Thus for this fault, thresholds will be exceeded in monitors 1, 3 and 4. The fact that all three of these monitors show that a fault has occurred suggests the hypothesis that the roll-rate gyro is malfunctioning.

One could make similar hypotheses for the remaining three measurements Φ, Ψ and r and hence detect and locate a fault - subject to the hypothesis being true. The first step to take to ensure that the hypothesis is a reliable one is to reduce the probability that a false-alarm fault signal will be generated, for example as a consequence of disturbances acting on the process itself or through parameter variations. This is the problem of robustness with respect to unkown inputs which has already been discussed by Frank and Wünnenberg in Chapter 3.

The aim of this Chapter is to develop the theme of the robust unknown input observer for fault detection (UIOD) into a direction making use of the *eigenstructure assignment* approach. Section 4.2 provides the essential theory of eigenstructure assignment and shows how to design a full-order observer using this approach. An example of the design of a discrete time observer is given to illustrate the method.

Section 4.3 gives a detailed study of the robustness properties of full-order state observers in the context of the theory of *invariant subspaces*. The problem of disturbance-decoupling is analysed as a background study for the application of the UIOD to a sensor fault detection problem with emphasis on eigenstructure assignment design.

Section 4.4 is then devoted to the problem of designing robust observers with application to fault detection. The approach used is that of eigenstructure assignment. This section makes use of the invariant subspace concepts laid down in Section 4.3.

Sub-section 4.4.3 provides a design study of the application of a robust UIOD to a sensor fault detection problem, namely the aircraft problem mentioned above. It is shown, in particular, that by using a discrete-time dead-beat observer approach based on eigenstructure assignment a very effective design can be achieved. A single UIOD monitor can be used to detect the faults in each of the p sensor channels independently. The example given has *five* state variables, *two* control inputs and *three* measurements (p = 3). Results are shown to illustrate the potential and robustness of the approach.

4.2 OBSERVER DESIGN USING EIGENSTRUCTURE ASSIGNMENT

4.2.1 Background theory

The method of eigenstructure assignment has been applied mostly to the problem of the design of a closed-loop modal structure to a controlled system which can be referred to as the *control* problem. The observer system can be thought of as a *dual* to the control problem and eigenstructure assignment can thus be extended to encompass the design of linear time-invariant observers.

Many methods are available to compute accurately the feedback gains for either the *full state feedback* or the *output feedback* control problems and these techniques can easily be extended to the observer design case. For the single input (single control signal) or single input measurement (observer case) system the eigenvalue assignment (pole-placement) approach has a unique solution for the feedback gain calculation. However, for the multi-input case (for both controllers and observers) the feedback gain matrix calculation is under-determined.

The consequence of this is that a number of difficulties can occur when the pole-placement approach is applied to a multi-input (multivariable) system. One difficulty often encountered is that the feedback design may not lead to a well-conditioned closed-loop matrix i.e. the eigenvalues of the closed-loop matrix may be sensitive to even slight numerical perturbations in the process system parameter.

A further problem is that, for the multivariable case the modal structure is not determined by the eigenvalues alone. Thus, by using pole-placement (by only assigning the eigenvalues with no consideration of eigenvector assignment) there are many possible closed-loop eigenvectors and some of these may not give rise to suitable closed-loop time or frequency response behaviour.

In order to construct a robust observer it is necessary to assign a suitable modal structure so that the state estimates converge quickly to the actual states of the process system whilst remaining insensitive to disturbance signals and parameter variations. It is considered very difficult to achieve this when the observer has a multivariable structure unless all the degrees of freedom in feedback design are used correctly.

It is therefore necessary to use as many degrees of freedom as possible in designing an observer in order to provide a mechanism for decoupling the observer's response from unknown disturbances. The degrees of freedom available in the design of an observer are related to the degrees of freedom of a (similar) dual control problem.

To gain an appreciation for the role of the eigenvalues and eigenvectors of both the controlled process and observer systems consider first the following controlled system in state-space form:

$$\left.\begin{aligned} \dot{\underline{x}} &= A\,\underline{x} + B\,\underline{u} \\ \underline{y} &= C\,\underline{x} + D\,\underline{u} \end{aligned}\right\} \tag{4-2.1}$$

where $\underline{x} \in R^n$, $\underline{y} \in R^p$ and $\underline{u} \in R^n$.

We can consider the following output feedback control law:

$$\underline{u} = -\,K_c\,C\,\underline{x} \tag{4-2.2}$$

such that the closed-loop state equations become:

$$\left.\begin{aligned} \dot{\underline{x}} &= F_c\,\underline{x} \\ \underline{y} &= (I_n - D\,K_c)\,C\,\underline{x} \end{aligned}\right\} \tag{4-2.3}$$

where $F_c = (A - B\,K_c\,C)$ and K_c is the output feedback gain matrix.

For full state feedback the matrix C in the control law (4-2.2) is replaced by the identity matrix such that the state equations become:

$$\left. \begin{array}{l} \dot{\underline{x}} = F_c \; \underline{x} = (A - B \; K_c) \; \underline{x} \\ y = C \; \underline{x} - D \; K_c \; \underline{x} \end{array} \right\} \qquad (4\text{-}2.4)$$

Now the observer system which reconstructs the entire state-space from a set of p independent measurements is given by:

$$\left. \begin{array}{l} \dot{\hat{\underline{x}}} = A \; \hat{\underline{x}} + B \; \underline{u} + K_o(y - \hat{y}) \\ \hat{y} = C \; \hat{\underline{x}} + D \; \underline{u} \end{array} \right\} \qquad (4\text{-}2.5)$$

Where K_o is an n x p observer feedback matrix and the $\hat{\underline{x}}$ and $\hat{y}$ vectors are the estimates of the process states and outputs, respectively. In order to reconstruct the system state it is necessary for the matrix pair [A,C] to be observable. It will also be assumed that [A,B] is a controllable pair in order to achieve maximum controller design freedom.

The error dynamics of the observer are given by:

$$\dot{\underline{e}} = (A - K_o \; C) \; \underline{e} \qquad (4\text{-}2.6)$$

with $\quad \underline{e} \triangleq \underline{x} - \hat{\underline{x}}$

Equation (4-2.6) can be re-written as:

$$\dot{\underline{e}} = A \; \underline{e} - K_o \; \underline{e}_y \qquad (4\text{-}2.7)$$

where $\quad \underline{e}_y = C(\underline{x} - \hat{\underline{x}})$

The vector $\underline{e}_y$ can be considered as a control signal applied to the observer estimation error.

On considering further the matrix $(A - K_o C)$ in equation (4-2.6) it follows that:

$$(A - K_o \; C)^T = A^T - C^T \; K_o^T \qquad (4\text{-}2.8)$$

Hence, the *dual* or *transposed* system with state vector $\underline{e}^*$ can be given as:

$$\dot{\underline{e}}^* = A^T \underline{e}^* - C^T K_O^T \underline{e}^* \qquad\qquad (4\text{-}2.9)$$

with $\underline{e}^* \in R^n$

It is clear then that equation (4-2.9) has the form of the state equation of the controlled process under state feedback. C^T can be compared with the matrix B and K_O^T can be compared directly with the matrix K_C of the control system.

For convenience of notation, here, let:

$$F_O = (A - K_O C) \qquad\qquad (4\text{-}2.10)$$

It is straight forward to show that the eigenvalues of the matrices F_O and F_O^T are identical, however their eigenvectors form *reciprocal* bases i.e. the *left* eigenvectors of F_O are the *right* eigenvectors of F_O^T and vice-versa.

If we consider these eigenvectors, then:

$$\left.\begin{aligned}
F_O \, \underline{m}_i &= \underline{m}_i \, \lambda_i \\
\underline{v}_i^T F_O &= \lambda_i \, \underline{v}_i^T \\
F_O^T \underline{v}_i &= \underline{v}_i \, \lambda_i
\end{aligned}\right\} \quad \text{for } i = 1, \ldots, n \qquad (4\text{-}2.11)$$

Where the $\underline{m}_i$ and the $\underline{v}_i$ are the n *right* and *left* eigenvectors, respectively of the observer problem, whilst the $\underline{v}_i$ and $\underline{m}_i$ are also the right and left eigenvectors of the *dual control* problem given by equation (4-2.9).

We can now re-write equation (4-2.11) as:

$$\left.\begin{aligned}
F_O \, M &= M \, \Lambda \\
F_O^T V &= V \, \Lambda
\end{aligned}\right\} \qquad\qquad (4\text{-}2.12)$$

where the matrices V and M are the appropriate n x n *modal* matrices containing the right eigenvectors (as columns) of the observer and dual control problems, respectively. The matrix Λ is the diagonal matrix containing the distinct eigenvalues, i.e.:

$$\Lambda = \text{diag} \, (\lambda_1, \ldots, \lambda_n) \qquad\qquad (4\text{-}2.13)$$

The use of the diagonal matrix is valid here as long as the eigenvectors are *linearly-independent*. In the case of repeated (non-distinct) eigenvalues the matrix Λ must be replaced by the Jordan block structure (Brogan, 1974). In the additional case in which the eigenvectors corresponding to repeated eigenvalues are identical (geometric multiplicity) the set of *generalised* eigenvectors must be used as a basis for the Jordan form.

The duality between the state feedback control and observer problems means that the observer eigenvalues and eigenvectors can be assigned to the appropriate dual control problem. This is the well known observer eigenstructure assignment problem as discussed by O'Reilly (1983).

It is necessary to use the dual control system form as the right eigenvectors of the observer are not directly assignable. On assigning the right eigenvectors to the dual control problem these eigenvectors become the left eigenvectors of the observer system. This is an important approach to the design of *robust* observers and will be used later in this Chapter in the design procedure for a robust fault detection system based on observers.

4.2.2. Eigenstructure assignment details

From the above it is clear that the observer feedback eigenstructure assignment problem can be handled by means of a transformation into the dual control form. If the pair (A,C) provides an observable realisation of the system, then, by duality (A^T, C^T) is also a controllable pair. The observer full state feedback matrix K_o can then be designed using maximum freedom and all n eigenvalues can be placed arbitrarily. For convenience of notation we will now drop the subscripts o and c and refer to the gain matrix as simply K.

In a now classical paper (Moore, 1976) the freedom available to assign eigenvectors for an arbitrary self-conjugate set of eigenvalues using state feedback has been described. Moore gives both necessary and sufficient conditions for a *general* state feedback matrix K to exist which satisfies the standard control problem:

$$(A - BK) \; \underline{v}_i = \lambda_i \; \underline{v}_i, \quad i = 1, \ldots, n \qquad (4\text{-}2.14)$$

The *dual control* problem can also be solved by substituting C^T for B,

A^T for A and K^T for K in equation (4-2.14).

In order to present Moore's result, associate with each λ_i a matrix $Q(\lambda_i) \in R^{n \times (n+m)}$ such that:

$$Q(\lambda_i) = [\lambda_i I - A : B] \tag{4-2.15}$$

and a compatibly partitioned matrix:

$$L(\lambda_i) = \begin{bmatrix} P(\lambda_i) \\ T(\lambda_i) \end{bmatrix} \tag{4-2.16}$$

Where the columns of $L(\lambda_i)$ form a basis for the null space of $Q(\lambda_i)$. The necessary and sufficient conditions to find a real feedback matrix K satisfying (4-2.14) are:

(a) The $\underline{v}_i \in C^n$ are linearly independent vectors
(b) $\underline{v}_i = v_i^*$, whenever $\lambda_i = \lambda_j^*$

*here denotes the complex-conjugate operation
Furthermore, if such a feedback matrix K exists and rank (B) = m then K is unique.

The robust observer design problem involves the use of full state feedback, however for the more general case of output feedback the basic result due to Moore has been extended by Srinathkumar (1978). The *generalised eigenstructure* problem has also been solved by Klein and Moore (1977) in which eigenvectors with multiplicity are assigned by computing the generalised eigenvectors.

Burrows and Patton (1988) describe a CTRL-C package for interactive design of feedback systems by eigenstructure assignment. The package enables the user to design the generalised eigenstructure of a system and also provides an indication of eigenvalue sensitivity, matrix conditioning, together with iterative robustness improvement. The CTRL-C package has been used as a design tool in this work.

In the more general control problem and from a practical standpoint, the necessity of full-state feedback in the eigenstructure assignment can be thought to be restrictive or undesirable as, for many problems only output measurements are available rather than states. For this general case consider now the control law in the form:

$$\underline{u} = -K \underline{y} = -K C \underline{x} \tag{4-2.17}$$

where it is assumed that:

$$\text{rank } [B] = m \tag{4-2.18}$$

and:

$$\text{rank } [C] = p \tag{4-2.19}$$

Using these assumptions the feedback problem can be stated as follows (Sobel and Shapiro, 1985).

Given a set of desired eigenvalues λ_i, $i = 1, \ldots, p$ and a corresponding set of eigenvectors $\underline{v}_i$, $i = 1, \ldots, p$, find a real matrix $K \in R^{m \times p}$ such that the closed-loop eigenvalues of A - BKC contain the λ_i as a subset. Furthermore, the corresponding eigenvectors are required to be close to the set of desired eigenvectors $\underline{v}_i$.

The following result due to Srinathkumar (1976, 1978) describes the freedom available to assign the eigenstructure in the above (control) problem.

Given the controllable and observable system described by the state-space triple (A, B, C) and using the assumptions that the matrices B and C are full rank, then a maximum of (m, p) closed-loop eigenvalues can be assigned and a maximum of (m, p) eigenvectors can be *partially* assigned with a minimum of (m, p) entries in each eigenvector arbitrarily chosen using constant output feedback.

For a particular problem it may be desirable to consider more than the minimum of (m, p) entries in a given eigenvector. In this case, the set of basis vectors which span the allowable subspace for the given eigenvector must be determined and a best possible achievable eigenvector assigned to the closed-loop system using, for example least-squares.

It should be noted that the freedom described above is based on output feedback when the rank of the output matrix C is full and given by p. For full state feedback the rank of the output matrix (Identity matrix) is clearly n and thus, for this case one must substitute the integer n instead of p above.

The observer design involves full state feedback and hence if (A, C) form an observable pair then n eigenvalues can be freely assigned together with n eigenvectors. The observer is driven by p independent measurements (equivalent to m controls in the *control problem*) and hence p elements can be assigned exactly in each eigenvector. The n left eigenvectors of the observer are assigned together with the n eigenvalues by assigning the eigenvectors as right eigenvectors to the dual control problem.

4.2.3 Computation of allowable subspaces

From the results due to Moore (1976) it has been seen that the choice of eigenvalue λ_i determines the allowable subspace in which any chosen eigenvector must reside. This section considers a computational method for constructing allowable eigenvector subspaces for the cases of real and complex eigenvalues. The details for the computation of subspaces for the case of real eigenvalues is given, however the reader is referred to the literature for details of the computation of allowable subspaces corresponding to the case of complex eigenvalues (Mudge and Patton, 1988).

If it is assumed that $\lambda_i \in R$, then the corresponding eigenvector $\underline{v}_i \in R^n$. Equation (4-2.15) may be rewritten as:

$$(\lambda_i I - A) \, \underline{v}_i + B K \underline{v}_i = 0 \qquad\qquad (4-2.20)$$

If $Q(\lambda_i)$ and $L(\lambda_i)$ are as defined by equations (4-2.15) and (4-2.16) respectively, then any arbitrary vector $\underline{l}_i$ which post-multiplies $L(\lambda_i)$ will give a resulting vector which lies in the null-space of $Q(\lambda_i)$. Thus:

$$[\lambda_i I - A : B] \begin{bmatrix} P(\lambda_i) \\ T(\lambda_i) \end{bmatrix} \underline{l}_i = 0 \qquad\qquad (4-2.21)$$

because $L(\lambda_i)$ is a basis for the null-space of $Q(\lambda_i)$. Expanding equation (4-2.21) yields:

$$(\lambda_i I - A) \, P(\lambda_i) \, \underline{l}_i + B \, T(\lambda_i) \, \underline{l}_i = 0 \qquad\qquad (4-2.22)$$

Comparison of equations (4-2.20) and (4-2.22) yields:

$$\underline{v}_i = P(\lambda_i) \, \underline{l}_i \qquad\qquad (4-2.23)$$

with $\underline{v}_i \in$ Span $[P(\lambda_i)]$ and $\underline{l}_i$ determining where in the allowable subspace $\underline{v}_i$ exists. It also follows from (4-2.20) and (4-2.22) that:

$$K \underline{v}_i = P(\lambda_i) \underline{l}_i$$

$$= \underline{u}_i \qquad\qquad (4\text{-}2.24)$$

which provides one method of computing the required feedback matrix K. The required null-space of $Q(\lambda_i)$ will be computed using the singular value decomposition (SVD) which is defined as follows (Golub and Van Loan, 1983):

If $Z \in R^{m \times n}$, then there exist orthogonal matrices:

$$U = [\underline{u}_1, \ldots, \underline{u}_m] \in R^{m \times m}$$

and:

$$V = [\underline{v}_1, \ldots, \underline{v}_n] \in R^{n \times n}$$

Such that:

$$Z = U \text{ diag } (\sigma_1, \ldots, \sigma_q) V^T \qquad\qquad (4\text{-}2.25)$$

where $q = \min (m, n)$ and $\sigma_1 \geqslant \sigma_2 \geqslant \ldots \geqslant \sigma_q \geqslant 0$.

This is the singular value decomposition of Z. The σ_i are the *singular values* of Z and the vectors $\underline{u}_i$ and $\underline{v}_i$ are the i-th left and the i-th right singular vectors, respectively.

Consider now the SVD of $Q(\lambda_i) \in R^{n \times (n+m)}$ defined in equation (4-2.16):

$$Q(\lambda_i) = [\lambda_i I_n - A : B]$$

$$= \begin{array}{c} U \\ \\ q \end{array} \begin{bmatrix} \sigma_1 & & & \overset{q}{:} & \overset{p}{\underline{0}} \\ & \sigma_2 & & : & \underline{0} \\ & & \ddots & : & \vdots \\ & & & \sigma_q & : & \underline{0} \end{bmatrix} V^T$$

$$= U [\text{diag}(\sigma_i) : \underline{0}] V^T \qquad\qquad (4\text{-}2.26)$$

By post-multiplying both sides of equation (4-2.26) by the orthogonal matrix V, it is noted that the last p columns of the $Q(\lambda_i)V$ product are null and so an appropriate null-space of $Q(\lambda_i)$ can be found by using the last p columns of V. For full-state feedback one must substitute the integer n instead of p. This method has the advantage that existing poles of the open-loop system or values near to them may still be chosen as eigenvalues of the closed-loop system. This would be important, for example, in the eigenvector assignment of a partially controllable system in which certain modes can be fixed or are invariant by feedback. Alternative methods which utilise the matrix $(\lambda I_n - A)^{-1}$ in the subspace calculation may thus lead to numerical problems.

The feedback gain K is determined from equation (4-2.24) as:

$$K V = U \tag{4-2.27}$$

Furthermore, if all the columns of V are linearly-independent, then it follows that:

$$K = U V^{-1} \tag{4-2.28}$$

The existence of the matrix V^{-1} is satisfied when the eigenvectors of the closed-loop system are not linearly-dependent (or even nearly so). For the degenerate case when repeated eigenvalues lead to linearly-dependent eigenvectors a chain of *generalised* eigenvectors must be generated as described in the next sub-section.

4.2.4 Multiple eigenvector assignment

Assignment of eigenvalues of multiplicity larger than m (where m is the number of inputs) is not possible according to the normal set of rules (Klein and Moore, 1977) and hence some modifications for this case are necessary. When eigenvectors of *geometric multiplicity* greater than zero are to be assigned (i.e. linearly-dependent eigenvectors) the same situation arises.

The theory of eigenstructure assignment can easily be extended by first denoting eigenvalues corresponding to linearly-dependent eigenvectors by λ_i and $\underline{V}_{i,j-1}$ for $j = 1, 2, \ldots, d_i$ (Klein and Moore, 1977). A chain of d_i *generalised* eigenvectors corresponding to the eigenvalue λ_i of the closed-loop matrix (A - B K) satisfy:

$$[\lambda_i I_n - A : B] \begin{bmatrix} P_j(\lambda_i) \\ T_j(\lambda_i) \end{bmatrix} l_i = P_{j-1}(\lambda_i) \, l_i \tag{4-2.29}$$
$$j = 1, 2, \ldots, d_i$$

By noting that:

$$\left. \begin{array}{l} \underline{v}_{i,j} = P_j(\lambda_i)\ \underline{l}_i \\ \text{and: } K\ \underline{v}_{i,j} = T_j(\lambda_i)\ \underline{l}_i \end{array} \right\} \qquad (4\text{-}2.30)$$

It follows that:

$$Q(\lambda_i) \begin{bmatrix} \underline{v}_{i,j} \\ F\ \underline{v}_{i,j} \end{bmatrix} = \underline{v}_{i,j-1} \qquad (4\text{-}2.31)$$

$$\text{for } j=1,2,\ldots,d_i$$

Thus using the SVD approach:

$$U[\text{diag}(\sigma_i) : 0]\ V^T \begin{bmatrix} \underline{v}_{i,j} \\ K\ \underline{v}_{i,j} \end{bmatrix} = \underline{v}_{i,j-1} \qquad (4\text{-}2.32)$$

By writing: $V = [V_1 : V_2]$

together with: $\Sigma = \text{diag}\ (\sigma_i)$

It then follows that:

$$\begin{bmatrix} \underline{v}_{i,j} \\ K\ \underline{v}_{i,j} \end{bmatrix} = V_1\ \Sigma^{-1}\ U^T\ \underline{v}_{i,j-1} + \text{span}\ \{V_2\} \quad (j \geqslant 1) \qquad (4\text{-}2.33)$$

where $\underline{v}_{i,o} \in \text{span}\ \{V_2\}$

Equation (4-2.33) shows how a string of *generalised* eigenvectors may be assigned by the calculation of a suitable null-space matrix V_2 via singular value decomposition.

4.2.5 Direct eigenstructure assignment

It has been shown that the eigenvector corresponding to a given eigenvalue must lie in an allowable subspace which is determined by the plant matrix A, the input matrix B and the eigenvalue itself. If information is available about a suitable weighting of the system states for each mode, it is then possible to choose a desired eigenvector specification. This will not necessarily be achievable because it may not lie within the prescribed allowable subspace. In such a case, an allowable eigenvector which is in some sense *closest* to the desired eigenvector must be selected.

For a given problem only a few components of a desired eigenvector, $\underline{v}_i^d$ are usually specified (Harvey and Stein, 1978). The remaining components may be chosen arbitrarily. To account for this a row-reordering operation is computed such that:

$$\text{RTRANS } \underline{v}_i^d = \begin{bmatrix} \underline{v}_i^\theta \\ x \\ \vdots \\ x \end{bmatrix} \tag{4-2.34}$$

Where RTRANS is a matrix which performs the row transformation operation and $\underline{v}_i$ is a subvector of specified eigenvector elements.

The desired form of equation (4-2.34) may not reside in the prescribed subspace required. If this is the case then:

$$\begin{bmatrix} \underline{v}_i^o \\ x \\ \vdots \\ x \end{bmatrix} = \text{RTRANS } P(\lambda_i) \ \underline{\varsigma}_i \tag{4-2.35}$$

$$= \begin{bmatrix} W \\ W' \end{bmatrix} \underline{\varsigma}_i \tag{4-2.36}$$

Where $\underline{v}_i^o$ is an allowable sub-vector which has components that are as close as possible to the desired form $\underline{v}_i^\theta$. A perturbation vector $\underline{\varsigma}_i$ is required corresponding to the projection of $\underline{v}_i^d$ into the allowable subspace. This is accomplished by minimising the error between the *desired* and *assigned* components in a least-squares sense:

$$\underline{\varsigma}_i = (W^TW)^{-1} \ W^T \ \underline{v}_i^\theta \tag{4-2.37}$$

For the case of complex-conjugate eigenvalues the procedure follows in a similar manner (Mudge and Patton, 1988).

4.2.6 Example of observer design using eigenstructure assignment

The eigenstructure assignment procedures described above have been based on the design of the feedback control problem given by equations (4-2.3) and (4-2.4). An example will be provided which makes use of this theory in the design of an observer.

As the eigenvectors of the continuous and discrete-time state-space equations are identical for a given system we will now go straight into the design of a discrete-time observer.

The discrete-time realisation of the observer estimation error system corresponding to a sample time ΔT is then given by:

$$\underline{e}_{n+1} = \Phi_{Cl} (\Delta T) \underline{e}_n \qquad (4-2.38)$$

where $\Phi_{Cl} (\Delta T) = \exp \{(A - K C)\Delta T\}$ \qquad\qquad (4-2.39)

The open-loop observer matrix is given by:

$$\Phi = \exp \{A \Delta T\} \qquad (4-2.40)$$

Consider the 5th order system state-space matrices A, B, C and D as follows:

$$A = \begin{bmatrix} 47.9162 & -4.0553 & 20.3064 & -159.7866 & 368.4026 \\ -9.4616 & 3.5458 & -0.7647 & -4.5779 & -97.1197 \\ -10.7088 & 4.2715 & -5.2909 & 13.4040 & -94.2993 \\ -19.8411 & 19.4585 & -4.2874 & -26.6870 & -246.7910 \\ -22.7281 & 9.0371 & -7.4134 & 37.0994 & -205.6052 \end{bmatrix} \qquad (4-2.41)$$

with the eigenvalue spectrum:

$$\{-2.765, -3.799 \pm j\ 1.46, -80.379, -95.379\}$$

together with:

$$C = \begin{bmatrix} 1 & 0 & 0 & 0 & 0 \\ 0 & 0 & 1 & 0 & 0 \\ 0 & 0 & 0 & 0 & 1 \end{bmatrix} \qquad (4-2.42)$$

and with $D = 0$ \qquad\qquad\qquad (4-2.43)

For $\Delta T = 0.001$ (1 msec.) the open-loop discrete-time observer matrix Φ is given by:

$$\Phi = \begin{bmatrix} 1.0464 & -0.0040 & 0.0197 & -0.1546 & 0.3584 \\ -0.0086 & 1.0031 & -0.0005 & -0.0055 & -0.0888 \\ -0.0101 & 0.0040 & 0.9949 & 0.0123 & -0.0884 \\ -0.0174 & 0.0182 & -0.0036 & 0.9708 & -0.2236 \\ -0.0214 & 0.0085 & -0.0070 & 0.0346 & 0.8062 \end{bmatrix} \qquad (4-2.44)$$

The closed-loop eigenvalues are *required* to be:

[0, 0, 0, 0.8869, 0.9094]

The choice of these numerical values has arisen from the sampling process. The equivalent continuous system eigenvalues are three at minus infinity together with real eigenvalues of -120 and -95, respectively. The eigenvectors corresponding to the zero-valued eigenvalues are required to be independent. Note that this requirement means that, although there are p repeated eigenvalues (where p = 3) it is not necessary to use a computation involving generalised eigenvectors. For this problem it also important that the left eigenvectors corresponding to the zero-valued eigenvalues of the observer should be as orthogonal as possible. The following eigenvectors corresponding to the zero-valued eigenvalues are required to be assigned as *left* eigenvectors of the observer matrix Φ_{Cl}.

$$v_1^T = [1.0 \quad 0 \quad 0 \quad 0.159 \quad 0]$$

$$v_2^T = [0 \quad 0 \quad 1.0 \quad -0.012 \quad 0] \qquad (4\text{-}2.45)$$

$$v_3^T = [0 \quad 0 \quad 0 \quad -0.035 \quad 1.0]$$

The remaining two eigenvectors corresponding to the required closed-loop eigenvalues [0.8869, 0.9094] can be assigned but are allowed to take on any directions as long as the eigenvalues together with v_1^T, v_2^T, v_3^T are assigned correctly.

In order to perform the assignment for the observer it is necessary to assign the right eigenvectors $\underline{v}_1$, $\underline{v}_2$, and $\underline{v}_3$ to the dual control problem i.e. to the discrete-time closed-loop matrix Φ_{Cl}^T. The matrix C^T becomes the B matrix for the dual control problem.

The assignment of the above structure to the dual system yielded the following achievable eigenvectors:

$$\lambda_i \rightarrow \quad 0 \qquad 0 \qquad 0 \qquad 0.8869 \quad 0.9094$$

$$V = \begin{bmatrix} 1.000 & 0.002 & 0.006 & 0.516 & 0.521 \\ 0.001 & -0.004 & -0.008 & -0.128 & -0.221 \\ 0.002 & 1.000 & -0.005 & -0.046 & 0.226 \\ 0.159 & -0.012 & -0.035 & 1.000 & 1.000 \\ 0.006 & 0.000 & 1.000 & -0.125 & 0.438 \end{bmatrix} \qquad (4\text{-}2.46)$$

$$\underline{v}_1 \qquad \underline{v}_2 \qquad \underline{v}_3 \qquad \underline{v}_4 \qquad \underline{v}_5$$

The *dual control* system feedback gain matrix is computed as:

$$K^T = \begin{bmatrix} 1.125 & 0.055 & -0.016 & -0.509 & -0.038 \\ -0.057 & 2.917 & 1.012 & 0.456 & 0.032 \\ 0.196 & 5.954 & -0.053 & 0.758 & 0.889 \end{bmatrix} \qquad (4\text{-}2.47)$$

The closed-loop *dual control* system matrix is:

$$\Phi_{Cl}^T = \begin{bmatrix} -0.078 & -0.064 & 0.006 & 0.492 & 0.017 \\ -0.004 & 1.003 & 0.004 & 0.018 & 0.009 \\ 0.076 & -2.922 & -0.017 & -0.459 & -0.039 \\ -0.155 & -0.006 & 0.012 & 0.971 & 0.035 \\ 0.163 & -6.043 & -0.035 & -0.981 & -0.083 \end{bmatrix} \qquad (4\text{-}2.48)$$

The corresponding *observer* closed-loop matrix Φ_{Cl} is thus given by:

$$\Phi_{Cl} = \Phi - KC \qquad (4\text{-}2.49)$$

$$\Phi_{Cl} = \begin{bmatrix} -0.078 & -0.004 & 0.076 & -0.155 & 0.163 \\ -0.064 & 1.003 & -2.922 & -0.006 & -6.043 \\ 0.006 & 0.004 & -0.017 & 0.012 & -0.035 \\ 0.492 & 0.018 & -0.459 & 0.971 & -0.981 \\ 0.017 & 0.009 & -0.039 & 0.035 & -0.083 \end{bmatrix} \qquad (4\text{-}2.50)$$

The left eigenvectors of the observer are given by (4-2.46) by transposing the matrix V. This example shows the design of a discrete-time observer with *three* measurement inputs. The assignment of zero-valued eigenvalues forms the basis of the design of a *dead-beat* observer. The dead-beat design problem will be referred to later in this chapter in the context of robust fault detection.

4.3 OBSERVER ROBUSTNESS ANALYSIS

4.3.1 Problem specification and robustness definition

The assumption made before designing an observer is that the process to which the observer is to be applied can be represented by a linear model; this is a poor assumption for most real applications. It is instructive to consider a representation of a real system which is subject to parameter variations and disturbances. In this way we can examine the effects of unmodelled disturbances acting on the observer and investigate under what conditions the observer can be made robust against unknown inputs.

Let the complete dynamics be represented by:

$$\dot{x} = \begin{bmatrix} x_o \\ --- \\ x_u \end{bmatrix} = \begin{bmatrix} A_o + \Delta A & A_c \\ ------------ \\ A_{u1} & A_{u2} \end{bmatrix} \begin{bmatrix} x_o \\ -- \\ x_u \end{bmatrix} + \begin{bmatrix} B_o + \Delta B \\ -------- \\ B_u \end{bmatrix} u + \begin{bmatrix} \gamma_o \\ -- \\ \gamma_u \end{bmatrix} \omega$$

$$y = [C_o \; C_u] \; [x_o \; x_u]^T$$

(4-3.1)

The matrices A_o, B_o form the linearised plant to be modelled in the observer. ΔA, A_c, A_{u1}, A_{u2}, ΔB, and B_u represent the parameter variations and plant uncertainty which describe the time-varying dynamics of the plant, the subscript u denoting unmodelled effects. The subscript c denotes coupling to unmodelled dynamics. γ_o and γ_u are the system disturbance distribution matrices, whilst ω is the disturbance vector. C_o is the pxn output matrix of the linearised plant.

The nominal linear system (A_o, B_o, C_o) is to be used in a linear observer with feedback gain K. The full-order form of the state observer is used as the linear (nominal) model is *assumed* to replicate the actual process. In practice, of course, there are modelling uncertainties.

When the observer is applied to the nominal system (denoted by the o subscripts) then the state estimate error is dependent only upon the initial error between the actual state and the observer estimate. The estimation error state equation corresponding to this idealised case has been given in Section 4.2. Equation (4-2.5) can now be extended to account for the effects of uncertainties and disturbance terms. When driven by the actual plant outputs, the estimation error dynamics can be represented as follows:

$$\dot{\underline{e}} = (A_o - KC_o)\underline{e} + \underbrace{[\Delta A | \Delta B | A_c - KC_u | D_o]}_{E} \underbrace{[\underline{x}_o | \underline{u} | \underline{x}_u | \underline{v}]^T}_{\underline{d}^T} \qquad (4\text{-}3.2)$$

where $\underline{e} = \underline{x}_0 - \hat{\underline{x}}_0 \in R^n$, and K is the $n \times p$ observer gain matrix. A_o is an $n \times n$ matrix and C_o is a $p \times n$ matrix. The estimation error equation now has an extra input vector which will have the effect of driving the error away from its zero steady-state during disturbances and plant parameter variations. The observer error dynamics may now be completely described by:

$$\dot{\underline{e}} = (A_o - KC_o)\underline{e} + E\underline{d} = D_o\underline{e} + E\underline{d} \qquad (4\text{-}3.3)$$

Where $\underline{d}$ is an *unknown* disturbance input vector and E is, in general, an *unknown distribution matrix*. It is important to use (at the design stage) frequency domain or sensitivity information about the disturbances acting on the observer in an attempt to make the estimation error insensitive to the effects of the unmodelled dynamics. It is often the case that the frequency bandwidth of the disturbance is known approximately. This information can be put to good use in designing an observer which is insensitive to these unknown input signals. Certain approaches can be applied to the design of a robust (disturbance-discriminating) observer in an attempt to decouple the effects of the disturbance term $E\underline{d}$.

It is important to note that, with the exception of very few special cases, neither the disturbance distribution matrix E nor the disturbance input vector $\underline{d}$ are known. In some special cases the directions defined by the columns of the matrix E may be constant and this case will be referred to as the case of *structured uncertainty*.

Whether the uncertain terms are structured or not a mechanism of disturbance decoupling is still required if the observer is to be applied to real non-linear and stochastic processes. Sufficient disturbance decoupling will thus permit accurate reconstruction of the state vector $\underline{x}$ even when the characteristics of the disturbance inputs are not known.

An observer which provides an accurate reconstruction of the state vector despite the effects of unknown inputs is termed a *robust observer* (UIO) in the literature (see Chapter 3 for further background details). Alternatively the slightly confusing term *unknown input observer* has been used by many authors since the pioneering work of Basille and Marro (1969). Work on this topic includes research by the following investigators: Mita (1976, 1977), Sundareswaran et al. (1977), Bhattacharya (1978), Wang and Davison (1978), Schumacher (1980), Kobayashi and Nakamizo (1982), Fairman et al. (1984), Viswanadham and Srichander (1987), Ge and Fang (1988) and Wünnenberg and Frank (1988) (see also Chapter 3 of this book).

Observers may have a low sensitivity to unknown inputs of a certain class or having prescribed structure, however, the defined sensitivity under consideration may increase suddenly according to varied operating conditions. In this case the observer is *locally insensitive* to disturbances and modelling errors but is not necessarily robust. In order to clarify the concept of the truly robust observer it is necessary to quote the following important definitions of parameter sensitivity and robustness according to Frank (1983):

Definition 1 (Parameter Sensitivity)

A system function (or system) is said to be sensitive to parameter variations if a sensitivity measure S is unlike 0. In the special case $S=0$ the system function (or system) is said to be *zero-sensitive*. If a measure of S is small, the system function (or system) is termed insensitive or having low sensitivity.

Definition 2 (Robustness)

A system is called robust if the system property of interest remains in a bounded region in the face of a class of finite bounded perturbations.

The essential difference between the above two definitions may be summarised as follows. Whereas sensitivity characterises the tendency of a system defined as a fixed quantity at nominal parameter values, robustness implies invariance of a system property defined on a set or region under a class of finite structural or parameter changes. Thus robustness is a *global* property of a system and sensitivity can be interpreted as robustness in the specific case of infinitessimal parameter variations. One must be careful therefore to clarify the context in which the robustness property applies. This clarification has hitherto been lacking in much of the literature on the robust observer.

It should now be clear that an observer is as *robust* as possible in the presence of unknown inputs if the time response of the estimation error remains either completely or almost completely decoupled from disturbance terms E$\underline{d}$ (referring to equation (4-3.3)). The decoupling must be invariant to all known operating conditions of the plant. A robust observer thus remains insensitive to unknown inputs. The presence of a sensor or actuator fault or generated system malfunction will, in turn, lead to a malfunction of the observer state estimation mechanism as the observer is not applied to a normally functioning plant. This distinction between unknown inputs, disturbances or parameter variation effects and actual faults is, of course, the basis of the robust observer approach to fault detection and isolation.

Robust observers can be designed in a number of ways. Most approaches are derived from the *Principle of Invariance* provided originally by Rozonoer (1963, (a) + (b)) and extended to the so-called {A,B}-invariance initially referred to by Basille and Marro (1969).

The in-built parameter insensitivity that a robust observer possesses can be understood in terms of controllability and observability concepts as discussed by Wang (1965), Cruz and Perkins (1966), Tomovic and Vukobratovic (1972), Wonham (1974) and Mita (1976) - to name only a few who have mentioned this important topic in the literature. The chapter will later show that by defining a vector sensitivity measure $\underline{S}$ for an observer the actual sensitivity to unknown inputs can be minimised by designing either uncontrollable or unobservable subspaces. The measure $\underline{S}$ can be used for fault detection and isolation. The next sub-section provides a tutorial introduction to the essential background behind the theory of invariant subspaces. Section 4.3.3 then specialises this theory within the context of robust observer design as a pre-cursor to the robust fault detection problem which is described in section 4.4.

4.3.2 Introduction to Robustness Analysis using invariant subspace theory

The problem of invariance consists of determining the conditions under which external disturbances do not effect the systems controlled quantities. Rozonoer (1963 (a) + (b)) used variational techniques to force the variations in the outputs due to unknown disturbances to vanish (or be zero). He introduced the concept of *complete invariance* which requires the invariancy of the system throughout the phase space in a manner which is independent of the system coordinate and time. Wang (1965) re-expressed the concepts of invariance in terms of structural properties of linear systems and generalised the results to multivariable systems. He expressed strong invariance as *output uncontrollability*. This can be easily derived using a frequency domain approach and forcing the numerator of the transfer function to be zero. Consider, for example, the case of the generalised state-space system:

$$\left.\begin{array}{c} \dot{x} = Ax + Bu + y\omega \\[2mm] y = Cx + Du \end{array}\right\} \qquad (4\text{-}3.4)$$

where:

$x \in R^n$ is the state vector
$\omega \in R^r$ is an *unknown* disturbance vector
$y \in R^p$ is the measurement vector

A state feedback control law can be defined as:

$$u = -Kx \qquad\qquad (4\text{-}3.5)$$

such that (4-3.4) can now be re-written as:

$$\left.\begin{array}{c} \dot{x} = Fx + y\omega \\[2mm] y = Cx + Du \end{array}\right\} \qquad (4\text{-}3.6)$$

where:

$$F = (A - BK) \qquad\qquad (4\text{-}3.7)$$

The Laplace-transformed transfer relationship between the outputs y and the disturbances ω is given by:

$$y(s) = C(sI - F)^{-1} y\, \omega(s) - DK\, x(s) \qquad (4\text{-}3.8)$$

$$y(s) = G(s) \ \underline{\omega}(s) - DK \ \underline{x}(s) \tag{4-3.9}$$

which can be re-written as:

$$y(s) = \sum_{i=1}^{\infty} g_i \ s^{-i} \ \omega(s) - DK \ \underline{x}(s) \tag{4-3.10}$$

for which it can be shown that:

$$g_i = CF^{i-1} \ y \quad , \quad i = 1, \ 2, \ \ldots, \ \infty \tag{4-3.11}$$

Equation (4-3.9) can be re-written as

$$y(s) = C \ y + CF \ y \ s^{-1} + CF^2 \ y \ s^{-2} +$$

$$\ldots \ - DK \ \underline{x}(s) \tag{4-3.12}$$

The transfer function numerator terms related to the unknown inputs $\underline{\omega}$ will vanish *iff*:

$$CF^i \ y = 0 \quad , \quad i = 0, \ 1, \ 2, \ \ldots, \ n-1 \tag{4-3.13}$$

which is the condition for complete (or strong) invariance i.e. so that the outputs y become completely decoupled from the inputs $\underline{\omega}$.

Cruz and Perkins (1966) re-expressed the concept of invariance in terms of insensitivity to disturbance and parameter variation. They interpreted *strong invariance* as defining an equivalence class of signal-invariant systems, some of which do not satisfy the conditions of parameter insensitivity.

Under the theory of invariant subspaces, Basille and Marro (1969) introduced the concepts of (A, B) - *invariant subspaces contained in the nullspace of C* denoted by N(C). They applied the concept in the synthesis of an unknown input observer.

The most general approach to the disturbance decoupling problem is obtained using equations (4-3.11) and (4-3.12), under the assumption of structured uncertainty referred to in section (4.3.1). This has been described above under *the principle of invariance* as an approach accomplishing complete invariance. One of the authors who has discussed disturbance decoupling using the frequency domain is Bhattacharyya (1978). An equivalent system for disturbance decoupling using output feedback has been discussed by Schumacher (1980) using the theory of invariant subspaces, in particular using (C,A,B)-pairs. The (C,A,B)-pairs is an extension of the so called (A,B)- and (C,A)-invariant subspace first introduced by Basille and Marro (1969).

It should be assumed that feedback is applied to achieve disturbance decoupling using a suitable feedback matrix K resulting in the closed-loop matrix F, given in (4-3.7).

In many cases, the disturbance matrix y is not known and cannot be estimated with any reasonable accuracy. When such unstructured uncertainties exist in a plant, the approach must use the available system and output map structures at the design stage. The necessary condition for disturbance decoupling is thus:

$$CF^i = 0 \quad , \quad i = 1, 2, \ldots, n \qquad (4\text{-}3.14)$$

This has the effect of nulling the outputs by using feedback to achieve *unobservability*. A similar but special case occurs in the discrete-time domain in dead-beat feedback systems. This has the property that the outputs become zero within at most n-steps and then have zero steady-state error.

By considering the case of a single input disturbance, the above implies that disturbances emanating from the subspace spanned by n-1 columns of the disturbance controllability matrix are nulled. The exception is the direct transmission term corresponding to the single column of the input matrix thus making $C \, y \neq 0$ in equation (4-3.12). The direct transmission term does not satisfy the unobservability criterion and hence is the only term that can give rise to non-zero outputs. This is the only way in which the unknown input can affect the outputs.

When the system under consideration has *structured uncertainty* i.e. the directions of the columns of the disturbance matrix y are known, then the structure of y can be used at the design stage in order to obtain a *zero sensitive* closed-loop system. This can be achieved by constraining the disturbance controllability subspace to lie in the nullspace of C. This can be expressed in terms of the so called maximal (A,B)-invariant subspace in N(C) (Fessas (1979)). Other authors who have used the (A,B)- and (C,A)-invariance principle in the synthesis of unknown input observers are: Basille and Marro (1969), Mita (1977), Bhattacharyya (1978), Schumacher (1980), Kobayashi and Nakamizo (1982) and Fairman et al. (1984). Wünnenberg and Frank (1988) have also used the unknown input *dead-beat* observer.

The concepts of invariance and (A,B)- and (C,A)-invariance are related to the definition of system zeros and the *output zeroing* principle according to the following background literature. Silverman and Payne (1971) introduced the concept of *zero output systems*. Bengtsson (1974) gave a definition of zeros as the poles of the maximal (A,B)-invariant subspace in N(C). Later Kouvaritakis and MacFarlane (1976), Karcanias and Kouvaritakis (1978, 1979) studied the *output zeroing problem* and its relation to (A,B)-invariant subspaces as well as the associated characteristic frequencies transmitted by zero output systems.

Having examined briefly the invariance properties of disturbance decoupled state feedback control systems our attention can now turn to the *robust observer problem*. The next section uses the above theory to develop the properties of the robust observer based on invariant subspaces. The robust observer so defined will then be used for fault diagnosis purposes as a mechanism for discriminating between disturbances and faults and as a monitor to locate sensor or actuator faults.

4.3.3 The Robust Observer Using Invariant Subspaces

In order to examine the invariant conditions applicable to an observer it is necessary to consider the complete estimation error dynamics given by equation (4-3.3) and re-written in the form:

$$\dot{\underline{e}} = A_o\underline{e} - K\,\underline{e}_y + E\,\underline{d} \qquad\qquad (4\text{-}3.15)$$

where $\underline{e}_y$ is the measured estimation error and can be looked upon as a control signal which drives the error state towards a desired invariant subspace. In general, any well defined manifold or invariant subspace will be reached (if reachable) through asymptotic dynamic behaviour.

Using as a basis for the approach the (C,A)-invariant subspace theory, a subspace is required that is orthogonal to the columns of the observer closed-loop matrix $F_o = (A_o - KC_o)$ in order to achieve a special case of the *output zeroing problem* (Karcanias and Kouvaritakis, 1979). This is made possible by using a left annihilator, H of the closed-loop matrix as follows:

Define a constant matrix $H \in R^{p \times n}$ of full rank p (where p is the number of observer measurements inputs). Furthermore, let $\underline{S}(t)$ be a vector function of time t such that:

$$\underline{S}(t) = H \, \underline{e}(t) \quad , \quad \text{for all time t} \tag{4-3.16}$$

The required nullspace or *invariant manifold* will be reached *iff* the estimation error $\underline{e}(t)$ satisfies the following intersection of p hyperplanes in n-dimensional space S:

$$S = \bigcap_{j=1}^{p} s_j = (\underline{e} : H \, \underline{e} = \underline{0}) \tag{4-3.17}$$

where the subscript (t) has been discarded for notational convenience.

The subspace S is the nullspace of H denoted by N(H). Thus if the null-space manifold is reached, the system phase trajectories satisfy the equation:

$$\underline{S} = H \, \underline{e} = 0 \tag{4-3.18}$$

This may be interpreted in terms of pole-zero cancellation by a weighting matrix thus driving the system to become unobservable. The unobservability achieved becomes clear by looking at the observability matrix θ of the error dynamics:

$$\theta = [\, H^T \quad (HF)^T \quad (HF^2)^T \quad \ldots \quad (HF^{n-1})^T \,]^T \tag{4-3.19}$$

Following on from the discussion above, if the rows of H are *left eigenvectors* corresponding to *zero-valued* eigenvalues it then follows that:

$$HF = 0$$

and also that:

$$HF^i = 0 \quad , \quad i = 1, 2, \ldots, n-1 \tag{4-3.20}$$

and

$$\text{rank } (\theta) = \text{rank } (H) = p < n \tag{4-3.21}$$

Thus when equation (4-3.18) is satisfied, certain poles of the impulse response of the observer are cancelled by zeros (decoupling zeros) only if the vector $\underline{S}$ is used as an output measure. In particular, if the rows of the matrix H are *left* eigenvectors of the closed-loop observer matrix then, as rank (H) = p, each element of the vector $\underline{S}$ will constitute an output (in the zero-input or free response case) which has one mode - the mode corresponding to the particular row of $\underline{S}$.

The reason for the degree of *decoupling* becomes clear on considering that a particular left eigenvector is always *orthogonal* to the right eigenvectors of the remaining eigenvalues (modes). The proof for this is well known but is given here for completeness.

Proof:

Consider the closed-loop observer system matrix:

$$F_o = A_o - KC_o$$

Now, for left eigenvectors $\underline{v}_i^T$ and right eigenvectors $\underline{m}_i$ of F_o, where $i = 1, 2, \ldots, n$, then:

$$\underline{v}_i^T F_o = \lambda_i \underline{v}_i^T \tag{4-3.22}$$

Post-multiplying both sides of equation (4-3.22) by $\underline{m}_j$

$$\underline{v}_i^T F_o \underline{m}_j = \lambda_i \underline{v}_i^T \underline{m}_j \tag{4-3.23}$$

But, as

$$F_o \underline{m}_j = \lambda_j \underline{m}_j \tag{4-3.24}$$

equation (4-3.24) can be rewritten as:

$$\lambda_j \underline{v}_i^T \underline{m}_j = \lambda_i \underline{v}_i^T \underline{m}_j \tag{4-3.25}$$

Hence, if $\lambda_i \neq \lambda_j$, the only solution to the above equation is the trivial solution and it thus follows that:

$$\underline{v}_i^T \underline{m}_j = 0, \qquad \text{for } i \neq j \tag{4-3.26}$$

i.e. the left and right eigenvectors corresponding to mutually distinct eigenvalues are orthogonal. This completes the proof.

The physical significance of the pole-zero cancellation

occurring if the p rows of H are given by p independent left eigenvectors is that in each element of the measure s_i, for i = 1, 2, ..., p n-1 modes become unobservable. As the complete response of the output vector $\underline{S}$ is given by the convolution of the impulse response with the disturbance terms in E $\underline{d}$ this has an important consequence on the sensitivity to unknown inputs.

In particular, when the motion of the estimation error vector $\underline{e}$ lies within the null-space manifold i.e. for $\underline{e} \in N(H)$ the measure $\underline{S}$ is only dependent on one mode for each s_i. In general these modes may be distinct or may have algebraic multiplicity.

It follows further that, if the modes corresponding to the rows of H are *small in magnitude* with negative real parts then the measure $\underline{S}$ will have *weak sensitivity* to the disturbance E $\underline{d}$. In the limit as the bandwidth of these range space modes tend to zero, the sensitivity of E $\underline{d}$ also tends to zero whilst the motion is *within* or *close to* the null space manifold.

In this way an (H, F_O)-invariant subspace can be created. Furthermore, if the structure of the disturbance distribution matrix E is known and is fixed (corresponding to the case of structured uncertainty) then the matrix H can be designed so that:

$$H E = 0 \qquad\qquad (4\text{-}3.27)$$

For this special case complete disturbance-decoupling is achieved and the invariant subspace becomes the *supremal invariant* subspace in the null space of H. An additional special case arises when:

$$F_O E = 0 \qquad\qquad (4\text{-}3.28)$$

i.e. the columns of E are the *right* eigenvectors of the closed-loop observer matrix F_O. This is a special case of the (A,B)-invariant subspace (or (F_O, E)-invariant subspace in our notation). This, of course occurs when the columns of E are independent eigenvectors corresponding to zero-valued eigenvalues of F_O. The (F_O, E)-invariant subspace is unrealistic for a continuous system design as it involves the assignment of r zero-valued eigenvalues to F_O (assuming E has maximum column rank r). The use of zero-valued eigenvalues for continuous system design is unrealistic as it leads to the implementation of a non-asymptotically stable system. Furthermore, numerical sensitivity problems may cause the assigned eigenvalues to be small but positive thus giving an unstable system. However, this latter approach is feasible in the discrete time and a discussion of this has been given in Chapter 3 of this book.

In this work attention is drawn to the (C,A)-invariant subspace or *output zeroing* approach. In consistency with our notation for the observer we will now use the term (H, F_o)-invariancy to avoid possible confusion.

The matrix H provides a weighting of the modal behaviour of the observer giving rise to a zero-valued vector $\underline{S}$, in the ideal case. H can also be thought of as an output weighting matrix and the vector $\underline{S}$ is a sensitivity measure, and thus a *sensitivity function*.

The idea of the measure $\underline{S}$ being a sensitivity function is embodied in the robustness problem for fault detection. If the function $\underline{S}$ remains insensitive to the normal operation of a plant it can be made sensitive to system faults e.g. sensor faults. If the parametric insensitivity is maintained throughout the range of operation of the process plant then the design can be said to be *robust*.

Having made some important definitions we are now in a position to examine the structure of a robust observer design in closer detail and then show how a robust observer can be designed for fault detection applications.

The motion near the invariant subspace has a low sensitivity to parameter variations occurring in the range space of H. Once the actual N(H) manifold is reached the resulting motion is insensitive to disturbance and parameter variations occurring in the measured plant. However, should there be a fault e.g. due to a sensor mal-function then the measure $\underline{S}$ is also *locally insensitive* to both the fault and the disturbances. The motion is, however, driven away from the null-space hyperplane intersection but, if the disturbance, or fault signals are bounded (for example by the thresholds) then the motion will return again towards N(H). It is not necessary, of course, for the action of a fault to be bounded but it is essential for response to a disturbance to be bounded from above by a suitably chosen threshold.

Consider now the time derivative of $\underline{S}$, by differentiating equation (4-3.16):

$$\dot{S} = H \, \dot{\underline{e}} = 0 \tag{4-3.29}$$

On combining equations (4-3.29) and (4-3.15) and using $\underline{e}_{yeq}$ to denote the *equivalent control* vector which would be effective in maintaining the response of the weighted estimation error $H \, \underline{e}(t)$ within $N(H)$, it follows that:

$$H(A_o \, \underline{e} - K \, \underline{e}_{yeq} + E \, \underline{d}) = 0 \tag{4-3.30}$$

If the matrix $H \, K$ is non-singular i.e. rank $(H \, K) = p$, then $\underline{e}_{yeq}$ can be determined from equation (4-3.30) to satisfy the condition:

$$H \, N = 0 \tag{4-3.31}$$

where:

$$N = (I - K(HK)^{-1} \, H) \tag{4-3.32}$$

The nullspace of H is of order $n-p$ *iff* rank $(H) = p$ and furthermore rank $(N) = n-p$ (by the dimension theorem). Thus the $n \times n$ matrix satisfying (4-3.31) and (4-3.32) has p zero-valued eigenvalues.

As has been shown in Section 4.2 the observer feedback gain K can be designed using the freedom available in the eigenstructure assignment of the closed-loop matrix F_o.

The matrix H can, in principle, be any full rank matrix satisfying equations (4-3.31) and (4-3.32).

On satisfying these equations the *idealised* or *equivalent* control vector $\underline{e}_{yeq}$ is given by:

$$\underline{e}_{yeq} = (HK)^{-1} \, H[A_o \, \underline{e} + E \, \underline{d}] \tag{4-3.33}$$

From equation (4-3.31) it also follows that:

$$\dot{\underline{e}} = (I_n - K(HK)^{-1} \, H) \, [A_o \, \underline{e} + E \, \underline{d}] \tag{4-3.34}$$

$$= N \, [A_o \, \underline{e} + E \, \underline{d}] \tag{4-3.35}$$

which, for convenience of notation can be re-written as:

$$\dot{\underline{e}} = F_{eq} \, \underline{e} + E_{eq} \, \underline{d} \tag{4-3.36}$$

The matrices F_{eq} and E_{eq} are termed *equivalent* system and disturbance matrices, respectively.

From (4-3.29) it can be seen that:

$$H \dot{\underline{e}} = \underline{0} \qquad (4-3.37)$$

together with:

$$H F_{eq} = \underline{0} \qquad (4-3.38)$$

and:

$$H E_{eq} = \underline{0} \qquad (4-3.39)$$

Thus the equivalent control signal in the observer operates to maintain the weighted error response $H \underline{e}$ (t) within the invariant manifold.

In general, for a full rank matrix H it can be seen that a non-linear form of feedback structure is required to achieve the closed-loop equivalent system F_{eq}. It is clear that, for:

$$H(A_0 - K C_0) = H F_0 = H F_{eq} = 0 \qquad (4-3.40)$$

i.e. for the system matrices F_0 and F_{eq} to be identical, the n-p independent columns of both F_0 and F_{eq} must span N(H) otherwise the trivial solution $A_0 = K C_0$ will result. The non-trivial solution to equation (4-3.40) is only possible for linear feedback design when the rows of H are the p independent *left* eigenvectors corresponding to p zero-valued eigenvalues, as, for this case:

$$\underline{h}_i^T F_0 = \underline{h}_i^T F_{eq} = \lambda_i \underline{h}_i^T = 0 \qquad (4-3.41)$$

$$\text{for } i = 1,2,\ldots,p$$

and where $\lambda_i = 0$.   (4-3.42)

Thus, by using equation (4-3.26) the right eigenvectors $\underline{m}_k$ for k = p+1,...,n corresponding to n-p *non-zero* valued eigenvalues are orthogonal to the rows of H F, such that:

$$H \underline{m}_k = \underline{0} , \quad k = p+1,\ldots,n \qquad (4-3.43)$$

It is clear that the above n-p eigenvectors $\underline{m}_k$ are the right eigenvectors of *both* F_0 and F_{eq}. These n-p eigenvectors are thus a *basis set* for the N(H) subspace. The corresponding null-space eigenvalues are termed the *invariant zeros* (McFarlane and Karcanias (1976)) of the observer system; the null-space eigenvectors are the $n-p$ *zero state directions*.

Now, turning again to the disturbance-decoupling problem it follows from equations (4-3.30) and (4.3.31) that:

$$H \ N \ E = H \ E_{eq} = 0. \tag{4-3.44}$$

Thus complete decoupling of disturbances occurs whilst the motion lies in the invariant subspace $N(H)$. It is important to note that, as the product $H \ N$ is a $p \times n$ null matrix all disturbances regardless of their independence are decoupled from the response of $H \ \underline{e}(t)$ *iff* $\underline{S} = 0$. This is, of course, the definition of a (C,A)-invariant or (H,F_o)-invariant subspace which is independent of inputs and thus exhibits a free rectilinear time response of the form:

$$\underline{e}(t) = \sum_{k=p+1}^{n} \alpha_k \ \underline{m}_k \ e^{-\lambda_k t} \tag{4-3.45}$$

The scalar parameters α_k arise from the *residues* of the response, whilst the $\underline{m}_k$ are the *null-space* eigenvectors corresponding to the eigenvalues λ_k. It can thus be said that, if the observer system defined above has p zero-valued eigenvalues, the system behaves as a lower dimensional system of order $n-p$ as long as the motion stays within the manifold of $N(H)$. Outside $N(H)$ the system takes on full-order n.

The insensitivity to disturbance in the range space dynamics is also of interest. From equation (4-3.35) it follows that disturbance decoupling can also be achieved in the range space motion of H, $R(H)$ *iff*:

$$N \ E = 0 \tag{4-3.46}$$

Now rank $(N) = n-p$ and hence the null-space of N i.e. $N(N)$ is of order p which means that, if E is known and is of full column rank r then a *necessary* condition for disturbance decoupling is:

$$r \leqslant p \tag{4-3.47}$$

This implies that the number of independent disturbance signals must not be greater than the number of independent measurements of the system.

There are very few situations where the disturbance matrix E is known apart from the case of structured uncertainty, however it is of interest to note that when perfect disturbance decoupling in the observer is achievable equation (4-3.46) can easily constitute an additional eigenvalue-eigenvector problem in which all the independent columns of E are right eigenvectors of the matrix N corresponding to zero valued eigenvalues.

In order to illustrate the insensitivity to disturbances an example is given that uses the maximal controllable subspace of the disturbance matrix contained in N(C) (Mita, 1976).

Consider the uncertain dynamic system given by equation (4-3.1). If it can be assumed that the uncertainty is structured such that the disturbance matrix has known columns, then disturbance decoupling is achievable through the use of the so called unknown input observer expressed as follows:

$$\dot{\underline{z}} = F_o \underline{z} + K \underline{y} + B_o \underline{u} + E \underline{d} \qquad (4\text{-}3.48)$$

where $\underline{z} = T \hat{\underline{x}}$ and $\hat{\underline{x}}$ is the estimated state vector which is not necessarily the same order as the state vector $\underline{x}$. The subscript o relates to the nominal linear system as discussed in Section 4-3.1. The estimated state vector can be expressed as:

$$\hat{\underline{x}} = M_1 \underline{z} + M_2 \underline{y} \qquad (4\text{-}3.49)$$

In order for the estimates $\hat{\underline{x}}$ to converge asymptotically to the system states $\underline{x}$, it is required that the observer matrices satisfy:

$$\hat{F}_o T = T(A_o - K C_o) = T F_o \qquad (4\text{-}3.50)$$

$$\hat{B}_o = T B_o \qquad (4\text{-}3.51)$$

$$\hat{K} = T K \qquad (4\text{-}3.52)$$

The above constitute the well known Luenberger design equations (Luenberger, 1971). For disturbance-decoupling it is necessary for the additional equation to be satisfied as discussed in Chapter 3:

$$\hat{E} = T E = 0 \qquad (4\text{-}3.53)$$

The above design equations constitute the most general approach to disturbance decoupling when the matrix E is known. However, in most cases as equations (4-3.1) and (4-3.2) show, the uncertain terms which make up the columns of E are difficult to determine. It should be noted that the transformation matrix T should only be used as a step in simplifying the decoupling design. It may, of course, be the identity matrix in which case the observer equations become t hose used in Section 4.2.6. When (4-3.53) is satisfied the observer is known equivalently as a *zero sensitive observer* or *unknown input observer* (UIO) as discussed by Mita (1976, 1977) and others (see for example Chapter 3). By designing a suitable matrix, K, the above conditions can be re-written equivalently as (Mita, 1977):

$$P \left\{ E \quad F_o E \quad F_o^2 E \quad \ldots \quad F_o^{n-1} E \right\} = \underline{0} \qquad (4\text{-}3.54)$$

for some matrix P. Equation (4-3.54) can be re-expressed as:

$$P \ F_o^i = 0, \qquad i = 0, 1, \ldots, n\text{-}1 \qquad (4\text{-}3.55)$$

and thus the subspace spanned by the rows of P can be described as a (P, F_o)-invariant subspace with the columns of E being contained in the nullspace of P. When the rows of P are formed by the left eigenvectors of F_o corresponding to zero valued eigenvalues ($P \equiv H$ as discussed early in this section), it can thus be described as an (H, F_o)-invariant subspace as discussed in this chapter.

4.4 ROBUST FAULT DETECTION USING EIGENSTRUCTURE ASSIGNMENT

4.4.1 Introductory background

Fault detection and isolation (FDI) has gained widespread interest in the research community since the pioneering work of Beard (1971) and Jones (1973). New techniques have been developed to increase the reliability and performance of electronic systems beyond the level offered by the classical method of hardware redundancy. Some of these techniques have been developed using analytical redundancy. The concept of analytical redundancy is based on combining groups of sensors into algorithms of known physical relations which facilitate fault detection. All fault detection systems are based on exploiting some form of analytical redundancy and the concept has been thoroughly discussed in a number of chapters of this book.

Many FDI techniques use *parity functions* to produce *residuals*. A parity function is an algebraic relationship involving the observed measurements from the system such that the measurement noise is neglected. The residual is the value of a parity function which is (in the ideal case) *zero* in the *absence* of a fault and *nonzero* in the *presence* of a fault. The latter nonzero value is called a *fault signature* and is generally a function of the failure magnitude, time of occurrence of the fault and the type of the fault. The simplest parity function occurs when two instruments are used to measure the same quantity, with values y_1 and y_2. The scalar parity function is then given by:

$$r(t) = y_1(t) - y_2(t) \tag{4-4.1}$$

Since both measurements are usually corrupted by additive random noise (assumed to affect both sensors equally) the parity function itself is a random process and the expectation of r is zero when neither of the signals is biased by a fault. In more general terms, a vector of linear parity functions is expressed in the form:

$$\underline{r}(t) = Z\,\underline{y}(t) \tag{4-4.2}$$

where $\underline{y}(t)$ is the measurement vector and Z is a coefficient matrix.

The above definitions can be applied to the following discrete time state-space system:

$$\left. \begin{array}{l} \underline{x}(k+1) = \Phi\,\underline{x}(k) + \Delta\,\underline{u}(k) \\[2mm] \underline{y}(k) = C\,\underline{x}(k) + D\,\underline{u}(k) \end{array} \right\} \tag{4-4.3}$$

where $\underline{x} \in R^n$, $\underline{u} \in R^m$ and $y \in R^p$ and k denotes the k-th sampling instant.

Chow and Willsky (1984) summarised redundancy relations for equation (4-4.3) as a linear combination of present and lagged values of $\underline{u}$ and $\underline{y}$ such that if no fault occurs, the following parity signal $r(k)$ is nulled (Himmelblau, 1986):

$$r(k) = \sum_{i=0}^{p} \underline{X}^T \left[y(k-p+i) - \sum_{j=0}^{i-1} C \, \Phi^{i-j-1} \, \Delta \, \underline{u}(k-p+j) - D\underline{u}(k-p+i) \right] \quad (4-4.4)$$

Equation (4-4.4) can be arranged so that one side of the resulting expression represents a synthetic or artificial measurement that can be directly compared with y(k). Another interpretation is that equation (4-4.4) is essentially an auto-regressive model for y(k). The parity signal grows in magnitude and in a particular direction associated with the faulty sensor.

Chow and Willsky (1984) have shown that all redundancy equations (direct, temporal, open and closed-loop types) can be categorised under the *generalised parity space* i.e. they form a subspace of the generalised parity space. Lou et al. (1986) re-defined the problem of robust residual generation in geometrical terms using an unconstrained criterion for maximising the angle between the parity space and the operational space. The redundancy relations are then specified mathematically as follows.   The parity space **X** of order n is the subspace spanned by (n + 1) p-dimensional vectors given by:

$$\Pi = \left\{ \mathbf{x} \mid X^T \left[\Theta_d \right] = \underline{0} \right\} \quad (4-4.5)$$

where: $\Theta_d = \left[c^T \ (C \, \Phi)^T \ \cdots \ (C \, \Phi^n)^T \right]^T$

and: $$r(k) = \underline{X}^T \left\{ \begin{bmatrix} y(k-n) \\ y(k-n+1) \\ \vdots \\ \dot{y}(k) \end{bmatrix} - \Gamma \begin{bmatrix} u(k-n) \\ u(k-n+1) \\ \vdots \\ \dot{u}(k) \end{bmatrix} \right\} \quad (4-4.6)$$

and where:

$$\Gamma = \begin{bmatrix} D & \underline{0} & \underline{0} & \underline{0} & \underline{0} & \underline{0} \\ C \, \Delta & D & \underline{0} & \underline{0} & \underline{0} & \underline{0} \\ C \, \Phi \, \Delta & C \, \Delta & D & \underline{0} & \underline{0} & \underline{0} \\ C \, \Phi^2 \, \Delta & C \, \Phi \, \Delta & C \, \Delta & D & \underline{0} & \underline{0} \\ \vdots & \vdots & \vdots & \vdots & D & \underline{0} \\ C \, \Phi^{n-1} \, \Delta & \vdots \ \cdots & \vdots \ \cdots & C \, \Phi \, \Delta & C \, \Delta & D \end{bmatrix} \quad (4-4.7)$$

This subspace is the *left* nullspace of the matrix Θ_d as opposed to the complementary unobservable subspace. It should be noted that $\underline{X}^T$ is a parity weighting row vector which spans the parity space X.

The parity relation can also be expressed as the polynomial vector in the left nullspace of the singular pencil (Lou (1982), Massoumnia (1986 (b), 1988)) as:

$$\Pi(\nu) = \left[\begin{array}{ccc} C, & \nu I & - \Phi \end{array}\right]^T \tag{4-4.8}$$

The residual generator for equation (4-4.7) is a finite impulse response filter (FIR). This has been interpreted as a *dead-beat observer* (Massoumnia (1988), Wünnenberg and Frank (1988)).

Massoumnia (1988) has reformulated the generalised parity relations from a transfer matrix description of the system, and obtains parity relations for which the length of the observation time window is minimal. Wünnenberg and Frank (1988) on the other hand use a decoupled optimal and robust dead-beat unknown input observer (UIO) using a minimisation of a performance index. The details of this method are given in Chapter 3.

In addition to considering the UIO for sensor fault detection, Viswanadham and Srichander (1987) proposed the use of the bank of UIOs for robust *actuator* fault detection. Ge and Fang (1988) also applied this approach to robust fault detection of the *components* of a dynamic system.

The method described in this particular chapter uses the eigenstructure approach in the design of the vector weighting matrix required to generate a form of parity space. When the ideal vectors are not achievable because they do not lie in the allowable subspace as defined in Section 4.2.3, a least-squares projection of the vector onto the allowable subspace is used to obtain the optimal vectors. The required vectors, $\underline{h}_i^T$ can be selected so that the fault due to one sensor is emphasised whilst faults from the other sensors are de-emphasised. This has the advantage of reducing the number of observers required for fault detection and, in turn, also reduces the complexity of the decision logic used for fault isolation. The reduction in computation increases the response time to the detection of faults. The design can be carried out both in the continuous and discrete-time domains. However, since the method requires p zero-valued eigenvalues, the best solutions are obtained in the discrete-time domain as discussed in Section (4-3.2). The residual signals obtained are those of a finite impulse response filter (FIR), as discussed by Massoumnia (1988).

4.4.2 The eigenstructure assignment approach

Many methods for fault detection and isolation (FDI) have been based on a discrete-time version of the linear plant model with the exception of the detection filters (Jones (1973)). Some of the discrete-time FDI methods are based on *robust residual generation* as the basis of fault detection and have the potential to specify fault directions in the same way as a detection filter. They use the generalised parity space discussed in the last section. The *parity space* can now be interpreted as an *invariant manifold* in which, when certain conditions are satisfied, robust residual generation is achieved. The invariant manifold motion arises as the ideal case corresponding to a null parity function in the absence of a fault. The corresponding continuous design will now be discussed:

Consider the *nominal* open-loop linear time invariant system of equation (4-3.4) in the absence of faults and disturbances:

$$\dot{\underline{x}} = A_0\underline{x} + B_0\underline{u}$$

$$y = C_0\underline{x} + D_0\underline{u}$$

with:

$$\dot{\underline{x}} \in R^n \quad , \quad y \in R^p \quad , \quad \underline{u} \in R^m$$

The pair (C_0, A_0) is assumed to be observable. A full order observer for the above system has the linear structure as given by equation (4-2.5)

$$\dot{\hat{\underline{x}}} = A_0\hat{\underline{x}} + B_0\underline{u} + K(\underline{y} - \hat{y})$$

with: $\underline{e} = \underline{x} - \hat{\underline{x}}$

and: $\dot{\underline{e}} = F_0\underline{e}$

where: $F_0 = (A_0 - K C_0)$

The presence of *sensor faults* can be modelled by an additional vector term $\underline{f}_s$ in the system dynamics as follows:

$$\dot{\underline{x}} = A_0\underline{x} + B_0\underline{x} \left.\vphantom{\begin{array}{c} \\ \\ \end{array}}\right\}$$

$$\underline{y} = C_0\underline{x} + D_0\underline{u} + \underline{f}_s \qquad (4\text{-}4.9)$$

where $\underline{f}_s = \Delta\, C\, \underline{x}$ denotes the *sensor fault vector*. The additional effect of sensor faults on the estimation error dynamics is then described by the following:

$$\dot{\underline{e}} = F_0\underline{e} + K\,\underline{f}_s$$

After the initial error transients have died away, assuming that F_0 corresponds to a stable system, the forced transient response due to sensor faults becomes:

$$\underline{e}(t) = \int_0^t e^{F_0(t-\tau)}\, K\,\underline{f}_s(\tau)\; d\tau \qquad (4\text{-}4.10)$$

The corresponding matrix $G(s)$ of transfer functions relating the residual error vector $\underline{e}$ to the sensor fault inputs $\underline{f}_s$ is given by:

$$G(s) = (sI_n - F_0)^{-1}\, K \qquad (4\text{-}4.11)$$

in which it can be noticed that the effect of the term $D_0\underline{u}$ has been cancelled through the assumption that the observer replicates the identical $D_0\underline{u}$ term in the process under investigation. Similarly, *actuator faults* are modelled by an additional term $B_0\underline{f}_a$ in the system model as follows:

$$\dot{\underline{x}} = A_0\underline{x} + B_0\underline{u} + B_0\underline{f}_a \left.\vphantom{\begin{array}{c} \\ \\ \end{array}}\right\}$$

$$\underline{y} = C_0\underline{x} + D_0\underline{u} \qquad (4\text{-}4.12)$$

The transfer function relating the residual error vector $\underline{e}$ to the actuator faults $\underline{f}_a$ which act as the forcing function is:

$$G(s) = (sI_n - F_0)^{-1}\, B_0 \qquad (4\text{-}4.13)$$

Now let us turn our attention again to the case where *sensor faults* only act on the system. From equation (4-4.11) the residual error vector in the s-domain due to sensor faults alone is given by:

$$\underline{e}(s) = (sI_n - F_o)^{-1} K \underline{f}_s \qquad\qquad (4\text{-}4.14)$$

which forms the basis for sensor fault detection. In order to design an effective fault detection scheme having sufficient design freedom, a weighting matrix H must be defined as in equation (4-3.17). Furthermore, assuming that the pair (C_o, A_o) is observable, the matrix H must satisfy the following conditions:

a) $H = \left\{ \mathbf{x} \mid \underline{x}^T (\nu I - F_o) = 0 \right\}$ \qquad\qquad (4-4.15)

   for some complex number ν

b) $H = W\, C_o$ \qquad\qquad (4-4.16)

   for some suitably designed matrix W

c) rank (H) = p \qquad\qquad (4-4.17)

The weighting matrix H defines an *invariant manifold* necessary for the zeroing of the observation error outputs in the absence of sensor faults. It is clear from equations (4-4.15) and (4-4.16) that not every left eigenvector can form the required weighting matrix H. After applying the weighting matrix, H to the error dynamics, the i-th sensor fault signal in the s-domain, $s_i(s)$ is related to the i-th fault by:

$$s_i(s) = \frac{\underline{h}_i^T \underline{k}_i}{s + \lambda_i} \cdot f_{si}(s) \qquad\qquad (4\text{-}4.18)$$

where $\underline{h}_i^T$ is the closed-loop eigenvector corresponding to the mode, λ_i, $\underline{h}_i^T$ is the i-th row of the matrix H, $\underline{k}_i$ is the i-th column of the gain matrix K and $f_{si}(s)$ denotes the Laplace Transform of the i-th sensor fault signal. Since the left eigenvector, when applied as the weighting has the effect of zeroing the remaining n-1 modes, then only the mode λ_i is transmitted through the observer system. It is clear that when $\underline{h}_i^T$ corresponds to a *zero-valued eigenvalue*, equation (4-4.18) becomes:

$$s_i(s) = \frac{\underline{h}_i^T K_i}{s} \cdot f_{si}(s) \qquad\qquad (4\text{-}4.19)$$

Equation (4-4.19) represents the direct transmission term in the transfer function due to excitation by the i-th column of the matrix K. This can easily be seen by writing the full transfer function (4-4.11) together with equation (4-4.16) as:

$$W\, G(s) = \frac{1}{s}\, H \left\{ I_n + \frac{F_o}{s} + \frac{F_o^{\,2}}{s^2} + \ldots + \frac{F_o^{\,n-1}}{s^{n-1}} \right\} K \qquad (4-4.20)$$

If the terms involving powers of F up to n - 1 in the expansion of equation (4-4.20) are nulled by the left annihilator matrix H then higher order terms are also nulled. In order for the null-space manifold to be reached by the observer error response the numerator terms in the transfer function matrix expansion must be nulled and this must hold true in the absence of either sensor or actuator faults.

For complete insensitivity to unknown inputs and sensor faults the following invariance conditions must be satisfied:

$$H\, F_o^{\,i} = 0 \qquad \text{for } i = 1,2,\ldots,n\text{-}1 \qquad\qquad (4-4.21)$$
$$H\, K = 0 \qquad\qquad (4-4.22)$$
and $$\quad H\, E = 0 \qquad\qquad (4-4.23)$$

For complete insensitivity to actuator faults the following additional condition must be satisfied:

$$H\, B_o = 0 \qquad\qquad (4-4.24)$$

Clearly, the above conditions cannot be completely satisfied if sensor and actuator fault detection is to be achieved. In order to sensitise the observer to *specific* faults it is necessary for the direct transmission terms H K and H B_o to be designed to provide a *window* for the transmission of sensor and actuator faults, respectively.

The window design can be achieved through eigenstructure assignment by making the rows of the product H K (or H B_o) sensitive to one sensor (or one actuator). The approach to achieve this is described as follows.

In order to achieve the above objectives it is necessary to turn to the discrete-time observer outlined in Section 4.2.6.

The z-domain transfer relationships for sensor and actuator faults can be written as:

$$\underline{e}(z) = (zI_n - \Phi_{cl})^{-1} K \underline{f}_s \qquad (4\text{-}4.25)$$

and $\quad \underline{e}(z) = (zI_n - \Phi_{cl})^{-1} \Delta_o \underline{f}_a \qquad (4\text{-}4.26)$

We can, therefore, replace equation (4-4.20) by its discrete-time equivalent as:

$$W\ G(z) = \frac{1}{z}\ H \left\{ I_n + \frac{\Phi_{cl}}{z} + \frac{\Phi_{cl}^2}{z^2} + \ldots + \frac{\Phi_{cl}^{n-1}}{z^{n-1}} \right\} K \qquad (4\text{-}4.27)$$

In the *absence* of faults the invariance conditions can be stated as:

$$\underline{h}_k^T\ \underline{e} = 0\ , \qquad \text{for } k = 1,\ldots,p \qquad (4\text{-}4.28)$$

and $\quad H\ \Phi_{cl}^i\ \underline{e} = 0\ , \text{ for } i = 1,2,\ldots,n\text{-}1 \qquad (4\text{-}4.29)$

As the matrix H comprises left eigenvectors corresponding to zero-valued eigenvalues of the closed-loop observer matrix Φ_{cl} condition (4-4.29) is completely satisfied whilst condition (4-4.28) is satisfied after a *finite* number of time steps in accordance with the dead-beat response (Leden, 1977). The input de-coupling arises as a consequence of the assignment of p zero-valued eigenvalues.

When condition (4-4.28) is satisfied for a given row $\underline{h}_k^T$ of H, the observer error state vector $\underline{e}$ will belong to a specific subspace spanned by the n - 1 range space eigenvectors orthogonal to $\underline{h}_k^T$. By designing the p $\underline{h}_k^T$ row vectors to be *mutually orthogonal* to each other, a special de-coupling structure arises in the sensor fault detection vector $\underline{S} = H\ \underline{e}$.

The p *parity* subspaces so formed then provide the independent insensitivity of each sensor fault channel. The parity subspaces are disjoint and allow the independent detection and localisation of sensor and actuator faults.

It is clear then that the product matrix H K (considering sensor fault detection) can be re-arranged by row operations to give a diagonally dominant structure. The inner product terms arising from the product H K then yield p independent *parity windows* corresponding to each of the p sensor channels, in turn.

A directly equivalent approach can be used for the design of m independent actuator fault detection parity spaces by using m of the p rows of H and by arranging the corresponding mxm product matrix to be diagonally dominant. For this, it is of course necessary for m ⩽ p.

The eigenstructure assignment technique outlined in Section 4.2 is used in a very straightforward manner to assign n eigenvalues to the discrete dual control problem. p of these eigenvalues are zero-valued (for sensor fault detection) and the corresponding p right eigenvectors are mutually orthogonal. On forming the observer system (by transposing the closed-loop matrix) the p right eigenvectors become p left eigenvectors for the observer which constitute the rows of the matrix H. As the p eigenvectors corresponding to zero-valued eigenvalues are orthogonal it is not necessary to use a generalised eigenstructure approach. It is necessary, of course to ensure that the observer closed-loop matrix Φ_{cl} is well-conditioned numerically and this can be ensured by minimising the condition number through a suitable choice of eigenvectors.

As a final comment it is important to note that the invariancy conditions described in this section yield the following matrix pencil formulation:

$$\underline{h}_k^T \, [\gamma \, I_n - \Phi_{cl}] = 0 \, , \quad \text{for } k = 1,\ldots,p \qquad (4\text{-}4.30)$$

In the present study γ is chosen to be zero-valued and is an eigenvalue of the closed-loop system matrix Φ_{cl}. Equation (4-4.30) thus shows the similarity between the generalised parity space described by Frank and Wünnenberg in Chapter 3 and Massoumnia (1988). This has also been referred to by equation (4-4.8) above. The eigenstructure assignment approach clearly falls into the same parity space observer class as discussed in Chapter 3.

4.4.3 Sensor fault detection example

The example is based on the lateral motion of a light aircraft, for this example application the system is considered to be fifth order (n = 5) with three measurements available (p = 3). The linearised (or nominal) system model is used in the design and the fault monitor is tested by applying it to the fully non-linear aircraft dynamic model.

Straight and level flight is not a particularly severe test for an IFD scheme applied to an aircraft, as false alarms are most likely whilst the aircraft is performing some manoeuvre (including significant parameter variations). The controller reference demands are changed during the simulated flight to replicate a varied range of pilotic inputs. For this application it was decided to limit the IFD application to the lateral motion of the light aircraft. The lateral motion state-space system for a trim flight setting is:

$$\dot{\underline{x}}_O = A_O \, \underline{x}_O + B_O \, \underline{u}$$

$$\underline{y}_O = C_O \, \underline{x}_O + D_O \, \underline{u}$$

(4-4.31)

with: $\underline{x}_O \in R$, $\underline{u} \in R^2$ and $\underline{y} \in R^3$

The system parameter matrices are:

$$A_O = \begin{bmatrix} -0.2770 & 0.0 & -32.9 & 9.81 & 0 \\ -0.1033 & -8.525 & 3.75 & 0.0 & 0 \\ 0.3649 & 0.0 & -0.639 & 0.0 & 0 \\ 0.0 & 1.0 & 0.0 & 0.0 & 0 \\ 0.0 & 0.0 & 1.0 & 0.0 & 0 \end{bmatrix}$$

$$B_O = \begin{bmatrix} -5.432 & 0.0 \\ 0.0 & -28.64 \\ -9.49 & 0.0 \\ 0.0 & 0.0 \\ 0.0 & 0.0 \end{bmatrix}$$

and:

$$C_O = \begin{bmatrix} 0 & 1 & 0 & 0 & 0 \\ 0 & 0 & 0 & 1 & 0 \\ 0 & 0 & 0 & 0 & 1 \end{bmatrix}$$

together with $D_O = 0$.

The state and control vectors have the following physical significance:

$\underline{x}^T$ = [v (side-slip velocity), p (roll-rate), r (yaw-rate),

   Φ (bank angle), Ψ heading angle)]

$\underline{u}^T$ = [ζ (rudder angle), ξ (aileron angle)]

Linear velocities are in metres per second, whilst angles and angular velocities have units of radians and radians per second, respectively.

Using a sampling period of ΔT = 0.01 (10 msec.), the discrete system matrices (correct to 4 decimal places) are given by:

$$\Phi = e^{A_o\Delta T} \quad \text{and} \quad \Delta = \int_o^{\Delta T} e^{A_o\tau} B \, d\tau \quad \text{with:}$$

$$\Phi = \begin{bmatrix} 0.9966 & 0.0005 & -0.3274 & 0.0979 & 0.0 \\ -0.0009 & 0.9193 & 0.0361 & 0.0 & 0.0 \\ 0.0036 & 0.0 & 0.9930 & 0.0002 & 0.0 \\ 0.00 & 0.0096 & 0.0002 & 1.0 & 0.0 \\ 0.0 & 0.0 & 0.01 & 0.0 & 1.0 \end{bmatrix}$$

$$\Delta = \begin{bmatrix} -0.0376 & 0.0 \\ -0.0017 & -0.2752 \\ -0.0947 & 0.0 \\ 0.0 & -0.0014 \\ -0.0005 & 0.0 \end{bmatrix}$$

The observer can be designed by assigning the eigenstructure of the *dual control* problem as discussed in Section 4.2.6. For this problem it is only necessary to assign *three* (p = 3) right eigenvectors to the dual problem, although all five eigenvectors are partially assignable. The three assigned eigenvectors then become left eigenvectors of the observer, as required, by transposition. As the problem has three inputs, three elements of each eigenvector can be assigned *exactly*, however, to satisfy equation (4-4.29) it is necessary to assign three non-zero elements of each row of H together with two zero (or very small magnitude) elements in each row.

   In order to achieve the correct structure of the matrix H the direct assignment approach outlined in Section 4.2.5 is used by means of the least-squares orthogonal projection. An additional requirement is that the rows of H must be mutually orthogonal (or nearly so) in order to provide the three independent detection spaces for the p (roll rate), Φ (bank angle) and Ψ (heading angle) sensor channels, respectively.

In order to achieve this the following weighting matrix structure is desirable:

$$H_d = \begin{bmatrix} 0 & 1 & 0 & a & b \\ 0 & c & 0 & 1 & d \\ 0 & e & 0 & f & 1 \end{bmatrix}$$

The discrete time design was chosen to have the following closed-loop eigenvalue spectrum:

$$\sigma(\Phi_{cl}) = \{\ 0,\ 0,\ 0,\ 0.8,\ 0.98\ \}$$

The rows of H_d i.e. the desired left eigenvectors were assigned as the right eigenvectors of the dual control problem. These eigenvectors correspond to the three zero-valued eigenvalues and the following achievable *right* (dual control problem) eigenvectors were obtained:

$$\begin{bmatrix} 0.0011 \\ 1.0 \\ -0.0388 \\ 0.0056 \\ 0.2783 \end{bmatrix} \quad \begin{bmatrix} 0 \\ -0.0052 \\ 0 \\ 1.0 \\ 0.0004 \end{bmatrix} \quad \begin{bmatrix} -0.0002 \\ -0.1879 \\ -0.0033 \\ 0.0014 \\ 1.0 \end{bmatrix}$$

$$\underline{h}_1 \qquad\qquad \underline{h}_2 \qquad\qquad \underline{h}_3$$

The parity matrix H is thus:

$$H = \begin{bmatrix} 0.0011 & 1.0 & -0.0388 & 0.0056 & 0.2783 \\ 0 & -0.0052 & 0 & 1.0 & 0.0004 \\ -0.0002 & -0.1879 & -0.0033 & 0.0014 & 1.0 \end{bmatrix}$$

It can easily be seen that this matrix approximates the required structure of H_d.

The resulting closed-loop observer matrix Φ_{cl} is:

$$\Phi_{cl} = \begin{bmatrix} 0.9966 & -0.7941 & -0.3274 & -0.0035 & -1.3623 \\ -0.0009 & -0.1924 & 0.0361 & -0.0010 & -0.0605 \\ 0.0036 & -5.3715 & 0.9930 & -0.0271 & -1.7218 \\ 0 & -0.0010 & 0.0002 & 0 & -0.0003 \\ 0 & -0.0539 & 0.0100 & -0.0003 & -0.0172 \end{bmatrix}$$

The *normalized* right eigenvectors of Φ_{cl} are:

$$\begin{bmatrix} 0.0236 \\ 0.0056 \\ 0.0540 \\ 1.0 \\ -0.0020 \end{bmatrix} \begin{bmatrix} -0.6254 \\ 0.2709 \\ 0.2987 \\ 1.0 \\ -0.6898 \end{bmatrix} \begin{bmatrix} 0.4900 \\ 0.1812 \\ 1.0 \\ 0.0009 \\ 0.0125 \end{bmatrix} \begin{bmatrix} 1.0 \\ 0.0182 \\ 0.5343 \\ 0.0001 \\ 0.0053 \end{bmatrix} \begin{bmatrix} 1.0 \\ 0.0006 \\ 0.0473 \\ 0 \\ 0.0005 \end{bmatrix}$$

$$\lambda_1 = 0 \qquad \lambda_2 = 0 \qquad \lambda_3 = 0 \qquad \lambda_4 = 0.8 \qquad \lambda_5 = 0.98$$

If the desired eigenvectors had been assigned exactly then HK would be diagonal and the three sensor fault parity (null) spaces would have been disjoint. The sparse structure of the matrix H arises from the sparseness of the matrix C_o.

To implement the fault detection observer it is only necessary to pre-multiply the *measured* estimation errors $\underline{e}_y$ by the matrix W given by:

$$W = \begin{bmatrix} 1.0 & 0.0056 & 0.2783 \\ -0.0052 & 1.0 & 0.0004 \\ -0.1879 & 0.0014 & 1.0 \end{bmatrix}$$

The required fault detection signals are then given by:

$$\underline{S}(t) = W \, \underline{e}_y(t)$$

with the three elements of the vector $\underline{S}(t)$ providing a specific sensitivity to the p (roll rate), Φ (bank angle) and Ψ (heading angle) measurement channels, respectively.

4.4.4 Aircraft system IFD design and simulation results

The fault detection and isolation (FDI) design approach discussed in this chapter has been based on the use of eigenstructure assignment to provide robust observer designs. The designs are completed in the discrete-time domain and give rise to special parity space de-coupling properties to enable an observer to be used for reliable fault detection *and* isolation.

A number of similar observers can be designed according to the scheme shown in Figure 4-1 at the beginning of the chapter. This is an application of the Generalised Observer Scheme (GOS) in which p + 1 observers are used with each observer being driven by a different subset of p out of p + 1 measurements.

In the example considered above, the system itself has *five* state variables whilst four of these are measured. The GOS scheme therefore consists of four observers with three measurement inputs applied in each case.

The combination of the four observers is sufficient to detect and isolate the faulty sensor uniquely by defining a fault status logical word. In this way *fault isolation* can be effected and in this case does not require statistical methods which would increase the computation complexity as well as the fault detection response time. Using the robust observer FDI technique it is possible to reduce the complexity further by using a single observer and p selected weighting operations for generation of residuals, detection and fault isolation in a single design step. The results of sample tests applied to a single observer as shown in Figures 4-2 to 4-5 will now be discussed.

Sample test signals were introduced into the system sensor measurements from a software fault generator. Representative fault signals in the form of consecutive step fault signals were injected into each sensor, in turn, with the simulated aircraft under normal trim flight and in the absence of gust disturbance. The consecutive step faults (shown in Figure 4-2) were applied on the roll-rate (p), the bank angle (Φ) and the yaw angle (Ψ), to demonstrate the ability to enhance faults due to one sensor whilst suppressing the faults due to the other sensors. Two fault-detect signals were used to achieve this (shown in Figure 4-3) and are given by the *zero-th order* scalar parity signals, s_k:

$$s_k = \underline{h}_k^T \underline{e} = \underline{w}_k^T \underline{e}_y \quad , \quad \text{for } k = 2, 3 \ldots, p$$

In Figure 4-3(a) the fault-detect signal $S_2(t)$ for isolating faults in the bank angle (Φ) sensor promptly announces the occurrence of a fault after 2.5 secs have elapsed. The effect of the other two faults and any other possible disturbances are completely nulled. The weighting vector, $\underline{h}_2^T$ used in forming the fault detect signal, $s_2(t)$ lies in the allowable subspace and therefore provides excellent decoupling. The other fault parity detect (parity) signal, $s_3(t)$ is shown in Figure 4-3(b) for enhancing faults in the yaw angle (Ψ) sensor. This signal also shows prompt detection at time t = 3.5 secs. The effects on $s_2(t)$ due to the faults in the roll rate (p) and bank angle (Ψ) sensors are almost nulled and any effect unlike zero is well below the threshold level. It should be noted that each fault-detect signal should have a different threshold level.

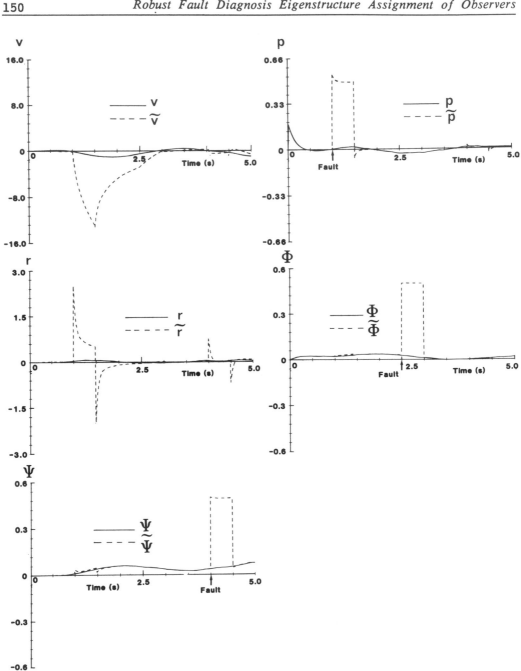

Figure 4-2: Effect of sensor faults on observer state estimates.
Three step faults on channels p (roll rate), Φ (bank
angle) and Ψ (heading angle) are clearly marked

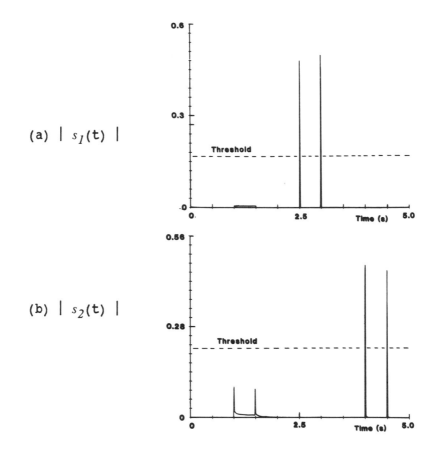

(a) $|\,s_1(t)\,|$

(b) $|\,s_2(t)\,|$

Figure 4-3: $s_1(t)$ and $s_2(t)$ fault detect signals corresponding to results shown in Figure 4-2. The diagram shows, in each case, a strong sensitivity to only *one* fault

The effect of sensor faults on the system states and the corresponding single fault-detect signal in the presence of gust disturbance is shown in Figure 4-4. It can be noted that the introduction of gust disturbances does not affect the fault monitor performance despite variations in the system state estimates of the sideslip velocity (v) and the yaw rate (r). The disturbances cause inaccuracy in the state estimation (particularly of the states not used for feedback) and can give rise to undesired high false alarm rate (FAR) in many techniques which are not based on disturbance de-coupling. Thus Figure 4-4 demonstrates clearly the power of the eigenstructure approach in minimising the effects of some sensor faults, disturbances and even model errors, whilst enhancing the effects of, say, one sensor fault. The power of the approach arises from the freedom in the assignment which has been utilised, whereas, on the other hand a pole-placement approach does not fully utilise this multivariable freedom.

Figure 4-5 shows the occurrence of a deadband fault at t = 3.0 secs on the Ψ (heading angle) sensor, represented by the dotted line in the state estimate. The fault detect signals show promptly the occurrence of a fault.

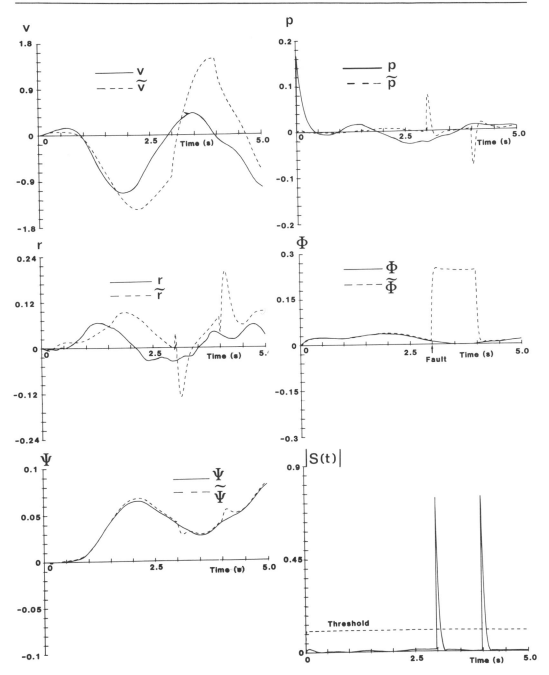

Figure 4-4: Fault monitor responses for lateral motion subsystem. The system is subjected to turbulence input and a single step fault on the Φ (bank angle) sensor

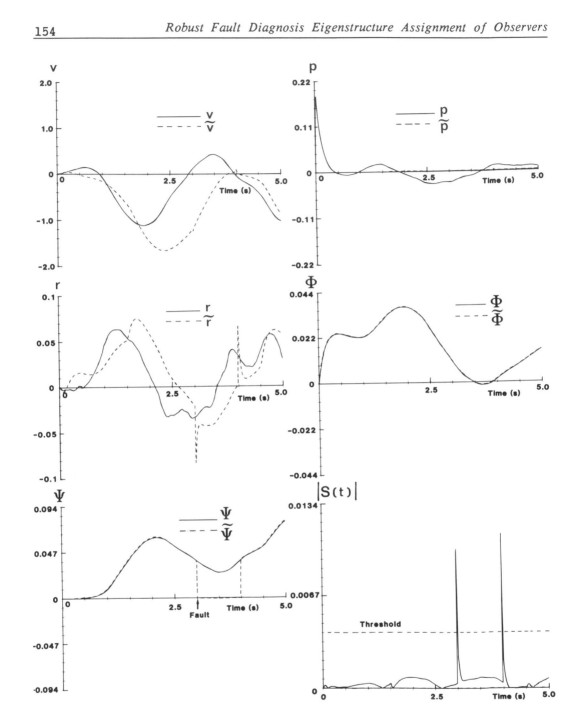

Figure 4-5: Fault monitor responses for the lateral motion subsystem. The system is subjected to turbulence input and a single dead-band fault on the Ψ (heading angle) sensor

Chapter Five

DIAGNOSABILITY OF SYSTEMS AND OPTIMAL SENSOR LOCATION

Shogo Tanaka

This Chapter analyses firstly the step-hypothesised generalised-like-lihood-ratio (GLR) method for fault detection in linear discrete dynamical systems and discusses the concept of detectability. It is shown that, in the case of component faults a certain null-space termed a weakly-diagnosable space can exist for component faults. The weakly-diagnosable space does not exist in the sensor fault case and this presents some problems. To avoid the difficulty two extended GLR methods are developed, one introducing a weight on all the GLR's calculated for different directions and the other taking into account the estimation of other variables than a bias based on the information about the control inputs and state vector of the system. Later the concept of detectability is utilised and, finally criteria for optimal sensor location are discussed based on the concepts of detectability, separability, and observability.

5.1 INTRODUCTION

Fault diagnosis is important in two aspects. The first aspect is an improvement in system availability, and the second, which is more important, is the protection from disasters. Using this point of view many fault detection techniques have been developed (see for example Willsky (1976)). In particular, the generalised-likelihood ratio (GLR) method is well-known for its speed in detection. However, its use has some limitations in that we must assume *a priori* several hypotheses which are likely to happen. Step and jump hypotheses have often been used for their convenience. However, the anomalous function which adds to the system has, in general, a very complicated form (time-varying and multi-dimensional) which cannot necessarily be modelled by a simple hypothesis such as, for example a step change.

Despite this, we cannot easily throw the step hypothesis away because of the several fascinating points such that it does not need many parameters to be estimated and that there are often the cases where it can be applied well (Willsky (1986)). But, as far as the author is aware, the details have not been strictly discussed.

So, in this Chapter, we first discuss the detectability of faults by the step-hypothesised GLR method by applying a discriminating measure *divergence* to the two innovations sequences of different conditions of normal and abnormal states, Willsky (1986), Tanaka et al. (1984) and Tanaka (1987), and show that if the system is observable and the anomalous function has a bias in some sense, not necessarily a bias in the sense of arithmetic mean, at some time, the anomaly or failure can, in principle, be detected. It is also shown that if the bias is included in what can be called a weakly-diagnosable space, it is difficult for the anomaly to be detected in the sense that the probability of missed alarm cannot be made smaller than a certain value, even if we used many observation data. Thus, the problem can never be overcome only by lengthening the width of window. Such a space can exist for dynamic faults rather than for sensor ones. In order to overcome the problem, we first develop a weighted step-hypothesised GLR method (*Detection Method I*) which gives a large weight on the GLR computed for the anomalous component in the direction of the weakly-diagnosable-space in order to minimise the probability of missed alarm.

But, even by this approach, every anomaly cannot always be effectively detected. This is because the anomalous function does not always have a bias of high level over a complete interval. Conversely speaking, the step-hypothesised method is effective only when the anomalous function has a bias over a long range or at least a balanced large bias if the range is short. Thus, another approach (*Detection Method II*) is proposed which makes the most use of the information about the state and the input of the system for constructing an adequate basis function to track the anomalous function and enables us to detect the general anomaly more accurately and rapidly, which cannot be necessarily detected by the step-hypothesised one. See Figure 5-1 for the relation of the proposed methods. We furthermore investigate the detectability by Detection Method II and show that if the input sequence is non-trivial and the fault affects the input distribution matrix, any fault can, in principle, be detected. Finally, we discuss the criteria for optimal sensor location based on the indices of *detectability*, *separability*, and *observability*.

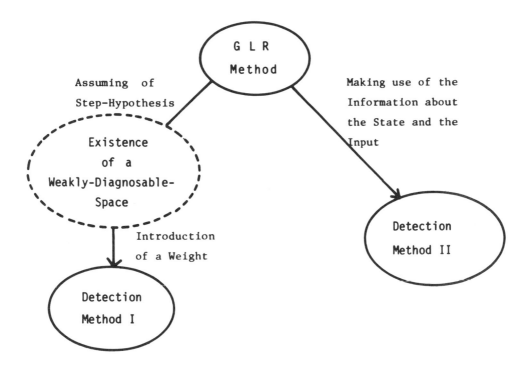

Figure 5-1: Proposed methods

5.2 PROBLEM FORMULATION

We consider here the *linear discrete* dynamical system defined by:

$$x(k+1) = Ax(k) + Bu(k) + \Gamma\omega(k) \tag{5-2.1}$$

$$y(k) = Hx(k) + v(k) \tag{5-2.2}$$

where $x(k) \epsilon R^n$: the state, $u(k) \epsilon R^r$: the input, $y(k) \epsilon R^m$: the observation, and $\omega(k) \epsilon R^p$ and $v(k) \epsilon R^m$ are mutually independent white Gaussian noises with zero means and covariances Q and R (positive definite). The initial state x(0) is assumed to be a Gaussian random variable with mean $\hat{x}(0/-1)$ and covariance $P(0/-1)$ independent of $\omega(k)$ and $\nu(k)$. The matrices A and H also assumed to be of full rank.

When a malfunction/fault occurs to the plant or the sensors, it generally causes changes ΔA, ΔB, and ΔH in the system matrices A, B, and H. This means that unexpected time-varying functions such as $\Delta Ax(k)$, $\Delta Bu(k)$, and $\Delta Hx(k)$ are included in the dynamics or measurement equations from fault occurrence time θ. Therefore, the malfunction or fault can be modelled by the following equations:

$$x(k+1) = Ax(k) + Bu(k) + \Gamma\omega(k) + \sigma(k+1,\theta_d)\ f_d\ (k+1) \qquad (5-2.3)$$

$$y(k) = Hx(k) + \nu(k) + \sigma(k,\theta_s)\ f_s\ (k) \qquad (5-2.4)$$

where $f_d(k+1)$ and $f_s(k)$ are unknown anomalous vector functions of appropriate dimensions and $\sigma(k,\theta)$ is the unit step function defined by $\sigma(k,\theta) = 0$ for $k < \theta$ and $\sigma(k,\theta) = 1$ for $k \geq \theta$. Clearly, the model is also valid for a pure bias type of fault.

We first consider here the step-hypothesised GLR method for fault detection and discuss the detectability aspects. Based on the discussion, we next consider some more refined alternatives for the fault detection and outline the optimal sensor location problem.

5.3 INNOVATIONS SEQUENCE

A. Behaviour of the innovations sequence

The optimal state estimation of the system (5-2.1) and (5-2.2) is computed from the well known Kalman filter equations:

$$\hat{x}(k/k-1) = A\hat{x}(k-1/k-1) + Bu(k-1) \qquad (5-3.1a)$$

$$\hat{x}(k/k) = \hat{x}(k/k-1) + K(k)\gamma(k) \qquad (5-3.1b)$$

$$\gamma(k) = y(k) - H\hat{x}(k/k-1) \qquad (5-3.1c)$$

$$P(k/k-1) = AP(k-1/k-1)A^T + \Gamma Q\Gamma^T \qquad (5-3.1d)$$

$$P(k/k) = P(k/k-1) - k(k)HP(k/k-1) \qquad (5-3.1e)$$

where:

$$K(k) = P(k/k-1)H^T V^{-1}(k) \qquad (5-3.2)$$

$$V(k) = HP(k/k-1)H^T + R \qquad (5-3.3)$$

$y(k)$ in (5-3.1c) is called the *innovations* and the sequence becomes white Gaussian noise with zero-mean and covariance $V(k)$ when the system is operating normally, Willsky (1976). We denote this optimal innovations sequence by $y^o(k)$. However, when a fault or anomaly has occurred to the system at time instant θ, the effect $\Delta y(k;\theta)$ caused by the anomaly appears superimposed on the innovations sequence $y(k)$ from instant θ onwards. For simplicity, the interest here is only with the dynamics (component) fault which are represented by the anomalous function $f(k)\sigma(k,\theta)$. Similar discussions are possible for other types of fault representations. Hence for this special case:

$$y(k) = \begin{cases} y^o(k) & ; \text{ no fault} \\ \Delta y(k;\theta) + y^o(k) & ; \text{ fault} \end{cases} \qquad (5-3.4)$$

where:

$$\Delta y(k;\theta) = G^o(k;\theta) * f(k)$$

$$\overset{\Delta}{=} \sum_{j=\theta}^{k} G^o(k;j)f(j) \quad \text{(convolutional sum)} \qquad (5-3.5)$$

Here, $G^o(k,\theta)$ is a fault signature matrix corresponding to a jump change in dynamics, Willsky and Jones (1976). The words *fault* and *anomaly* are used here interchangeably.

B. A Discriminating measure 'divergence'

When an anomaly or fault has occurred in the system, the innovations sequence $y(k)$ behaves differently from the way it does in the fault-free case as seen from (5-3.4). Hence, whether the anomaly detection can be easily made or not depends simply on the difference between the two innovations sequences arising from the no-fault and fault conditions. We adopt here the *divergence* as a measure of the distance between the two sequences and first discuss the detectability by the step-hypothesised GLR method. Of course, there are cases, as we mentioned earlier, for which this step-hypothesis does not work well. For such cases, we propose that other approaches given in this book should be examined.

Definition: For the two probability density functions $p_1(y)$ and $p_2(y)$ ($y \in Y$), the *divergence* is defined by Kullback (1959)

$$D(1;2) = \int_Y p_1(y) \ln \left(\frac{p_1(y)}{p_2(y)}\right) dy + \int_Y p_2(y) \ln \left(\frac{p_2(y)}{p_1(y)}\right) dy \qquad (5\text{-}3.6)$$

Properties: We have the following:

(i) $D(1;2) \geq 0$, and $D(1;2) = 0$ if and only if $p_1(y) = p_2(y)$.

(ii) $D(1;2)$ is invariant under non-singular transformations.

The divergence is calculated for $y(k)$ of (5-3.4) and it becomes:

$$D(H_0;H_1) = \sum_{j=\theta}^{k} \Delta y^T(j;\theta) V^{-1}(j) \Delta y(j;\theta) \qquad (5\text{-}3.7)$$

where H_0 and H_1 represent the states of no fault and a fault condition, respectively.

5.4 DETECTABILITY BY USUAL STEP-HYPOTHESISED GLR METHOD

We first discuss the detectability when the usual step-hypothesised GLR method is applied to the fault detection. To simplify the subsequent discussion, we now introduce the inner product defined by:

$$< a(k), b(k) > \triangleq \sum_{j=\theta}^{k} a^T(j) V^{-1}(j) b(j) \qquad (5\text{-}4.1)$$

(valid throughout this Chapter) where $a(k)$ and $b(k)$ are vector functions of dimension m. Using the notation, (5-3.7) can be rewritten as:

$$D(H_0;H_1) = < \Delta y(k;\theta), \Delta y(k;\theta) >$$

$$= \| \Delta y(k;\theta) \|^2 \qquad (5\text{-}4.2)$$

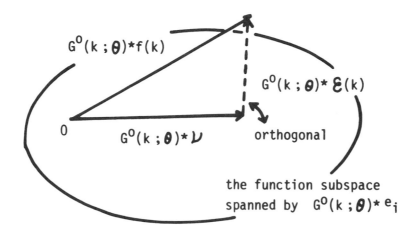

Figure 5-2: Geometrical explanation

We now decompose the anomalous function $f(k)$ into a constant $\nu \in R^n$ and a time-varying component $\epsilon(k)$ such that:

$$< G^o(k;\theta)*e_i, \ G^o(k;\theta)* \ \epsilon(k) > = 0 \quad , \ 1 \leq i \leq n \qquad (5\text{-}4.3)$$

where e_i is the i-th natural basis vector in R^n. Note that ν is not necessarily the arithmetic mean $\bar{\nu}$ of $f(k)$, but a vector which minimises $\| \ G^o(k;\theta)*(f(k)-\nu) \ \|^2$ with respect to ν (see Figure 5-2). This is in explicit form:

$$\nu = \bar{\nu} + \bar{\Delta}^{-1} \ \bar{\epsilon} = \bar{\Delta}^{-1}\bar{f} \qquad (5\text{-}4.4)$$

where $\bar{\Delta}$ is the n x n matrix whose (i,j) element is defined by $< G^o(k;\theta) *e_i, \ G^o(k;\theta)*e_j >$, and $\bar{\epsilon}$ and $\bar{f}$ are the n-dimensional vectors whose i-th components are respectively given by $< G^o(k;\theta)*e_i, \ G^o(k;\theta)*(f(k)-\bar{\nu}) >$ and $< G^o(k;\theta)*e_i, \ G^o(k;\theta)*f(k) >$. From (5-3.5), (5-4.2), and (5-4.3), we obtain:

$$D(H_o;H_1) = \| \ G^o(k;\theta)*f(k) \ \|^2$$

$$= \| \ G^o(k;\theta)*\nu \ \|^2 + \| \ G^o(k;\theta)* \ \epsilon(k) \ \|^2$$

$$= \| G(k;\theta)\nu \|^2 + \| G^o(k;\theta)* \epsilon(k) \|^2$$

$$= \nu^T C(k;\theta)\nu + D(H_1(\nu); H_1) \tag{5-4.5}$$

where:

$$G(k;\theta) \triangleq \sum_{j=\theta}^{k} G^o(k;j) \tag{5-4.6}$$

$$C(k;\theta) \triangleq \sum_{j=\theta}^{k} G^T(j;\theta)V^{-1}(j)G(j;\theta) \tag{5-4.7}$$

We see that $G(k;\theta)$ is a *fault signature matrix* of a dynamics step and $C(k;\theta)$ becomes the information matrix concerning the constant vector ν. $D(H_1(\nu); H_1)$ represents the divergence between the two innovations sequences of true fault H_1 and the *dynamic* step fault $\nu\sigma(k,\theta)$ (we denote it by $H_1(\nu)$). The first term $\nu^T C(k;\theta)\nu$ in the right-hand side of (5-4.5) represents a measure of the ease in detecting the fault by the step-hypothesis approach and the rate of the quantity occupying in $D(H_o; H_1)$ expresses the validity of adopting it rather than other hypotheses. The quantity is also equivalent to the divergence between the no fault and dynamics step fault $\nu\sigma(k,\theta)$, i.e.,

$$D(H_o; H_1(\nu)) = \nu^T C(k;\theta)\nu \tag{5-4.8}$$

If the system is observable, $C(k;\theta)$ becomes a positive definite matrix and so the quantity (5-4.8) takes a positive value for any anomaly or fault such that $\nu \neq 0$. Furthermore, the information matrix satisfies the relation $C(k;\theta) \leq C(k+1;\theta)$ for any $k \geq \theta$. This means that we can detect the anomaly by a dynamic step hypothesis, only if ν takes a non-zero value and we wait to judge till (5-4.8) takes an appropriate large value. In fact, by calculating the maximum likelihood estimate (MLE) of ν and the generalised likelihood ratio (GLR), we obtain:

$$\hat{\nu}(k;\theta) = C^{-1}(k;\theta) \sum_{j=\theta}^{k} G^T(j;\theta) V^{-1}(j)\gamma(j) \tag{5-4.9}$$

$$\mathit{l}(k;\theta) = \mathit{ln}\left[\frac{p(\gamma^k/ H_1(\nu), \theta, \hat{\nu}(k;\theta))}{p(\gamma^k/ H_o)} \right]$$

$$= (1/2)\, \hat{\nu}^T(k;\theta)C(k;\theta)\hat{\nu}(k;\theta) \tag{5-4.10}$$

Substituting $y(j) = G(k;\theta)\nu + G^o(j;\theta)*\epsilon(j) + y^o(j)$ into (5-4.9), we find that the MLE $\hat{\nu}(k;\theta)$ obeys the Gaussian distribution with mean ν and covariance $C^{-1}(k;\theta)$ and thus the probability density function of the GLR $\mathit{l}(k;\theta)$ becomes a non-central Chi-squared one with non-centrality parameter: $2 T$

$$\delta = \nu \, C(k;\theta)\nu \qquad\qquad (5-4.11)$$

This quantity is nothing but (5-4.8) and this fact shows that the foregoing statement is correct.

From these considerations, we can define the detectability of the fault $f(k)$ by the divergence $D(H_o;H_1(\nu))$, when the step-hypothesised GLR method is applied. The similar approach is also taken by Willsky (1986). The divergence generally increases with k, because of the relation $C(k;\theta) \leqq C(k+1;\theta)$. However, we will see from the following theorem that it depends on the system and the anomaly.

Theorem 1: We assume the stationary state. Then, we let $\pi \overset{\Delta}{=} A$ (I_n-K*H), where K* is a Kalman gain in the stationary state and I_n is the n x n identity matrix. Then, if the following condition:

$$H(I_n - \pi)^{-1}\nu = 0 \qquad\qquad (5-4.12)$$

is satisfied, the divergence $D(H_o;H_1(\nu))$ is saturated by increasing k, i.e., the divergence is bounded.

Proof: Using the recursive equations on the fault signature matrix $G^o(k;\theta)$ (see Appendix 5.1), we can obtain the effect of the dynamic step ν on the innovations by:

$$\Delta y(k;\theta) = \sum_{j=\theta}^{k} G^o(k;j)\nu$$

$$= H [I_n + \sum_{j=\theta}^{k-1} (\pi_{k-1}\pi_{k-2}...\pi_j)] \nu \qquad\qquad (5-4.13)$$

where:

$$\pi_j \overset{\Delta}{=} A(I_n - K(j)H) \qquad\qquad (5-4.14)$$

Thus, in the stationary state, we have:

$$\Delta y(k;\theta) = H \sum_{j=0}^{k-\theta} \pi^j\nu \qquad\qquad (5-4.15)$$

Since we are now assuming the stationary state, Π is a stable matrix. This is easily verified by considering that the recursive equation of the estimation error $e(k/k) = x(k) - \hat{x}(k/k)$ is given by:

$$e(k/k) = (I_n - K(k)H)Ae(k-1/k-1)$$
$$+ (I_n - K(k)H)\Gamma\omega(k-1) - K(k)\nu(k) \qquad (5\text{-}4.16)$$

and that the matrices $(I_n-K(k)H)A$ and $A(I_n-K(k)H)$ are similar to each other. Combining the fact and (5-4.2),(5-4.15), we can conclude that under the condition (5-4.12) the divergence $D(H_0;H_1(\nu))$ is saturated by k.

This theorem means that for the anomaly satisfying (5-4.12) we cannot make the probability of missed alarm smaller than a certain value, even if we make use of a large quantity of observation data. We call the space constructed by such a vector ν a *weakly-diagnosable-space*. It should be clear that obviously the null space of the matrix $H(I_n-\Pi)^{-1}$ and the dimension of the space becomes $(n-m)$. The detectability inside the space depends also on the direction of ν in the space.

Corollary 1: If (I_n-A) is non-singular (this is the usual case), the weakly-diagnosable-space is coincident with the null-space of $H(I_n-A)^{-1}$.

Proof: Since (I_n-A) and A are both non-singular, the commutativity between $(I_n-A)^{-1}$ and A holds. This is easily verified by seeing that:

$$(I_n-A)^{-1}A = [\, A^{-1}(I_n-A)\,]^{-1}$$
$$= [(I_n-A)A^{-1}]^{-1} \;=\; A(I_n-A)^{-1} \qquad (5\text{-}4.17)$$

Thus, we have:

$$[\, I_m + HA(I_n-A)^{-1}K*\,]\,H = H + HA(I_n-A)^{-1}K*H$$
$$= H + H(I_n-A)^{-1}AK*H$$
$$= H(I_n-A)^{-1}\,[(I_n-A) + AK*H]$$
$$= H(I_n-A)^{-1}(I_n-\Pi) \qquad (5\text{-}4.18)$$

Here, both H and (I_n-A) are of full rank by assumption and Π is a stable matrix as mentioned earlier in Theorem 1. Hence, the matrix

[$I_m + HA(I_n-A)^{-1}K*$] of the left side of (5-4.18) becomes non-singular. Premultiplying (5-4.18) by the inverse of the matrix and postmultiplying it by $(I_n-\Pi)^{-1}$, we get:

$$H(I_n-\Pi)^{-1} = [I_m + HA(I_n-A)^{-1}K*]^{-1} H(I_n-A)^{-1} \qquad (5-4.19)$$

The use of equation (5-4.19) together with Theorem 1 leads to the proof of the corollary.

As a special case which satisfies the condition of Corollary 1, we can consider the case where A is a stable matrix. Then, we see that for any element v of the weakly-diagnosable-space the effect of the anomaly on the observation y(k) vanishes meanwhile as time passes. This fact agrees with our natural intuition.

Similar discussions are possible for other faults such as sensor ones and a composite one causing biases in the sensors and the dynamics. Firstly let us investigate the former case. The following theorem can be stated for this case:

Theorem 2: If the matrix $H(I_n-\Pi)^{-1}(I_n-A)$ is not of full rank (even if the system is observable), a weakly-diagnosable-space exists, and the space is given by the range space of Ω by H, i.e., $H\Omega$, where Ω is the null space of $H(I_n-\Pi)^{-1}(I_n-A)$.

Proof: In a similar manner to Theorem 1, we can obtain the effect of the sensor step v' on the innovations as follows:

$$\Delta y(k;\theta) = [I_m-H(AK(k-1) + \sum_{j=\theta}^{k-2} \Pi_{k-1}\Pi_{k-2}...\Pi_{j+1}AK(j))]v' \qquad (5-4.20)$$

So, in the stationary state, we have:

$$\Delta y(k;\theta) = [I_m -H(\sum_{j=0}^{k-\theta-1} \Pi^j)AK*]v' \qquad (5-4.21)$$

and furthermore by making (k-θ) appropriately large we obtain:

$$\Delta y(k;\theta) = [I_m-H(I_n-\Pi)^{-1}AK*]v' \qquad (5-4.22)$$

We now investigate the following matrix identity.

$$[I_m-H(I_n-\Pi)^{-1}AK*] H = H-H(I_n-\Pi)^{-1}AK*H$$

$$= H(I_n-\Pi)^{-1} [(I_n-\Pi)-AK*H]$$

$$= H(I_n-\Pi)^{-1}(I_n-A) \qquad (5-4.23)$$

Here, Π is a stable matrix and H is a matrix of full rank. So, if (I_n-A) is non-singular (although this is the usual case), the right side of (5-4.23) is of full rank and thus we have a non-singular matrix [$I_m-H(I_n-\Pi)^{-1}AK*$]. This means that $\Delta y(k;\theta)$ takes a non-zero value for any non-zero vector v'. However, when (I_n-A) is singular, it is possible for the right side matrix $H(I_n-\Pi)^{-1}(I_n-A)$ to become a matrix of non-full-rank. If it is not full rank (we let the rank be m', $m' < m$), the matrix $[I_m-H(I_n-\Pi)^{-1}AK*]$ no longer becomes a non-singular one and thus $\Delta y(k;\theta)$ in (5-4.22) can take the value zero even for a non-zero vector v'. Such a vector v' is given by the form $H\omega$, where ω is any vector in the null space of $H(I_n-\Pi)^{-1}(I_n-A)$. This completes the proof. Attention must be drawn to the dimension of the weakly-diagnosable-space. As the null space of H is included in that of $H(I_n-\Pi)^{-1}(I_n-A)$, the range space $\{H\omega : \omega \in \Omega\}$ has the dimension $(m-m')$. Hence, the weakly-diagnosable-space has the dimension $(m-m')$ and it does not exceed, as can be shown, the multiplicity of eigenvalue 1 of the system matrix A.

From Theorem 2, we obtain the following corollary immediately.

Corollary 2: If the system is observable and the matrix (I_n-A) is non-singular, the sensor fault condition has no weakly-diagnosable-space.

Proof: This follows in a straightforward manner from Theorem 2.

As the condition of the corollary is usually satisfied, Corollary 2 is more important than Theorem 2.

Next, we consider the composite fault case which causes dynamics and sensor biases to occur simultaneously. First, a corollary will be presented.

Corollary 3: If the system is observable, then the vectors

$$H \sum_{j=0}^{p} A^j v + v' \quad ; \quad 0 \leq p \leq n$$

do not all vanish for any non-zero vector $v \triangleq (v^T, v'^T)^T$, where $v \in R^n$ and $v' \in R^m$.

Proof: The corollary is easily proved by a contradiction.

Thus, the following theorem can be stated:

Theorem 3: If the system is observable, any cancellation of effects by dynamics and sensor steps on the innovations sequence cannot happen (and thus any fault causing biases on both the dynamics and the sensors can be detected, in principle, by the step-hypothesised GLR method).

Proof: Let the dynamics and the sensor steps (in the similar sense as in (5-4.3)) be $v \epsilon R^n$ and $v' \epsilon R^m$, respectively. Denoting now the observations in normal and abnormal states, respectively, by $y^o(k)$ and $y(k)$, then we have:

$$y(k) = y^o(k) + \Delta y(k;\theta)$$
$$= y^o(k) + [H \sum_{j=0}^{k-\theta} A^j v + v'] \sigma(k,\theta) \qquad (5\text{-}4.24)$$

By Corollary 3, we see that $\Delta y(k;\theta)$ does not always vanish for $k \geq \theta$. Furthermore, since the innovations sequence is obtained by a non-singular linear transformation of the observations (as easily seen from the recursive relation (5-3.1)-(5-3.3)), we can conclude that the effect of the dynamics and the sensor step faults on the innovations does not always vanish to zero. This completes the proof.

Theorem 3 means that the divergence:

$$D(H_o;H_1(\tilde{v})) = \tilde{v}^T \tilde{C}(k;\theta)\tilde{v} \qquad (5\text{-}4.25)$$

is positive for any non-zero vector $\tilde{v}$, where $\tilde{C}(k;\theta)$ is an information matrix for the vector $\tilde{v}$. But, a weakly-diagnosable-space can exist for the composite fault case.

Theorem 4: The condition under which the weakly-diagnosable-space exists is that there exists a non-zero vector $\tilde{v} \triangleq (v^T, v'^T)^T$ such that:

$$H(I_n-\Pi)^{-1}v + [I_m-H(I_n-\Pi)^{-1}AK^*]v' = 0 \qquad (5\text{-}4.26)$$

Proof: The condition follows by combining Theorems 1 and 2.

It is clear that the weakly-diagnosable-space is constructed by such a vector $\tilde{v}$. The dimension of the space is at least n. If (I_n-A) is non-singular (although this is the usual case), the space becomes equal to the null space of $[\,H(I_n-A)^{-1}\!\mid\! I_m\,]$. This is easily verified by (5-4.19) together with the relation:

$$[\,I_m - H(I_n-\Pi)^{-1}AK*\,] = [\,I_m + HA(I_n-A)^{-1}K*\,]^{-1} \tag{5-4.27}$$

which is also derived by using (5-4.19).

For reference, in Appendix 5.2 we show the behaviour of the estimation error of the state for the dynamic step fault included in the weakly-diagnosable-space.

5.5 PROPOSED DETECTION METHOD

A. Detection method I

In section 5.4, we have seen that there exists a weakly-diagnosable-space especially for dynamic faults. Of course, such a space can exist for a composite fault case. But, the discussions are quite similar and the case of dynamic malfunction is more essential. So, we consider here only the former case for simplicity.

As we have seen, the dimension of the weakly-diagnosable-space for the dynamic fault case is $(n-m)$. This means that the information matrix $C(k;\theta)$ has $(n-m)$ eigenvalues which are bounded by increasing k. The remaining ones, i.e., m other eigenvalues increase with the time k. We consider now a k such that $(k-\theta)$ is appropriately large. Then, the $(n-m)$ smallest eigenvalues take almost their convergence values. We denote the corresponding eigenvectors by $s_1, s_2, \ldots, s_{n-m}$ and the other eigenvectors by $s_{n-m+1}, \ldots, s_n$.

If in general we assume that the dynamics step v takes the form $v=S\beta$, where S is an $n \times q$ matrix and β is a q-dimensional vector, then the GLR can be calculated as:

$$\ell(k;\theta) = (1/2)\hat{v}^T(k;\theta)C(k;\theta)\hat{v}(k;\theta) \tag{5-5.1}$$

where $\hat{v}(k;\theta)$ is the MLE of the dynamics step v and given by:

$$\hat{v}(k;\theta) = S\,[S^T C(k;\theta)S\,]^{-1}\,(S^T \sum_{j=\theta}^{k} G^T(j;\theta)V^{-1}(j)\gamma(j)\,) \tag{5-5.2}$$

Letting now $S \overset{\Delta}{=} [s_1, s_2, \ldots, s_n]$ and noting that $s_1, s_2, \ldots, s_n$ are orthogonal eigenvectors of $C(k;\theta)$, it can easily be shown that:

$$\ell(k;\Theta) = \sum_{j=1}^{n} \ell_j(k;\Theta) \tag{5-5.3}$$

where:

$$\ell_j(k;\Theta) = (1/2)\frac{1}{s_j^T C(k;\Theta)s_j} < G(k:\Theta)s_j, \gamma(k)>^2 \tag{5-5.4}$$

The quantity $\ell_j(k;\Theta)$ represents the GLR for the component of the dynamic step v on the direction of s_j.

We have usually adopted the following detection law:

$$\ell(k;\Theta) = \begin{cases} < \epsilon \; ; \; \text{no fault} \\ \geq \epsilon \; ; \; \text{fault} \end{cases} \tag{5-5.5}$$

We see from (5-5.3) that this detection law is equal to:

$$\sum_{j=1}^{n}\ell_j(k;\Theta) = \begin{cases} < \epsilon \; ; \; \text{no fault} \\ \geq \epsilon \; ; \; \text{fault} \end{cases} \tag{5-5.6}$$

So, this detection law assumes implicitly a uniform detectability on all directions of $s_1, s_2, \ldots, s_n$. However, as has already been shown, the vector v belonging to the (n-m) dimensional subspace spanned by (n-m) vectors $s_1, s_2, \ldots, s_{n-m}$, i.e. the weakly-diagnosable-space, has a weak sensitivity against the fault detection. This means that by the detection law (5-5.5) the probability of missed alarm for the dynamics step v appearing from near the space becomes very large compared to the other directions and thus the total probability of missed alarm becomes also large. Thus, it would be better to take this fact into account in the fault detection procedure. One approach to be considered is to introduce a weight on each GLR $\ell_j(k;\Theta)$ as follows:

$$\sum_{j=1}^{n} c_j\ell_j(k;\Theta) = \begin{cases} < \epsilon \; ; \; \text{no fault} \\ \geq \epsilon \; ; \; \text{fault} \end{cases} \tag{5-5.7}$$

where $c_1, c_2, \ldots c_n > 0$. The first (n-m) weighting coefficients $c_1, c_2, \ldots, c_{n-m}$ must, of course, be chosen larger than the remaining ones. We can consider the following criteria to assist in determining the coefficients.

$$\begin{array}{ll} \underset{\|v\|_E = \alpha}{\text{Max}} \quad \text{(Probability of Missed Alarm)} \rightarrow \underset{\{c_j\}}{\text{Min}} & (5\text{-}5.8) \end{array}$$

$$\int_{\|v\|_E = \alpha} \text{(Probability of Missed Alarm)} dv \rightarrow \underset{\{c_j\}}{\text{Min}} \qquad (5\text{-}5.9)$$

where $\|v\|_E$ represents the Euclidean norm of v defined by $(v^T v)^{1/2}$ and α means an appropriate level of anomaly, for example, the minimum level which must be detected. The criteria must, of course, be performed with the probability of false alarm fixed as a certain value. But, because of the difficulty in obtaining the optimal solution analytically, we will have to rely on a numerical computation method in the determination.

In the equations below, we show how to evaluate the probabilities of false and missed alarms. We first rewrite (5-5.4) as follows:

$$\ell_j(k;\theta) = (1/2) < \frac{G(k;\theta)s_j}{\|G(k;\theta)s_j\|} , y(k) >^2 \qquad (5\text{-}5.10)$$

This becomes clear on recalling the definitions of the norm $\|\cdot\|$ and the information matrix $C(k;\theta)$. We also see that the following relations hold:

$$< \frac{G(k;\theta)s_i}{\|G(k;\theta)s_i\|} , \frac{G(k;\theta)s_j}{\|G(k;\theta)s_j\|} > = \delta_{ij} \quad (1 \leqq j \leqq n) \qquad (5\text{-}5.11)$$

where δ_{ij} is the Kronecker delta. This equation is easily derived by noting that $s_1, s_2, \ldots, s_n$ are orthogonal eigenvectors of the information matrix $C(k;\theta)$.

We now define the symbol:

$$\epsilon_j = \frac{G(k;\theta)s_j}{\|G(k;\theta)s_j\|} \qquad (1 \leqq j \leqq n) \qquad (5\text{-}5.12)$$

Then, we find that $\epsilon_1, \epsilon_2, \ldots, \epsilon_n$ become orthonormal vectors of the vector space whose inner product is defined by (5-4.1) and that the detection law (5-5.7) can be rewritten as:

$$\sum_{j=1}^{n} c_j < \epsilon_j, y(k) >^2 = \begin{cases} < 2\epsilon & ; \quad \text{no fault} \\ \geqq 2\epsilon & ; \quad \text{fault} \end{cases} \qquad (5\text{-}5.13)$$

This means that the probability of false alarm can be obtained by re-placing $y(k)$ by $y^o(k)$ and collecting all the events which allow the judgement of *fault or failure*. That is,

$$P_F(c_1,c_2,\ldots,c_n,\epsilon) = \text{Prob.}\left\{ y^o(k); \sum_{j=1}^{n} c_j < \epsilon_j, y^o(k) > 2 \right.$$
$$\left. \underset{=}{\geq} 2\epsilon \right\} \tag{5-5.14}$$

After some calculations, we have:

$$P_F(c_1,c_2,\ldots,c_n,\epsilon) = 1 - \int_{\Omega_F} p(\eta)d\eta$$

$$, \eta \overset{\Delta}{=} (\eta_1,\eta_2,\ldots,\eta_n)^T \tag{5-5.15}$$

where $p(\eta)$ and Ω_F represent, respectively, the Gaussian probability density function with zero-mean and covariance I_n and the n-dimensional hyperellipsoid defined by:

$$\Omega_F \overset{\Delta}{=} \left\{ \eta ; \sum_{j=1}^{n} c_j\eta_j^2 \underset{=}{\leq} 2\epsilon \right\} \tag{5-5.16}$$

The probability of missed alarm for the anomaly vector v can be similarly obtained by:

$$P_M(c_1,c_2,\ldots,c_n,\epsilon,v) = \text{Prob.} \left\{ y^o(k) ; \sum_{j=1}^{n} c_j(< \epsilon_j, y^o(k) > \right.$$

$$\left. + d_j)^2 \underset{=}{\leq} 2\epsilon \right\} \tag{5-5.17}$$

where:

$$d_i \overset{\Delta}{=} < \epsilon_i, G(k;\theta)v > \quad (1 \underset{=}{\leq} i \underset{=}{\leq} n) \tag{5-5.18}$$

We easily see that (5-5.17) reduces to:

$$P_M(c_1,c_2,\ldots,c_n,\epsilon,v) = \int_{\Omega_M} p(\eta)d\eta \tag{5-5.19}$$

where $p(\eta)$ and Ω_M represent, respectively, the same probability density function as described above and the n-dimensional hyperellipsoid defined by:

$$\Omega_M \overset{\Delta}{=} \left\{ \eta ; \sum_{j=1}^{n} c_j(\eta_j+d_j)^2 \underset{=}{\leq} 2 \epsilon \right\} \tag{5-5.20}$$

The only remaining task is to adjust the parameters $c_1,c_2,\ldots,c_n$ and

ϵ such that the probability of false alarm P_F takes a specified value and that at the same time the probability of missed alarm P_M takes also a desirable one for an appropriately defined anomaly level as shown in (5-5.8) and (5-5.9).

B. Detection method II

The bias ν does not always occupy the dominant part of the anomalous function $f(k) = \Delta Ax(k) + \Delta Bu(k)$. It is just for the case where $f(k)$ has a large bias on a round interval of an appropriate length that the step hypothesised GLR method has a great effect in the fault detection. But, practically, there are often the cases where $f(k)$ has not such a bias. So, we cannot necessarily extract all the useful information in the fault detection by the dynamics step hypothesis. This means that we can expect a much more accurate and rapid fault detection by taking into account also the estimation of dominant time-varying modes other than a bias which are contained in the anomalous function $f(k)$. If we consider appropriate basis functions $\Phi_{ij} = e_i \Psi_{ij}(k;\theta)\sigma(k;\theta)$ $(1 \leq j \leq p_i, \ 1 \leq i \leq n)$ to express the anomalous function $f(k)$, where e_i denotes the i-th natural basis on R^n, the fault detection is achieved by estimating the anomalous function $f(k)$ by a linear combination of these basis functions. About the choice of these basis functions we mention later.

Now by introducing the functional subspace S spanned by the functions $\epsilon_{ij} = G^0(k;\theta)*\Phi_{ij}(k)$ $(1 \leq j \leq p_i, \ 1 \leq i \leq n)$, we see that the GLR can be expressed as:

$$l(k;\theta) = (1/2) \ [\ \| \ y(k) \ \|^2 -$$

$$\min_{\{c_{ij}\}} \| \ y(k) - G^0(k;\theta)*(\sum_{i,j} c_{ij}\Phi_{ij}(k)) \ \|^2 \ ]$$

$$= (1/2) \ [\ <y(k),y(k)> - <y(k),y(k)> \ ]$$

$$= (1/2) \ < y*(k),y*(k) > \tag{5-5.21}$$

where $y*(k)$ is the component of $y(k)$ projected orthogonally to the function subspace S and $y(k)$ is the orthogonal component of $y(k)$ to the subspace S (see Figure 5-3). Some additional calculations then yield:

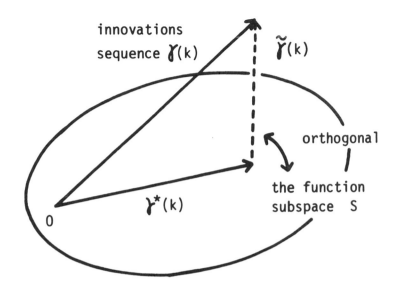

Figure 5-3: Orthogonal projection of innovations sequence

$$y*(k) = \sum_{j=1}^{N} c_j^{o*} \epsilon_j^o \tag{5-5.22}$$

and

$$\ell(k;\theta) = (1/2) \sum_{i=1}^{N} \sum_{j=1}^{N} c_i^{o*} c_j^{o*} < \epsilon_i^o \, , \, \epsilon_j^o >$$

$$= (1/2) \, [c_1^{\ast o}, c_2^{\ast o}, \ldots, c_N^{\ast o}] \left[\left[< \epsilon_i^o, \epsilon_j^o > \right] \right] \begin{bmatrix} c_1^{o*} \\ c_2^{o*} \\ \vdots \\ c_N^{o*} \end{bmatrix} \tag{5-5.23}$$

where:

$$
\begin{bmatrix} c_1^{o*} \\ c_2^{o*} \\ \vdots \\ c_N^{o*} \end{bmatrix} = \begin{bmatrix} < \epsilon_i^o, \epsilon_j^o > \end{bmatrix}^{-1} \begin{bmatrix} < \epsilon_1^o, y(k) > \\ < \epsilon_2^o, y(k) > \\ \vdots \\ < \epsilon_N^o, y(k) > \end{bmatrix} \qquad (5\text{-}5.24)
$$

Here, $N = p_1 + p_2 + \ldots + p_n$ and the renumbering of the coefficients and the basis functions is done for clarification. The $N \times N$ matrix $[<\epsilon_i^o, \epsilon_j^o>]$ represents the information matrix for the coefficients $\{ c_j^o \}$. This matrix is the Gram matrix corresponding to the basis functions $\{ \epsilon_j^o \}$ and it gives an information about the linear independence of the functions.

An important point in applying the method is that we should not provide more basis functions than are truly needed to track the anomaly. This is particularly important from the viewpoint of decreasing the probability of false and missed alarms. Redundant basis functions can be *a priori* removed by checking the strength in the linear independence of $\{\epsilon_{ij}\}$. However, as a second stage, we can further select truly important ones from the finally provided basis functions adaptively while using the method. This selection is made by checking the increase and decrease in the GLR. Anyway, when q basis functions are used in the calculation of the GLR, the probability density of the GLR in the no-fault case obeys the Chi-squared one with q degrees of freedom. So, we must take into account the number of used functions in determining the threshold for the GLR. The usual dynamics step hypothesis consists of nothing but estimating the anomalous function $f(k)$ by a linear combination of $e_i \sigma(k, \theta)$ $(1 \leq i \leq n)$. So, if the above basis functions ϕ_{ij} $(1 \leq j \leq p_i,\ 1 \leq i \leq n)$ contain the step functions $e_i \sigma(k, \theta)$ $(1 \leq i \leq n)$, the detection method proposed here includes the step-hypothesised GLR method as the special one.

We next consider the choice of the basis functions. Although we can consider smooth functions like triangular or spline functions as the basis functions to express the anomalous function $f(k) = \Delta A x(k) + \Delta B u(k)$, we can also make use of the information about the state $x(k)$ and the input $u(k)$ of the system because they are constituent

elements of the function f(k). So, we can consider the basis
functions $e_i u_j(k)\sigma(k,\theta)$ $(1 \leq j \leq r, 1 \leq i \leq n)$ and $e_i m_j(k)\sigma(k,\theta)$
$(1 \leq j \leq n, 1 \leq i \leq n)$ to express the anomalous function f(k), where
$u_j(k)$ is the j-th component of the input u(k) and $m_j(k)$ is the j-th
mode function of the closed loop system: $x(k+1)=(\Lambda-BF)x(k)$, where F
is a feedback coefficient matrix. In other words, we can consider the
basis functions $G^O(k;\theta)*e_i u_j(k)$ $(1 \leq j \leq r, 1 \leq i \leq n)$ and
$G^O(k;\theta)*e_i m_j(k)$ $(1 \leq j \leq n, 1 \leq i \leq n)$ for the subspace S in order to
track the anomaly effect $\Delta y(k;\theta)$. This approach is also valid for
sensor faults with a gain change.

In addition to the basis functions to express the term $\Delta Ax(k)$, we
can also consider the observation signal y(k). This approach is
effective, especially when some state variables can be resolved directly only by the observation at each time step. Since the observation is usually corrupted by noise, it is better to use a low-pass
filter for smoothing. Letting these functions be $t_j(k)$ $(1 \leq j \leq \alpha)$,
we have then the basis functions $G^O(k;\theta)*e_i t_j(k)$ $(1 \leq j \leq \alpha, 1 \leq i \leq$
n) to express the anomaly effect $G^O(k;\theta)*\Delta Ax(k)$ by the term $\Delta Ax(k)$.
As $G^O(k;\theta)$ works as an impulse response of a low-pass filter, we can
also use such an observation y(k) directly instead of the smoothed
functions $\{t_j(k)\}$ because the convolution sum of the noise and the
fault signature matrix $G^O(k;o)$ takes a small value compared with that
of $G^O(k;\theta)$ and the state variables and thus the noise effect
vanishes. This approach is powerful for the case where we cannot
obtain reliable information about the state x(k) and the input u(k)
of the system.

5.6 SOME COMMENTS ON THE DETECTION AND DETECTABILITY

A. Actuator faults

When a bias is added to the input of the i-th actuator, i.e. the case
where u(k) takes the form of u(k) + $\alpha e_i \sigma(k,\theta)$, the step-hypothesised
approach is directly applicable, because the fault can be exactly
modelled by the dynamic step $\alpha b_i \sigma(k,\theta)$, where b_i represents the i-th
column of B. So, if the system is such that b_i takes a vector near
the weakly-diagnosable-space, the fault is difficult to detect. In
such a case, *Detection Method I* has an advantage to be exploited.

Other faults corresponding to the i-th actuator can be modelled by the addition of the anomalous function $\Delta b_i u_i(k) \sigma(k;\theta)$ to the dynamic equation, where Δb_i is a change in b_i and $u_i(k)$ is the i-th component of $u(k)$. Then, we have:

$$\Delta y(k;\theta) = G^o(k;\theta) * [\Delta b_i u_i(k)]$$

$$= [G^o(k;\theta)*u_i(k)] \Delta b_i$$

$$= G*(k;\theta) \Delta b_i \qquad\qquad (5\text{-}6.1)$$

This means that we can detect the anomaly by a dynamic step-hypothesis on B by introducing the new fault signature matrix [$G^o(k;\theta)*u_i(k)$] and that the diagnosability of the fault is governed by the information matrix:

$$C*(k;\theta) \overset{\Delta}{=} \sum_{j=\theta}^{k} G*^T(j;\theta)V^{-1}(j)G*(j;\theta) \qquad\qquad (5\text{-}6.2)$$

We easily obtain the following theorem.

Theorem 5: If $u_i(k)$ is non-trivial, i.e., not constantly zero and time-varying, the actuator fault has no weakly-diagnosable-space for the parameter change Δb_i.

The extension to a multi-actuator-fault case is straightforward. We can also suggest the dual control which provides not only the state control, but also a rapid and more accurate fault detection.

B. Plant failures

We presented a general fault detection method in section 5.5B. Here, we limit our interest to fault detection only by the use of the input sequence.

Defining an appropriate matrix $\Delta B' \in R^{n \times r}$ such that:

$$< G^o(k;\theta)*(\Delta A x(k) - \Delta B' u(k)), G^o(k;\theta)*(\Delta B + \Delta B')u(k) > = 0, \qquad (5\text{-}6.3)$$

we obtain:

$$D(H_o;H_1) - D(H_1(\Delta B'');H_1) = \| G^o(k;\theta)*(\Delta B'' u(k)) \|^2$$

$$= \| [G^o(k;\theta)*u(k)] \, column(\Delta B'') \|^2 \qquad (5\text{-}6.4)$$

where $\Delta B'' \overset{\Delta}{=} \Delta B + \Delta B'$ and column $(\Delta B'')$ is the vector composed of the elements of $\Delta B''$. (5-6.4) is just the divergence $D(H_o;H_1(\Delta B''))$. The detectability of the method is governed by the information matrix

defined for the new fault signature matrix [$G^O(k;\theta)*u(k)$] and we
see that there exists no weakly-diagnosable-space for the change $\Delta B''$,
if $u(k)$ is non-trivial. Therefore, under the non-triviality of the
input sequence, we can detect any fault, in principle, by the
method.

C. Hierarchical failure diagnosis

In the section 5.5, we presented two extended detection methods. That
is, the weighted step-hypothesised GLR method (*Detection Method I*) and
the GLR method based on the information about the state and the input
of the system (*Detection Method II*). But, these methods are essentially
for the fault detection in dynamics (plant and actuators) and
sensors. So, in order to obtain more detailed information about the
fault, another diagnosis method like a multi-hypothesis test must be
performed. This multi-hypothesis test is such that GLR is computed
for each physical parameter and by comparing the GLR's the details of
the fault is obtained.

Of course, any fault can be detected and at the same time the
isolation of the fault and the estimation of the fault level are also
achieved by implementing the multi-hypothesis test from the
beginning. But, it generally requires much computation time to be
served for an on-line computer implementation. Thus, the detection
methods proposed here should be regarded as the method for the
second-layer of a hierarchical fault diagnosis system which has the
function of only informing the higher level, that is the first layer,
of the fault occurrence and an appropriate multi-hypothesis test of
the first-layer takes over the work in order to obtain more detailed
information about the fault and to consider how to cope with the
situation. See Figure 5-4 for the block diagram of a hierarchical
fault diagnosis system. In Figure 5-4 two detection methods are shown
corresponding to *Detection Methods I* and *II* as the method for the
second-layer. But, of course, we can use only one. There is also the
case where the fault detection in the second-layer alone is
sufficient such as the case where we can replace the faulty sensor
with a new one without knowing the form of the fault.

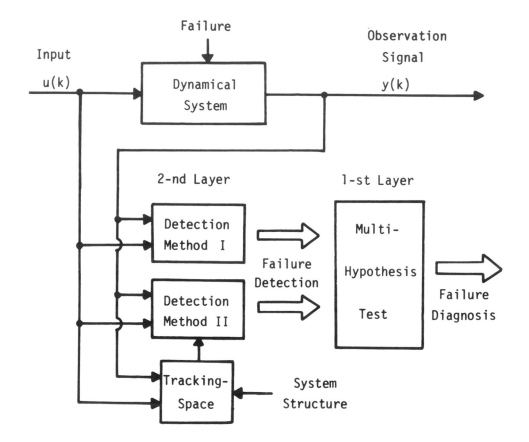

Figure 5-4: Block diagram of the hierarchical fault diagnosis system

5.7 OPTIMAL SENSOR LOCATION

A. Detectability and separability

We have seen that an information matrix plays an important role in detectability. We define here some detectability and separability indices to describe the ability of the system in detecting and separating the faults. We first consider the fault detection by a bias v appearing in the anomalous function $f(k)$. For the bias v, for example, appearing from the entire space R^n, we can define the detectability index by:

$$J(k;\theta) \stackrel{\Delta}{=} \min_{\| \nu \|_E = 1} D(H_0; H_1(\nu))$$

$$= \lambda_{min} [C(k;\theta)] \qquad (5\text{-}7.1)$$

and for the case where ν appears only from a subspace spanned by $s_1, s_2, \ldots, s_q$, we can similarly define:

$$J'(k;\theta) \stackrel{\Delta}{=} \lambda_{min} [S^T C(k;\theta) S] \qquad (5\text{-}7.2)$$

where $S \stackrel{\Delta}{=} [s_1, s_2, \ldots, s_q]$ and $\lambda_{min}[.]$ denotes the minimum eigenvalue of the matrix. We can also consider the case where the bias ν can occur from two subspaces S_1 and S_2, where $S_1 \cap S_2 = \{0\}$. Then, it becomes often a problem to judge from which subspace the anomaly vector ν has occurred. Thus, we need to define a quantity to describe the easiness of the separation. We can define the separability index by a function of the minimum angle ω between the two functional subspaces $\Delta y(k;\theta)$ ($\nu \in S_1$) and $\Delta y(k;\theta)$ ($\nu \in S_2$), i.e.,

$$s(k;\theta) \stackrel{\Delta}{=} (\sin \omega)^2$$

$$= \lambda_{min} [(S_1^T C(k;\theta) S_1)^{-1/2} G (S_1^T C(k;\theta) S_1)^{-1/2}] \qquad (5\text{-}7.3)$$

where S_1 and S_2 are matrices composed of the basis vectors of S_1 and S_2, respectively, i.e., $S_1 \stackrel{\Delta}{=} [s_{11}, s_{12}, \ldots, s_{1q1}]$, $S_2 \stackrel{\Delta}{=} [s_{21}, s_{22}, \ldots, s_{2q2}]$, and G is the Gram matrix of the orthogonal components of $G(k;\theta)s_{1j}$ ($1 \leq j \leq q_1$) to the subspace spanned by $G(k;\theta)s_{2j}$ ($1 \leq j \leq q_2$). The separability index between the dynamics and the sensor faults can be also defined in the same way.

Extension to a general case is also possible. For example, as the detectability index characterising the ability of the system in detecting the specified anomalous functions $g_j(k)\sigma(k,\theta)$ ($1 \leq j \leq \alpha$), we can define:

$$J''(k;\theta) \stackrel{\Delta}{=} \min_{1 \leq j \leq \alpha} \| G^0(k;\theta) * g_j(k) \|^2 \qquad (5\text{-}7.4)$$

where $\{g_j(k)\}$ are n-dimensional vector functions. This index serves for obtaining a sensor location which is especially sensitive to the faults characterised by the anomalous functions $\{g_j(k)\}$. If we take into account also the separation between those faults, we can consider:

$$J'''(k;\theta) \triangleq \lambda_{min} [< G^o(k;\theta)*g_i(k) , G^o(k;\theta)*g_j(k) >] \qquad (5-7.5)$$

We see that (5-7.1) is a special case of (5-7.5) and is obtained by letting $g_j(k) = e_j$ and $\alpha = n$.

B. Optimal Sensor Location

An effectiveness of the observation system in the failure detection can be evaluated by the detectability and separability indices. Of course, in the evaluation, observability index must be also taken into account, because the primary objective of observation systems is in knowing the state of the system as exactly as possible. So, desirable criteria for sensor location are generally such that which maximise the three indices. For example, for the case $v \in S_1$ or S_2, where $S_1 \cap S_2 = \{0\}$, we can consider:

$$c_1[\min_{S_1,S_2} J'(k;\theta)] + c_2 s(k;\theta) + c_3 \lambda_{min}[P^{-1}(k/k)] \rightarrow \max_H \qquad (5-7.6)$$

or, as an alternative,

$$c_1 \left(\frac{\min_{S_1,S_2} J'(k;\theta)}{\max_H \min_{S_1,S_2} J'(k;\theta)} - 1 \right)^2 + c_2 \left(\frac{s(k;\theta)}{\max_H s(k;\theta)} - 1 \right)^2$$

$$+ c_3 \left(\frac{\lambda_{min} [P^{-1}(k/k)]}{\max_H \lambda_{min} [P^{-1}(k/k)]} - 1 \right)^2 \rightarrow \min_H \qquad (5-7.7)$$

where c_1, c_2, $c_3 \geqslant 0$. These criteria have an effect for the systems which have flexibility in the location of sensors such as distributed parameter systems or the lumped ones which require an improvement in detectability and/or separability.

We mentioned here only min-max approaches. But, the other approaches such as taking a trace or a determinant of the information matrix instead of minimum-eigenvalue operation are also possible. See Tanaka et al. (1984), Müller and Weber (1972) for the details and the physical interpretation.

5.8 NUMERICAL EXAMPLE

As the philosophy behind the Detection Method I has been clarified, we demonstrate here only the effectiveness of Detection Method II. We apply the detection method to the anomaly detection of a linear motor car system. We consider here a simplified model defined by (see Breinl and Müller (1982) for example):

$$\dot{x}(t) = A(L)x(t)+b(L)u(t)$$

with:

$$x(t) = (\ z(t), \dot{z}(t), z(t) \ )^T$$

$$A(L) = \begin{bmatrix} 0 & 1 & 0 \\ 0 & 0 & 1 \\ \dfrac{K_s R}{mL} & \dfrac{K_s}{m} & \dfrac{K_i K_s}{mL} - \dfrac{R}{L} \end{bmatrix} \ , \quad b(L) = \begin{bmatrix} 0 \\ 0 \\ -\dfrac{K_i}{mL} \end{bmatrix}$$

(see Figure 5-5) where m=16 Kg, R=8 Ω, K_s=57000 N/m, K_i=K_s=114 N/A, L=0.5 Vs/A are assumed. We furthermore assume the observation:

$$y(t) = \begin{bmatrix} 1 & 0 & 0 \\ 0 & 0 & 1 \end{bmatrix} x(t)$$

and the feedback control:

$$u(t) = -F\hat{x}(t) = -f_1\hat{x}_1(t) - f_2\hat{x}_2(t) - f_3\hat{x}_3(t)$$

where f_1= -9123.5 V/m, f_2= -470.2 Vs/m, and f_3= -7.1 Vs^2/m [8].

By assuming a sampling time of ΔT=0.015s, x(0)=(0.01, 0.05, 0.0)^T, $\hat{x}$(0/-1)= x(0), P(0/-1)=$10^{-4}I_3$, $\Gamma=I_3$, Q=$9\times10^{-8}I_3$, and R=$9\times10^{-8}I_2$, we obtain the discrete system (5-2.1) and (5-2.2). Let now the anomaly correspond to the variation of the inductance value L from 0.5 to 0.54 (+8% variation) at the discrete time 15 (i.e., Θ=15). Then, the application of *Detection Method II* gives us the following results, that is, 76% by the method based on only the information about the variation modes of the state, 82% by that based on the information about the input sequence, and lastly 88% by that based

on the information about the state by the use of the observation
signal. In the applications, we selected three most dominant
functions adaptively from the given basis functions.

For comparison, we next applied a Chi-squared test and the
step-hypothesised GLR method and obtained the detection rates 40% and
64%, respectively. We see that the proposed methods are much superior
to the conventional ones in the detection rate. In the simulations,
we assumed 100 different noise sequences for each method and adopted
the window-length of 11 and the threshold allowing a false alarm of
5%.

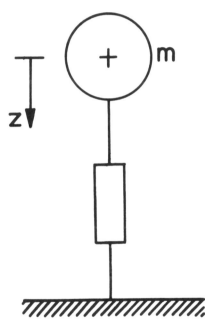

Figure 5-5: Simple model of a linear motor car

For reference, we show in Figures (5-6) to (5-8) the distribution of
the detection time only for the correct alarm for the Chi-squared
method, the dynamic fault step-hypothesised method, and the proposed
one based on the information about the state by the observation
signal. We see that the proposed methods are also superior to the
conventional ones in the detection speed. Figure (5-9) shows an
evolution of the GLR by the proposed method using the input sequence.
We can see a rapid rise in the GLR after the occurrence of the
anomaly.

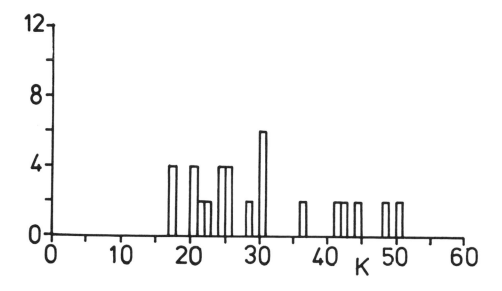

Figure 5-6: Detection time for chi-squared test

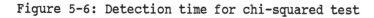

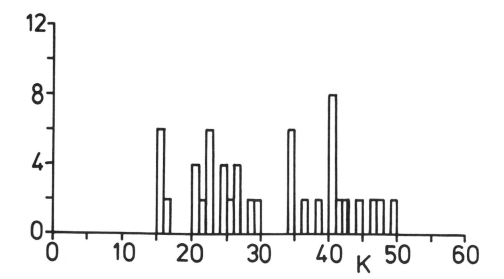

Figure 5-7: Detection time for dynamics step-hypothesised GLR method

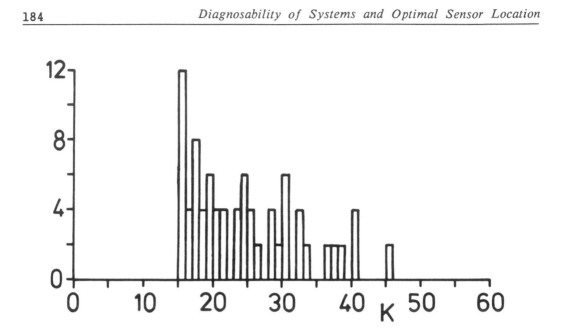

Figure 5-8: Detection time for proposed method making use of the observations

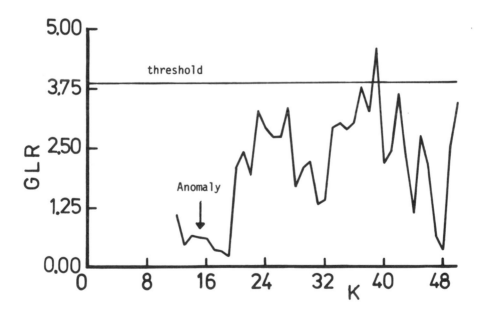

Figure 5-9: An Evolution of GLR for proposed method (making use of the input

In the application of the proposed method, we used the available information about the state and the input separately and also limited the number of actually used basis functions to n (=3) for simplicity. But, if we try to combine the informations and also change the number adaptively corresponding to the situation, we may be able to expect much higher detection rates. But, the details must be determined by a trade-off between the detection rate and the computational time.

In addition to the computation time *detection method II* needs extra time compared with the conventional ones for selecting some functions from the given basis functions and computing the GLR. So, if it will make a problem in on-line use (of course, it depends on the system), we can consider a slight modification such that the fault detection should be performed only at appropriately-spaced times, but not at all times.

5.9 CONCLUSIONS

The paper analysed first the step-hypothesised GLR method for a general fault in linear dynamical systems and discussed the detectability. By the analysis, we found that a weakly-diagnosable-space can exist for dynamic system faults rather than for sensor ones and it is beyond the problem of window-length.

To overcome the difficulty, two extended methods were developed. The first method is an introduction of a weighted step-hypothesised GLR method which can decrease the probability of missed alarm only at the sacrifice of delaying the detection time for the failures of directions of high sensitivity. Another method is the GLR one based on the estimation of other quantities than a bias by the use of the available information about the state and the control input of the system.

Of course, the fault diagnosis may also be achieved by a multi-hypothesis testing, but it generally requires a lot of computational time to be served as an on-line use by a small computer which is attached to the system. Thus, the detection methods proposed here should be regarded as the convenient methods for the second-layer of a hierarchical fault diagnosis system and the detailed information about the fault should be gained by an appropriate multi-hypothesis testing only after the judgement 'fault' by the methods described above. Continuously, in this chapter, criteria for optimal sensor location were discussed based on the detectability, separability, and observability indices, and finally the effectiveness of the proposed detection methods was verified by a simulation compared with other conventional ones.

As for constructing the basis functions, we can also consider another approach such as that based on the learning, that is the approach making the basis functions adaptively by processing the innovations. This remains as future work.

ACKNOWLEDGMENT

I would like to thank Professors P. C. Müller and P. M. Frank of Wuppertal and Duisburg universities (FRG), respectively, for their helpful comments on this research and Mr. Hahn for his assistance in gathering simulation data. I also acknowledge research support from Alexander von Humboldt Foundation in Federal Republic of Germany (FRG).

APPENDIX 5.1

We derive the equation (5-4.13). The fault signature matrix $G^O(k;\theta)$ for a dynamics jump satisfies the following recursive relation, Willsky (1986).

$$G^O(k;\theta) = H[A^{k-\theta} - AF^O(k-1;\theta)\ ] \tag{5-A1}$$

$$F^O(k;\theta) = AF^O(k-1;\theta) + K(k)G^O(k;\theta)\ \ ; k=\theta,\ \theta+1,.. \tag{5-A2}$$

where $G^O(\theta;\theta)=H$ and $F^O(\theta;\theta)=K(\theta)H$. Letting $k=\theta+1$, then we have:

$$G^O(\theta+1;\theta) = HA(I_n-K(\theta)H) \tag{5-A3}$$

and:

$$F^O(\theta+1;\theta) = AK(\theta)H+K(\theta+1)HA(I_n-K(\theta)H) \tag{5-A4}$$

Next, letting $k=\theta+2$, similarly:

$$G^O(\theta+2;\theta) = HA(I_n-K(\theta+1)H)A(I_n-K(\theta)H) \tag{5-A5}$$

Generally, we have:

$$G^O(k;\theta) = HA(I_n-K(k-1)H)A(I_n-K(k-2)H) \; x \; ...$$

$$...x \; A(I_n-K(\theta)H) \tag{5-A6}$$

The proof can be made more rigorously through *mathematical induction*.

APPENDIX 5.2

We consider how the dynamics step included in the weakly-diag-nosable-space affects the estimate of the state. For the dynamic step $v\sigma(k,\theta)$, we have:

$$y(k) = y^O(k) + G(k;\theta)v \tag{5-A7}$$

$$x(k) = x^O(k) + \sum_{j=0}^{k-\theta} A^j v \tag{5-A8}$$

$$\hat{x}(k/k) = \hat{x}^O(k/k) + F(k;\theta)v \tag{5-A9}$$

where $G(k;\theta)$ and $F(k;\theta)$ satisfy the following recursive equations:

$$G(k;\theta) = H \; [\; \sum_{j=0}^{k-\theta} A^j - AF(k-1;\theta) \;] \tag{5-A10}$$

$$F(k;\theta) = AF(k-1;\theta) + K(k)G(k;\theta) \quad ; \; k=\theta, \; \theta+1,... \tag{5-A11}$$

where $G(\theta;\theta)=H$ and $F(\theta;\theta)=K(\theta)H$. Furthermore, $\hat{x}(k/k)$ and $\hat{x}^O(k/k)$ represent, respectively, the estimate of $x(k)$ by a Kalman filter assuming no fault and the optimal estimate under no fault.

Since $G(k;\theta)v$ tends to zero as k increases for such a v described above, we have the following relation by combining the above fact with (5-A10) and (5-A11).

$$G(k;\theta)v = H \; [\; \sum_{j=0}^{k-\theta} A^j - AF(k-1;\theta) \;] \; v$$

$$= H \; [\; \sum_{j=0}^{k-\theta} A^j v - F(k;\theta)v \;] \; \rightarrow \; 0 \tag{5-A12}$$

We see from (5-A8), (5-A9), and (5-A12) that the quantity:

$$\sum_{j=0}^{k-\theta} A^j v - F(k;\theta)v$$

is the estimation error in the state $x(k)$ caused by the dynamics step failure and is always included in the null-space of the observation matrix H. We also see that $F(k;\theta)$ does not necessarily converge to a constant matrix, whereas $G(k;\theta)$ does if the system has a stationary state. More precisely speaking, $F(k;\theta)$ converges only when all the eigenvalues of the system matrix A are real and stable. Otherwise, $F(k;\theta)$ oscillates or diverges.

Chapter Six

BAND-LIMITING FILTER APPROACH
TO FAULT DETECTION

J.G. Jones and M.J. Corbin

A method is described whereby faults in control systems may be detected and identified on the basis of the generation and cross-comparison of band-limited signals from dissimilar sources. Whilst signals measured at different points in the system may differ significantly when viewed over a wide bandwidth, over limited pass-bands there may be shown to exist simple relationships which can be verified by means of appropriate filtering and comparison logic. The band-limiting filters employed are in effect used to monitor the propagation of information, originating in the structure of external inputs, through the system.

6.1 INTRODUCTION

One way to achieve fault-tolerance in safety-critical control systems is by means of multiple lanes of identical hardware, including sensors, actuators and computers. Two disadvantages of this approach are the weight penalty paid (particularly important in flight control) and the possibility of a common-mode (design) fault of the system. The latter is most relevant in the context of the current trend towards digital implementation of control laws, in which the possibility of software-design errors arises. A tool that is applicable to both of these problems is *analytical redundancy*, an expression denoting the use of additional computing to monitor system health in terms of the mutual consistency of dissimilar signals. As described in other chapters of this book, this approach makes use of analytical relationships among sets of signals, such as outputs from different sensors, or a sensor output and a computer output (such as an actuator-demand signal), derived from theoretical models of system behaviour.

Previous studies of analytical redundancy have generally made extensive use of Kalman filters and associated state-space models of the system, together with Gaussian models of external disturbances and minimum-mean-square (MMS) design criteria. The alternative approach to be described stems from the view that the criteria inherent in the design of Kalman filters tend to sacrifice sharpness of response to significant transient inputs (in the context of aircraft control, these may consist of pilot demands or sudden gusts) in favour of good noise suppression. In contrast, the method proposed here places less emphasis on noise suppression and more on good resolution of the transient effects of fluctuating inputs occurring during normal system operation.

The underlying theory concerns the information content of fluctuating signals. The manner in which *information* is defined in this context goes back to Gabor (1946) and is based on the use of *minimum-uncertainty cells* or components to represent an arbitrary signal. The band-limiting filters employed are in effect used to monitor the propagation of information, originating in the structure of external inputs, through the system. The technique employed to design the filters makes use of a minimum-uncertainty decomposition of system inputs.

In common with other approaches to analytical redundancy, the basis of the present method may be interpreted in terms of signal *reconstruction*. Some, generally unobserved, signal occurring during system operation is reconstructed independently from dissimilar sources of measurement; the assessment of system health is then based on the compatibility of these independent reconstructions. However, whilst many existing methods are based upon the reconstruction of system state variables, the method described here is based upon the reconstruction, over limited passbands, of system inputs. The band-limiting filters employed are thus designed such that the comparison signals are not only constrained to have prescribed common passbands in the frequency plane but are also tailored to have common timeplane signatures in response to identifiable features of external inputs such as pilot demands or gusts. As a result of these constraints, dissimilar signals recorded at different points in the system, such as pilot control-stick transducer, rate-sensor, accelerometer or actuator-demand, lead after filtering to sets of derived band-limited signals sufficiently closely matched to be used for cross-comparison purposes.

The implementation of the cross-comparison logic involves combining, in the form of weighted sums or differences, band-limited signals from dissimilar sources to form a *null* signal, whose exceedance of a prescribed threshold is used to indicate the occurrence of possible system failure. The simplest case, in which the null signal comprises just the difference between two matched band-limited channels, is illustrated in Figure 6-1.

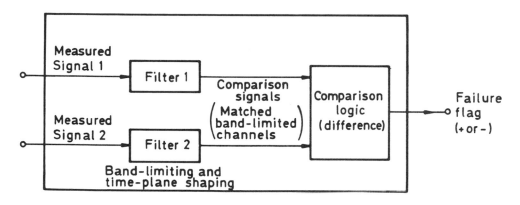

Figure 6-1: Basic principle

More generally, by employing an overall system-performance monitor in which various combinations of channels are checked in this way, patterns of threshold-exceedance flags can, in principle, be generated, from which may be derived not only an indication of possible system failure but also a pointer to the likely failed component.

A simple example of the concept of matched band-limited channels, taken from a ground-based simulator study of a manoeuvring combat aircraft, is illustrated in Figure 6-2. This illustration is idealised, in the sense that no gust disturbances were simulated, but shows the basic principle. Trace (a) shows the longitudinal control-stick input. Trace (b) is the aircraft pitch-rate response, equivalent to the idealised output of a pitch-rate sensor, after filtering with a simple first-order high-pass filter (ie washout). Trace (c) is the result of passing trace (a) through a band-pass filter chosen so that (b) and (c) correspond to matched channels, in a sense to be defined in Section 6.2. These signals may be seen to fluctuate in a closely related manner. Moreover, the common peaks correspond to the occurrence of identifiable features in the pilot stick input, in the form of increments (or changes in level) over a particular range of time intervals. The role of signal increments as a central feature of the design method is discussed in the following section.

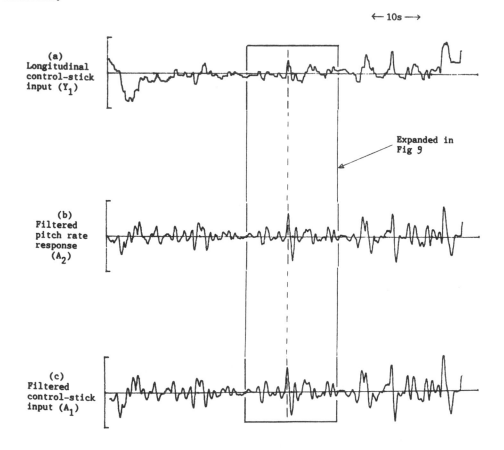

Figure 6-2: Illustration of matched band-limited channels (traces (b) and (c)) derived from pitch-rate response and control-stick input

6.2 BASIS OF DESIGN METHOD

We have mentioned state-space models and Kalman filters as concepts underlying much of the previous work on analytical redundancy. The proposed alternative approach stems from a quite distinct mathematical background which concerns the information content of fluctuating signals. The problem of how to represent mathematically the fluctuations in a time-dependent process was discussed by Gabor (1946) who described a method of signal decomposition into discrete elements (which he called *logons*) each of which is bounded in both time and frequency, the duration Δt and bandwidth Δf of an elementary fluctuation being constrained by an *uncertainty relation* or minimum-area condition:

$$\Delta f \times \Delta t \doteq {}^1/_2 \tag{6-2.1}$$

Brillouin (1962) subsequently discussed the role of Gabor's discrete elements as *cells of information* in the time-frequency plane. The significant discrete events which can be detected by band-limiting filters take the form of changes or increments in input-signal intensity. In modelling such events we are thus concerned with the gradient dx/dt of the input signal $x(t)$, rather than the overall magnitude of $x(t)$. We are thus led to a particular form of the Gabor logon which we refer to as a smooth increment.

Definition: A *smooth increment* (Figures 6-3 and 6-4) is a signal element satisfying the following conditions:

(a)   the time history $x(t)$ contains a smooth transition in amplitude over interval Δt, and

(b)   the significant *gradient energy* $\int \left(\dfrac{dx}{dt}\right)^2 dt$ associated with the signal gradient is constrained within intervals Δt and Δf satisfying equation (6-2.1).

It follows from condition (a) that Δt corresponds approximately to one half-cycle at the centre-frequency f_0 and hence that $2\Delta t = (f_0)^{-1}$ and (using condition (b) together with equation (6-2.1)) $\Delta f/f_0 \doteq 1$.

The *octave-width* constraint, ie $\Delta f/f_0 \doteq 1$, implies that the gradient of the signal takes the form of a broadband wave-packet

(Figure 6-3(a)). A smooth-increment profile is obtained by integration (Figure 6-3(b)). A more convenient practical approximation for engineering design purposes is obtained by suppressing side-bands (Figure 6-4(a)) leading to a smooth-ramp profile (Figure 6-4(b)). The effect of suppressing side-bands in practical applications is small.

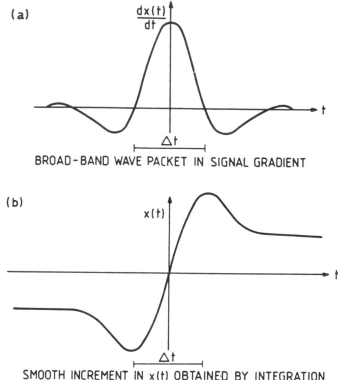

(a)

$\dfrac{dx(t)}{dt}$

BROAD-BAND WAVE PACKET IN SIGNAL GRADIENT

(b)

$x(t)$

SMOOTH INCREMENT IN x(t) OBTAINED BY INTEGRATION

Figure 6-3: Wave-packet representation of smooth increment

The band-limiting filters to be employed, in the proposed implementation of analytical redundancy, act as smooth increment detectors (for convenience we will refer to them simply as *increment detectors*) and are designed as matched filters for input patterns of the form illustrated in Figure 6-4(b). Such filters have been used previously in the detection of discrete gusts and wind-shears and in the localisation of sharp gradients in grey-level intensity (edges) in visual image processing.

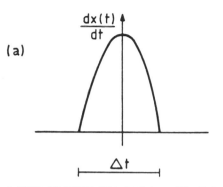

(a)

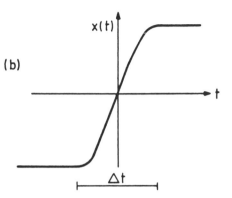

(b)

WAVE PACKET (IN SIGNAL GRADIENT)
WITH SIDE BANDS SUPPRESSED

SMOOTH INCREMENT IN x(t)
OBTAINED BY INTEGRATION

Figure 6-4a & b: Smooth-ramp profile derived from wave-packet with side-bands suppressed

The detection process is essentially as illustrated in Figure 6-5. On the occurrence of an input in the form of a smooth increment of prescribed duration, the matched filter is required to respond with an output fluctuation in the form of a single smooth pulse, ideally with little overswing of the datum level. Such an output may be regarded as an approximate reconstruction of the gradient of the input. This requirement on the shape of the response pulse imposes a band-pass constraint such that the filter bandwidth is of the order (but rather in excess of) one octave. The occurrence of an input of the prescribed form may then be associated with an output peak whose amplitude exceeds some threshold level (chosen with reference to the ambient background noise). In the more general situation a set of such threshold levels may be used to detect input increments of differing intensities.

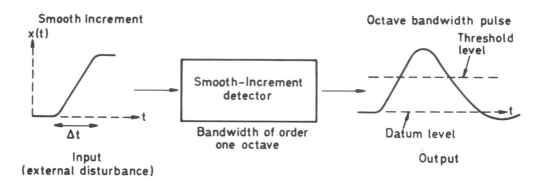

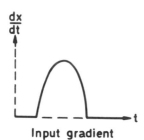

Figure 6-5: Ideal response smooth-increment detector to matched input

The sense in which the filter is matched to a smooth increment of given duration is illustrated in Figure 6-6. Here the response-peak amplitudes $y(\Delta t)$ are shown as a function of Δt for a class of inputs whose amplitudes w take the form $w = \Delta t^k$, where $0<k<1$. The value $k=1/2$ may be shown to give a well-defined peak in the function $y(\Delta t)$. However, other values of k, not too different from 1/2, give equally satisfactory definition; the value $k=1/3$ has been used in many of the examples to be presented as the associated inputs, in the situation where the disturbances are atmospheric gusts, may then be given the statistical interpretation of having equal probability, Jones (1980, 1986(b)). However, this statistical interpretation of the method, when applicable, is an added bonus which enhances the quantitative assessment of system performance but is not essential to the basic system-design method to be described, which is in fact insensitive to the precise value chosen for k.

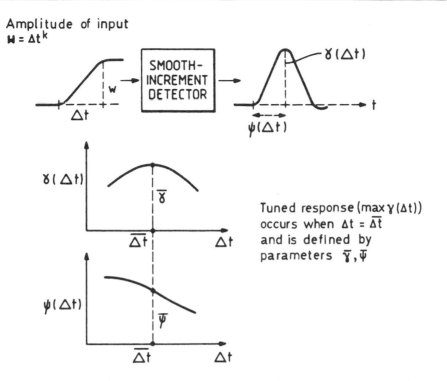

Figure 6-6: Smooth-increment detector for increment of duration $\Delta t = \overline{\Delta \epsilon}$

The condition $0 < k < 1$ implies that, whilst the amplitude w increases with increasing Δt, the mean gradient $w/\Delta t$ decreases. As a result the maximum output $\overline{y}$ for a given band-limiting filter (Figure 6-6) occurs in response to an increment whose duration equals some intermediate tuned time-interval $\overline{\Delta t}$.

Whilst a fluctuation, above threshold level, in the response of a single filter carries information concerning the occurrence of an event in the system input, the joint responses of two such filters matched to the same input carry in addition information concerning the proper functioning or health of the signals in which the band-limited channels are embedded. Two such filters will be said to be *matched channels* if they satisfy the relationships:

$$\overline{\Delta t}_1 = \overline{\Delta t}_2, \quad \overline{y}_1 = \overline{y}_2, \quad \overline{\Psi}_1 = \overline{\Psi}_2 \; . \tag{6-2.2}$$

The third condition above requires (see Figure 6-6) that, in response to their common tuned input, the output peaks occur at the same instant.

Although the design constraints on two matched channels, equations (6-2.2), refer specifically to response peaks, the bandwidth condition (ie $\Delta f / f_0 \doteq 1$) implies that signals having matched peaks will also be closely related over their whole amplitude range. This result has been discussed by Marr et al. (1979), a paper on the information content of band-limited signals in the context of visual information processing, and has been confirmed by the illustrative examples to be described.

6.3 FIRST STEPS IN IMPLEMENTATION

Figure 6-7(a) illustrates a situation in which the external input is measured (Y) by an ideal sensor and is thus directly available for processing by the increment detector I. Whilst this is generally the case for pilot inputs (through a transducer on the pilot's control) it is not so for gust inputs, as angle-of-incidence sensors measure a combined effect of gust and aircraft motion.

An alternative situation is illustrated in Figure 6-7(b). Here the measured signal X is related to the external disturbance through a transfer function S representing some aspect of system response. Examples of such signals are outputs from sensors measuring normal acceleration or pitch rate, or alternatively outputs from the control computer in the form of actuator-rate demand signals. The requirement in this case is to design a complementary filter F such that the combination of S and F in series (dashed line in Figure 6-7(b)) has the desired properties of an increment detector, as outlined in section 6.2.

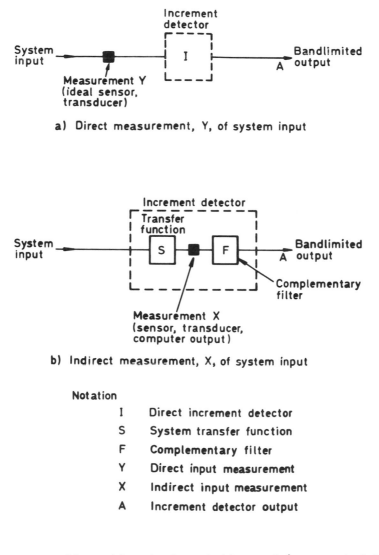

a) Direct measurement, Y, of system input

b) Indirect measurement, X, of system input

Notation

I	Direct increment detector
S	System transfer function
F	Complementary filter
Y	Direct input measurement
X	Indirect input measurement
A	Increment detector output

Figure 6-7: Alternative implementations of increment detector

Figures 6-8 and 6-12 illustrate simple examples of the possibilities open for system monitoring purposes using the above techniques. It is assumed at this stage that flight is in calm air (the influence of gusts is introduced in Section 6.4).

A. Matched channels incorporating direct measurement of input
In a controlled system, signals will be present at many points in the

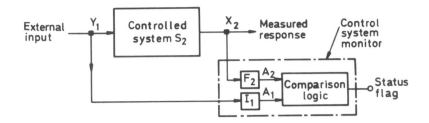

a) Controlled system with measured external input and output

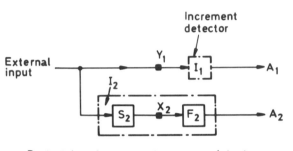

Dashed box denotes an increment detector
—■— Denotes a measurement point

b) Schematic diagram of Fig 8a

Figure 6-8: Example of monitor using measured external input

system, but access for measurement for a control system monitor will only be available at various points in the system. Examples in Figure 6-8(a) are Y_1 and X_2, being respectively a measured external input and a measured response of the controlled system S_2. The response signal X_2 may be fed into other parts of the controlled system, or provided for monitoring only.

A signal from measurement point Y_1 is received by the control-system monitor (Figure 6-8(a)) and band-limited by increment detector I_1 to provide an output A_1. A signal from measurement point X_2 is processed by filter F_2 to provide an output A_2. Outputs A_1 and A_2 are processed by a logic element (in accordance with Figure 6-1) to produce a status flag. Note that the measurement X_2 cannot be directly compared with the measurement Y_1, but is an indirect measurement of the input signal having been acted upon by a transfer function S_2. The combination of filter F_2 and transfer function S_2 acts as an increment detector (Figure 6-8(b)).

As a practical example, X_2 could represent the output of a rate sensor and S_2 the transfer function of the total closed-loop system, including aircraft dynamics and control-system effects, to a pilot input. Note that disparities between the signals A_1 and A_2 may be associated either with faults of the sensor itself or with faults of the control-law implementation in the computer (eg due to a software design error). The latter situation however, as will be seen subsequently, may in general be detected more rapidly by a monitor of the actuator-demand signal. The complementary filter F_2 is first designed so that the combination of S_2 and F_2 in series (dashed box, Figure 6-8(b)) has the desired properties of an increment detector. A first-order lowpass filter (washout) may be sufficient for F_2. I_1, acting directly on the control-stick transducer signal Y_1, is then designed as a matched increment detector (output A_1). Possible failures are flagged when (A_1-A_2) exceeds some prescribed threshold.

The example taken from a ground-based simulator study and presented in Figure 6-2 illustrated such a case. As already mentioned, these results are idealised, in the sense that no gust disturbances were simulated, and are intended only to illustrate basic principles. The incorporation of gust effects is treated in section 6.4.

Trace (a) (Figure 6-2) shows the longitudinal control-stick input, and corresponds to signal Y_1 in Figure 6-8. Trace (b) corresponds to signal A_2 in Figure 6-8, obtained by passing the output of a pitch-rate sensor (X_2) through a first-order high-pass filter (F_2) in the form of a washout. This filter was implemented simply by substracting a smoothed version of the raw signal from the signal itself, where the smoothing took the form of a running average over the preceding 1.2 seconds. This time constant was chosen to be approximately three times the *risetime* to a peak in the step response of the unfiltered signal. Trace (c) corresponds to signal A_1 in Figure 6-8. It is the result of passing trace (a), ie Y_1, through a band-pass filter I_1 chosen so that A_1 and A_2 correspond to matched channels in the sense defined in Section 6.2 (equation (6-2.2)).

The band-pass filter I_1 was implemented as a smoothed increment function in the form:

$$\bar{Y}_T(t) = \int_{t-T}^{t} Y_T(\tau)g_T(t - \tau)d\tau \qquad (6-3.1)$$

where:

$$Y_T(t) = x(t) - x(t - T) \tag{6-3.2}$$

is the increment function acting on input $x(t)$, and $g_T(t)$ is a weighting or smoothing function which essentially averages $Y_T(t)$ over an interval of order T.

The function $\overline{Y}_T(t)$ is clearly a band-pass filter, $Y_T(t)$ providing the high-pass, and $g_T(t)$ the low-pass, constituents. In subsequent sections more efficient forms of band-pass filter are described, that would be recommended for practical use in a performance monitor. However, the filter described by equations (6-3.1) and (6-3.2) allows us conveniently to illustrate the relationship between peaks in the filter output and the design concept summarised in Figures 6-5 and 6-6. A large increment in the input $x(t)$, over interval T, clearly will cause a large peak value to occur in $Y_T(t)$ and hence also in $\overline{Y}_T(t)$. Thus $\overline{Y}_T(t)$ may be seen to act as a smooth-increment detector in the sense of section 6.2.

In order to match the signal A_1 to A_2 (Figure 6-2), the time constant T was chosen so that the first of equations (6-2.2) was satisfied. The value required was in fact $T = 0.73$ second. Note that the second of equations (6-2.2) simply requires an appropriate amplitude factor and the third a simple time delay. Comparison (Figure 6-2) of the matched channels A_1 and A_2 shows a close relationship between peaks in the fluctuations. The form of equations (6-3.1) and (6-3.2) shows, moreover, that these peaks may be associated with identifiable features, in the form of increments over a time interval of approximately $T = 0.73$ second, in the control-stick input (as illustrated in Figure 6-9).

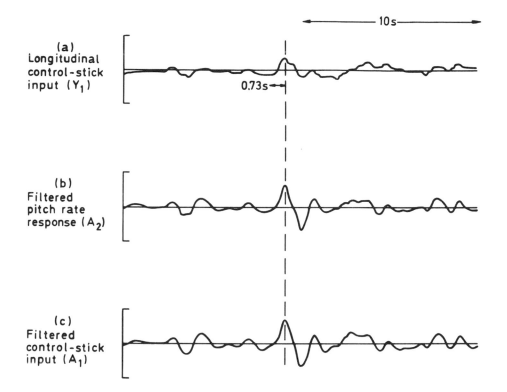

Figure 6-9: Association of peaks in matched band-limited channels
with a particular increment in the control-stick input

Figure 6-10 illustrates a similar pair of matched band-limited
channels, where the measured sensor output X_2 (Figure 6-8) is now
taken to be normal acceleration. The matched-channel outputs A_1 and
A_2 were generated exactly as in the case of Figure 6-2, except that
here the washout time constant was taken to be 2.9 seconds. The
difference between the time constants chosen for the two cases
reflects the differing frequency content of the energy in the unfiltered pitchrate and normal-acceleration channels. In each case the
washout time constant was chosen to be approximately three times the
rise time to a peak in the step response of the unfiltered signal.

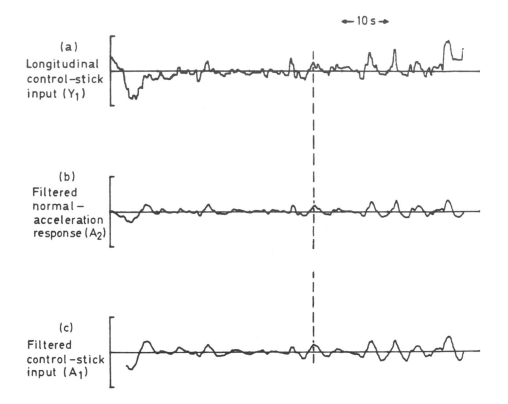

Figure 6-10: Illustration of matched band-limited channels (traces (b) and (c)) derived from control-stick input and normal-acceleration response

Whilst this time constant of 2.9 seconds is a natural choice in terms of the freqency-dependance of the energy in the unfiltered normal acceleration channel, for the purpose of speedy detection of system failures, matched channels based on a shorter time constant would probably be both practicable and preferable.

In the above examples, X_2 represents a measured sensor output. In an alternative interpretation of Figure 6-8, the signal X_2 describes the actuator-position or actuator-rate demand signal (from the control computer) resulting from a pilot input to the system. Note that, as the control computer derives the actuator-demand signal from sensor responses as well as the pilot input, the transfer function S_2 again involves aircraft dynamic response in addition to control-system effects. In practice, the optimum pass-band for monitoring an actuator-demand signal (particularly actuator-rate) tends to be at higher frequencies than that for monitoring sensor signals. Increment detectors based on actuator-rate demand thus have small time constants, making them attractive as a possible basis for monitoring for failures in the control computer (including software functions).

In Figure 6-11, the simulation study is used to illustrate this possibility. The signal X_2 (Figure 6-8) is now taken not to be a sensor signal but the actuator-demand signal from the control-system computer to the elevator. The matched-channel outputs A_2 and A_1 were generated by means of washout filtering of X_2 and band-limiting filtering of the pilot input Y_1 (based on the smoothed increment function, equations (6-3.1) and (6-3.2), as in the previous cases). The mean energy in the unfiltered actuator-demand signal is concentrated at significantly higher frequencies than in the previous (sensor output) cases and the washout filter time constant chosen was correspondingly short (0.2 second).

It is noteworthy that, whilst the external input, Y_1, is identical in Figures 6-2, 6-10 and 6-11, the filtered control-stick signals, A_1, correspond to different bandwidths and bear little relation to one another. However, each of the filtered signals A_1 follows closely the fluctuations in the matched channel A_2 derived from either a sensor response or an actuator demand. Band-limiting filters may thus be seen to provide a means of extracting relevant information, lying in an overlap region in the frequency content of dissimilar signals, which may be used as a basis for the validation of signal compatibility.

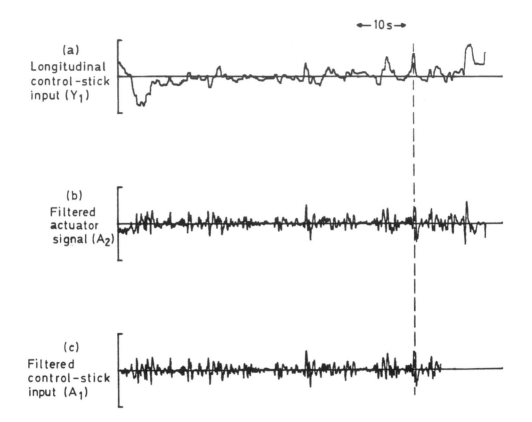

Figure 6-11: Illustration of matched band-limited channels (traces (b) and (c)) derived from control-stick input and actuator-demand signal

B. Matched channels based on indirect measurements of input

A different situation is illustrated in Figure 6-12a, in which the input to the controlled system is not amenable to direct measurement, but two indirect measurements X_3 and X_4 are available, having been acted upon by transfer functions S_3 and S_4, respectively. These measurements are filtered and subjected to comparison logic to produce a status flag. The combinations of transfer functions with filters (S_3 with F_3 and S_4 with F_4) are designed to form matched increment detectors (Figure 6-12(b)).

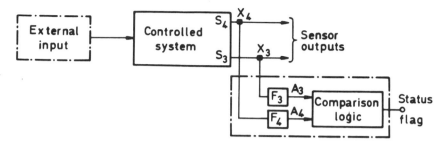

a) Controlled system with unmeasured external input

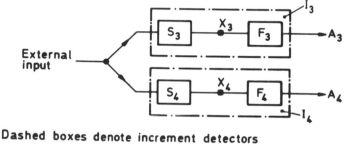

Dashed boxes denote increment detectors
—●— Denotes a measurement point

b) Schematic diagram of Fig 12a

Figure 6-12: Use of two measurements, each indirectly related to an
unmeasured input for monitoring

In a practical situation, X_3 and X_4 (Figure 6-12) could be the
outputs from two sensors, such as pitch rate and normal acceleration.
The transfer functions S_3 and S_4 describe the sensor responses though
the total closed-loop system. Comparison of signals A_3 and A_4, on the
basis of the difference signal (A_3-A_4), may then be used to check for
mutual compatibility of sensor outputs over the common pass-band.

An illustration from the simulator study is shown in Figure 6-13.
Trace (a) shows the control-stick input to the system. Trace (b)
corresponds to signal A_3 in Figure 6-12, obtained by passing the
output of the pitch-rate sensor (X_3) through a first-order high-pass

filter (F_3) in the form of a washout. Trace (c) of Figure 6-13 corresponds to signal A_4 in Figure 6-12, and is obtained by passing the output of the normal-acceleration sensor (X_4) through a first-order high-pass filter (F_4) in the form of a washout. The time constant of the latter filter was chosen so that A_3 and A_4 correspond to matched channels in the sense described in section 6.2. It may be seen that the filtered signals A_3 and A_4 fluctuate in a closely related manner.

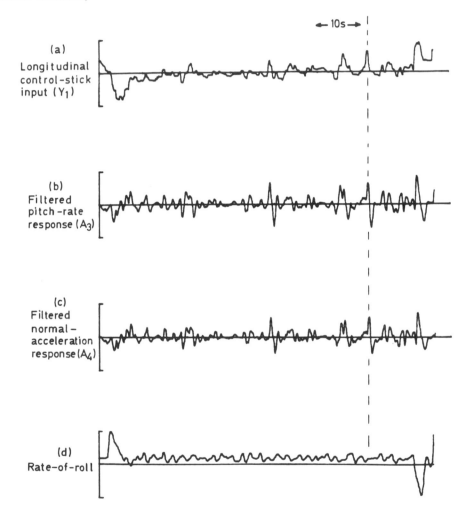

Figure 6-13: Illustration of matched band-limited channels (traces (b) and (c)) derived from pitch-rate and normal acceleration signals

An alternative interpretation of Figure 6-12 is also possible, in which the output X_3 from S_3 is a sensor signal and the output X_4 from S_4 is an actuator-demand signal. For instance it is possible to design complementary filters F_3 and F_4 such that a pitch-rate sensor signal and a tailplane-actuator demand signal are compared for mutual consistency over a suitably chosen common pass-band.

The design of the band-limiting filters in the above examples was entirely based on linearised small-perturbation equations for the aircraft longitudinal motion. The possibility of matched channels, such as A_3 and A_4 in Figure 6-13, differing from one another on account of either large-perturbation effects or cross-coupling from lateral motion needs in practice to be taken into account, leading perhaps to filter augmentation involving band-limited cross-feeds from other sensors. Such possibilities underline the need for full six-degree-of-freedom assessments of monitor performance, particularly as regards the specification of appropriate threshold levels for fault detection.

In the case of matched channels based on the outputs of pitch-rate and normal-acceleration channels (as in Figure 6-13), reference to the full six-degree-of-freedom equations indicates that, provided the normal-acceleration sensor output is corrected to give acceleration at the aircraft centre-of-gravity, the only effect that could be significant over the passband employed derives from a term involving the product of roll rate and sideslip angle. In the manoeuvre in Figure 6-13 the aircraft performed a banked turn, the entry and exit being marked by relatively large peaks in rate-of-roll (trace (d)). As can be verified from Figure 6-13, no significant disparities between A_3 and A_4 (traces (b) and (c)) occurred in association with roll-rate peaks.

6.4 SYSTEM WITH BOTH MEASURABLE AND NON-MEASURABLE INPUTS

In Section 6.3 we introduced possible monitor arrangements (Figures 6-8 and 6-12) for a situation in which the system was subjected to a single external input. The case of two inputs, for example pilot and gust inputs, may be treated by means of a straightforward extension of the same methods.

A possible monitor arrangement is shown in Figure 6-15, with subsystems illustrated independently in Figure 6-14. Figure 6-14(a)
illustrates the case of gust inputs alone, and is an exact analogue
of Figure 6-12(b). X_{G2} and X_{G3} are measured signals, for instance
outputs from two sensors such as pitch-rate and normal acceleration.
Alternatively X_{G2} or X_{G3} might be an actuator-demand signal. The
transfer functions S_{G2} and S_{G3} describe the responses of X_{G2} and X_{G3}
to gust inputs, through the total closed-loop system. The
complementary filters F_{G2} and F_{G3} are designed so that the
combinations of S_{G2} and F_{G2}, and of S_{G3} and F_{G3}, are matched
increment detectors for gust inputs. In the absence of pilot inputs,
comparison of signals A_{G2} and A_{G3}, on the basis of the difference
signal $(A_{G2}-A_{G3})$, could be used to check for mutual compatibility of
the signals X_{G2} and X_{G3} over the common pass band.

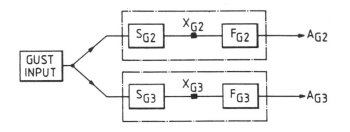

a) Comparison of outputs A_{G2} and A_{G3} derived from
   indirect measurements X_{G2} and X_{G3} of gust input
   (compare Fig 12)

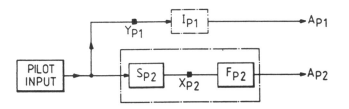

b) Comparison of output A_{P1} derived from direct
   measurement Y_{P1} with A_{P2} derived from indirect
   measurement X_{P2} (compare Fig 8)

Figure 6-14: Elements of monitor with both pilot and gust inputs

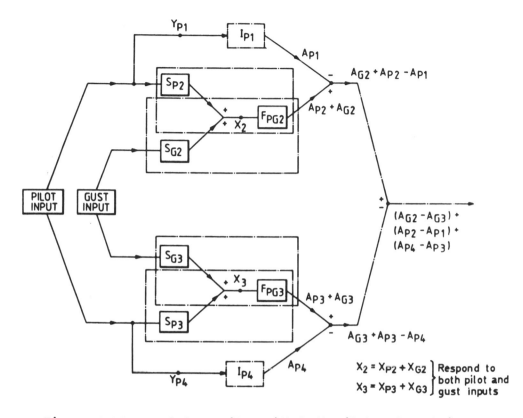

Figure 6-15: Complete monitor with both pilot and gust inputs

Figure 6-14(b) illustrates the case of pilot inputs alone, compare Figure 6-8(b). The measured signal X_{P2} is the response to a pilot input through a transfer function S_{P2}. A_{P2} is the result of passing X_{P2} through a complementary filter F_{P2}. The filter I_{P1} acts directly on the control stick transducer signal Y_{P1} to give an output A_{P1} matched to A_{P2}.

In fact, the measured signals, X_2 and X_3 in Figure 6-15, respond to both pilot and gust inputs and take the form:

$$X_2 = X_{P2} + X_{G2}, \; X_3 = X_{P3} + X_{G3} \qquad (6-4.1)$$

The system in Figure 6-15 incorporates both subsystems shown in Figure 6-14, with a common filter F_{PG2} playing the role of filters F_{P2} and F_{G2}. Experience has shown that it is in general possible to design a single filter to play such a dual role, the design process being iterative with attention to both pilot and gust inputs so that the frequency-plane and time-plane constraints of an increment detector are met adequately by both $S_{P2} \times F_{PG2}$, for increments in pilot input, and $S_{G2} \times F_{PG2}$, for increments in pilot input (see overlapping dashed boxes in Figure 6-15).

In response to simultaneous pilot and gust inputs the matching of A_{P1} to A_{P2} (see in Figure 6-14(b)) ensures that the difference $(A_{P2} - A_{P1})$ is small and hence that the output of the upper half of Figure 6-15 is predominantly the gust component A_{G2}. Similarly, in the lower half of Figure 6-15, $(A_{P3} - A_{P4})$ is small and the output is predominantly A_{G3}. But A_{G2} and A_{G3} are themselves matched (as in Figure 6-14(a)) so that the overall difference signal

$$(A_{G2} - A_{G3}) + (A_{P2} - A_{P1}) + (A_{P4} - A_{P3}) \qquad (6-4.2)$$

comprises a sum of components each of which remains small during normal system operation. In conjunction with prescribed threshold amplitudes, this signal may thus be used to flag possible system failures on the basis of the effects of input signals that occur during normal system operation, without requiring any special test procedures.

6.5 DESIGN CRITERIA FOR MATCHED INCREMENT DETECTORS

Consider the particular case where it is required to generate a pair of matched band-limited signals whose difference is the *null* signal used to indicate possible system malfunction.

Two basic, but conflicting, requirements for such band-limited signals arise. Since it is advantageous to detect any failure as rapidly as possible, signals with small time constants are desirable. According to the uncertainty relation expressed by equation (6-2.1) this implies wide bandwidths. On the other hand, over wide bandwidths pairs of signals measured at arbitrary points in the system would require complex filtering operations to achieve the necessary signal reconstructions, described in Section 6.1. The optimum compromise has been found to be a bandwidth of approximately, but rather more than, one octave.

Successive steps in the process of achieving matched signal

reconstructions are illustrated in Figure 6-16. Suppose the input to be a smooth increment, whose gradient takes the form of a band-limited pulse with bandwidth satisfying approximately the octave-width requirement. The input is assumed to be unmeasurable, measured signals being obtained, as in Figures 6-12(b) and 6-14(a), only after the input has passed through subsystems of the controlled system, designated by A and B in Figure 6-16. As the system is time invariant, the effects of A and B may be interpreted as the actions of *convolution filters*.

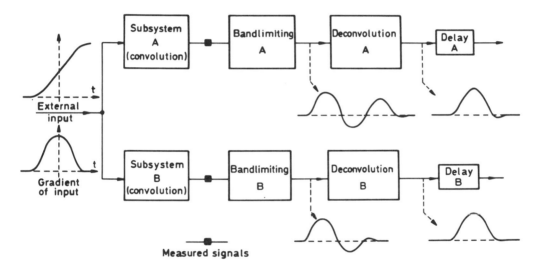

Figure 6-16: Matched reconstructions of signal gradient

The first step in the signal reconstruction process is band-limiting. However, even after filtering to a bandwidth of approximately one octave the resulting signals will in general not take the desired *reconstructed* form of a simple pulse but will, rather, contain *reverberant* overswings (Figure 6-16). Thus, in general, bandlimiting will be followed by a further stage of *deconvolution* in which the overswings are removed. The final step is to delay one or other of the two signal reconstructions so that the signal maxima occur at the same instant. Suitably scaling the amplitudes and differencing then leads to the required null signal.

Although the above process has been described for the situation where the input takes the idealised form of a smooth increment, the bandwidth constraints are such that the channels are in practice adequately matched for arbitrary inputs. In any particular system this result would be verified by numerical simulation, using typical fluctuating inputs, and error margins assessed. However, such an error analysis may also be performed theoretically on the basis of statistical models of the respective inputs (for details please refer to Appendix 6.1 at the end of this chapter).

6.6 EXAMPLES OF DESIGN PROCESS

The design of band-limiting filters is illustrated below by three examples. The first two are for the situation shown in Figure 6-8 involving measurement of the input. The third is for the situation shown in Figure 6-12 and relates two responses caused by the same input. The system being monitoring in these examples is a fly-by-wire aircraft pitch-control system combined with a linear second-order aircraft model described in Appendix 6.2. The system is of tenth-order overall. Block diagrams of the three monitor configurations are shown in Figure 6-17.

A. Elevator estimator (Figure 6-17(a))
The first example is based on the structure of Figure 6-8, using the pilot's pitch stick input Y_1 to obtain a band-limited estimate A_1 of elevator demand, and compares the result with the actual band-limited elevator signal A_2. Figure 6-18 shows a block diagram of the filter used. This is of only second order, though the response comes from the system S_2 (Figure 6-8) of tenth order. This reduction in order can mainly be attributed to the limited frequency band over which the comparison is performed.
The main stages in the design were as follows:

(a) A high-pass filter F_2 on the elevator signal was chosen to allow through responses of primary interest, while ensuring that the response to a pilot ramp input approximated closely to a single pulse with minimal overswing. No low-pass filter was needed, as the response is already attenuated at high frequencies.
(b) The form of the increment detector I_1 was chosen with regard to the response to be matched. The filter I_1 has three parameters, two time constants and a gain.

(c) A preliminary choice of these parameter values was made by ensuring that the peak responses of the two signals occur for the same length of pilot input ramp - the tuned ramp criterion.

(d) More accurate values were then found by minimising the envelope of the error between the responses over the whole useful range of pilot input ramps.

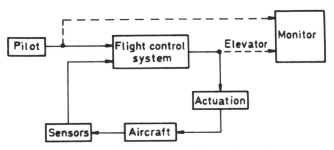

a) Pilots stick to control demand monitor

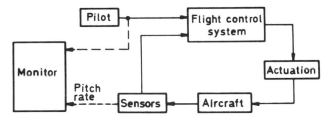

b) Pilots stick to aircraft motion monitor

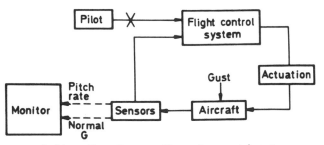

c) Aircraft motion monitor using gust inputs

Figure 6-17: Block diagram of control system with example monitors

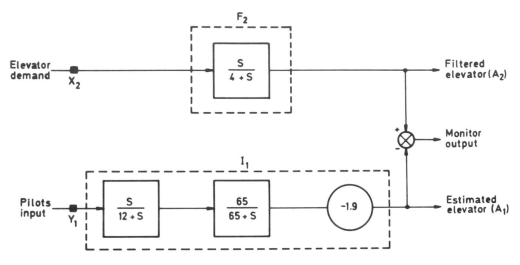

Note: S denotes the laplace transform

Figure 6-18: Block diagram of elevator demand monitor

The final responses are shown in Figure 6-19(a)&(b), in which the maximum monitor output is about 15% of the peak response in the filtered elevator output. In conjunction with a statistical model (see Appendix 6.1) of the pilot input, this monitor output could be used to set a threshold for fault detection at an acceptable false-alarm rate. Since the responses come from a closed-loop control system, the monitor will be able to detect any changes in the control system or aerodynamics which have an effect over the frequencies of interest. Figure 6-20 shows the effect of an error occurring in the control system, modelled by a reduction by two thirds in system gain, while Figure 6-21 shows the effect of a change in the M_w aerodynamic derivative.

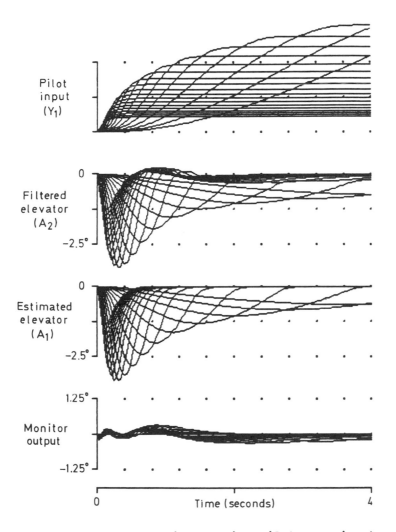

Figure 6-19 a: Elevator estimator with pilot ramp inputs

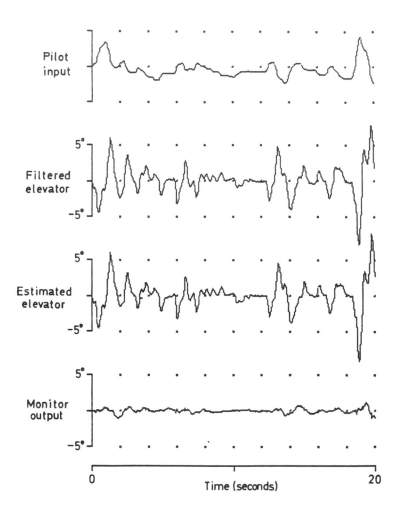

Figure 6-19 b: Elevator estimator with pilot input recorded on a
             simulator

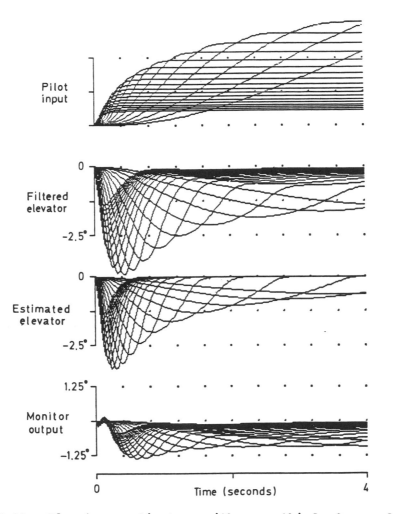

Figure 6-20a: Elevator estimator with one third of normal control
            gain

B. Pitch rate estimator (Figure 6-17(b))

This uses the same pilot's stick signal to estimate the pitch rate response, q, of the aircraft over a limited frequency band. Since the q response is much slower than the elevator response, a much lower frequency band has been selected for monitoring purposes. The form of the band-limiting filter I_1 (Figure 6-8) is also different (see Figure 6-22), having one time constant, a gain and a time delay as parameters. The time delay was found to be necessary in order to ensure time-plane matching between the two signals (see section 6.5). The most likely cause for the apparent appearance of a pure time

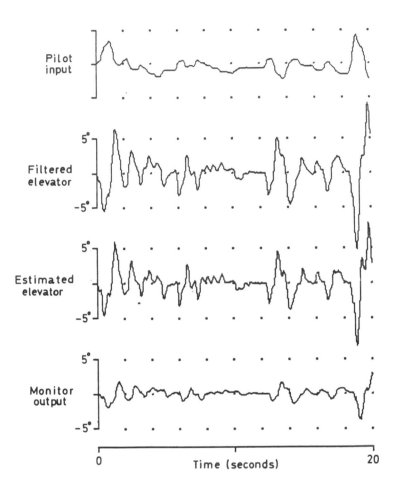

Figure 6-20b: Elevator estimator with one third of normal control
gain and recorded pilot input

delay, in an otherwise continous-time system, is the combined effects
of the high frequency lags in the control system, plus the actuator
and elevator power-control lags. In this case, we are using a very
simple component, the pure time delay, to model the effect of a very
high order system.

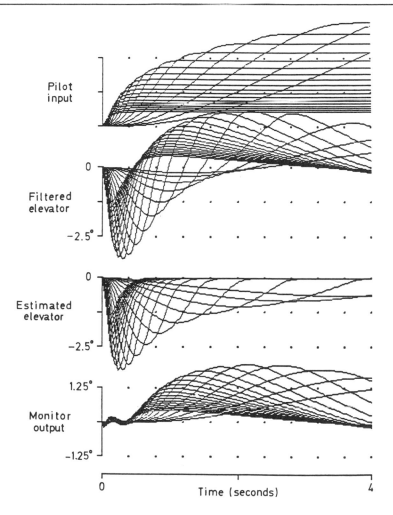

Figure 6-21a: Elevator estimator with changed aerodynamics

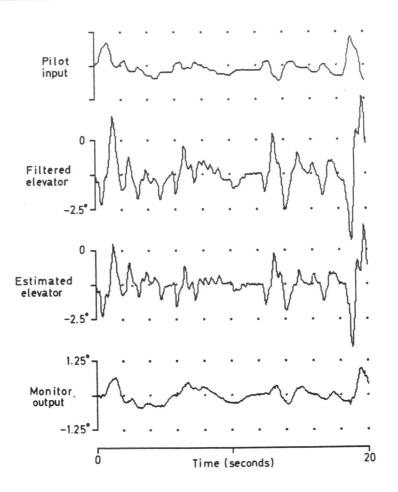

Figure 6-21b: Elevator estimator with changed aerodynamics and recorded pilot input

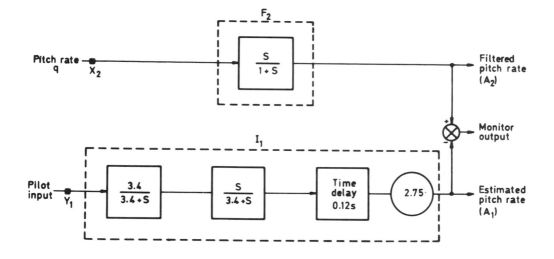

Figure 6-22: Block diagram of pitch rate monitor

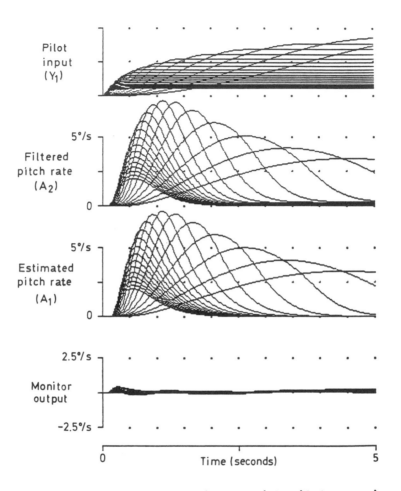

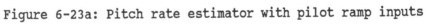

Figure 6-23a: Pitch rate estimator with pilot ramp inputs

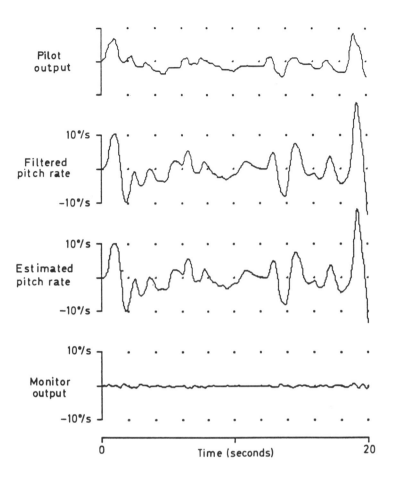

Figure 6-23b: Pitch rate estimator with recored pilot input

Figure 6.23(a)&(b) show the responses obtained in the absence of any faults, while Figures 6-24 and 6-25 show the effect of faults modelled as changes in the control system and aerodynamics, respectively. In the absence of faults the matching is correct to within 10%.

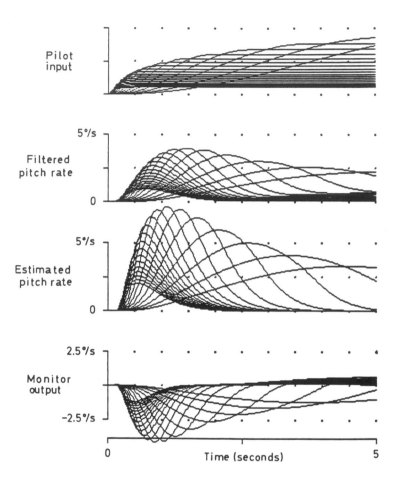

Pilot
input

Filtered
pitch rate

5°/s

0

Estimated
pitch rate

5°/s

0

Monitor
output

2.5°/s

-2.5°/s

0   Time (seconds)   5

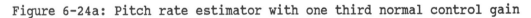

Figure 6-24a: Pitch rate estimator with one third normal control gain

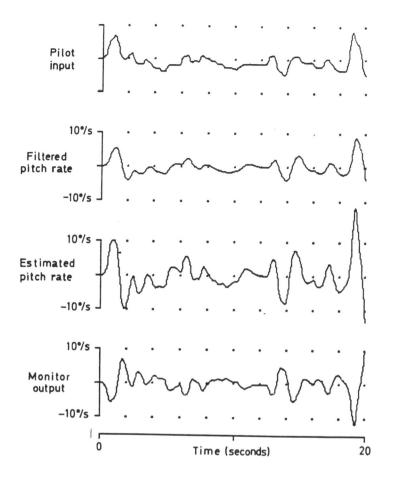

Figure 6-24b: Pitch rate estimator with recorded pilot inputs and one third normal gain

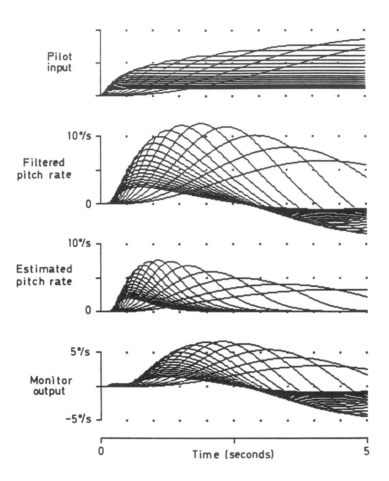

Figure 6-25a: Pitch rate estimator with changed aerodynamics

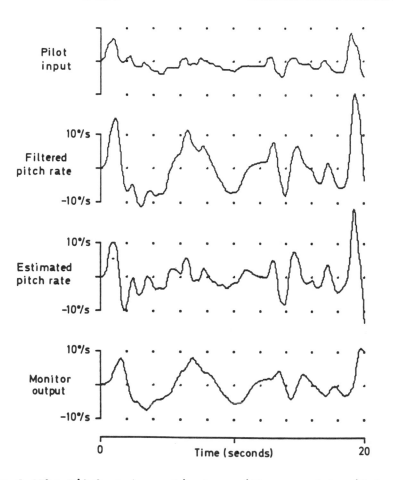

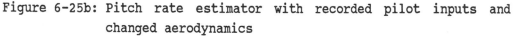

Figure 6-25b: Pitch rate estimator with recorded pilot inputs and changed aerodynamics

C. Gust monitor (Figure 6-17(c))

As it is not possible to obtain a direct measurement of the gust input, this monitor is based on the structure of Figure 6-12 and compares the pitch rate and normal acceleration responses resulting from the same vertical gust. Note that in this situation (Figure 6-12) the filters F_3 and F_4 are combined, respectively, with transfer functions S_3 and S_4, which include aircraft dynamics, to give overall increment detectors. The pitch rate response was already band-limited, and only needed a gain (F_3) applied to it (Figure 6-26). A second-order filter F_4 sufficed to make the normal acceleration match it to within 1%. Monitor outputs are thus too

small to be discernable in the responses without the faults as shown
in Figure 6-27.

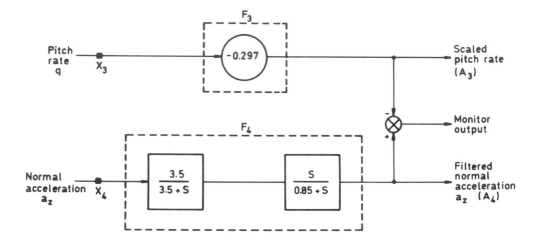

Figure 6-26: Block diagram of pitch rate (q) and normal acceleration
                (a$_z$) monitor using gust input

The signals A$_3$ and A$_4$ (Figure 6-27(a)) have a 20% overswing. A
disadvantage of such overswing is that it significantly increases the
time constant of the response. The overswing could only be removed by
increasing the complexity of the filters in the monitor. The next
section describes methods for removing such overswings. The
cost-effectiveness of introducing such additional complexity in order
to reduce the effective time constant of the monitor will depend upon
the requirements of the specific application. Responses of the
monitor, shown in Figure 6-26, to simulated faults are illustrated in
Figures 6-28 and 6-29.

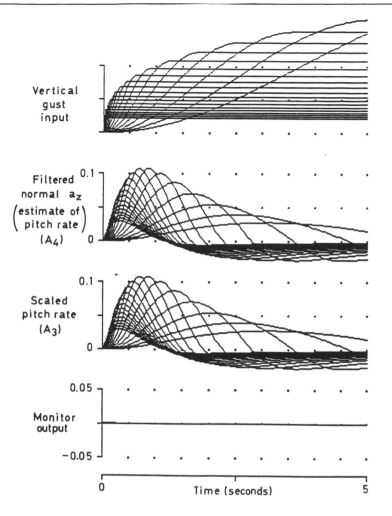

Figure 6-27a: q and a_z monitor with gust input ramps

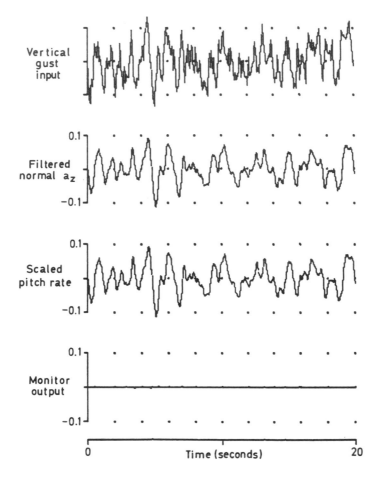

Figure 6-27b: q and a_z monitor with random vertical gusts

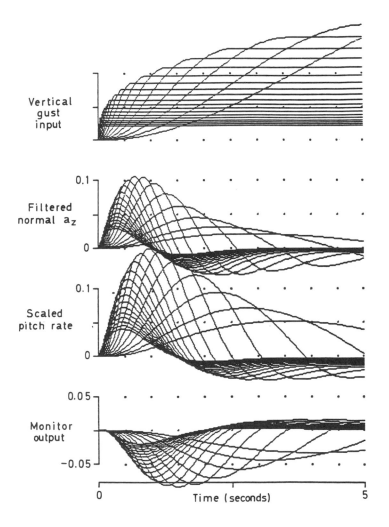

Figure 6-28a: q and a_z monitor with one third normal control gain

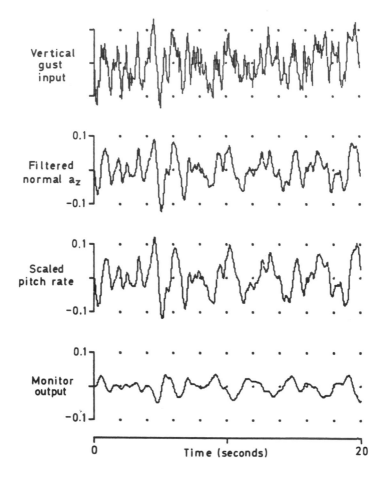

Figure 6-28b: q and a_z monitor with one third normal control gain and
random vertical gust input

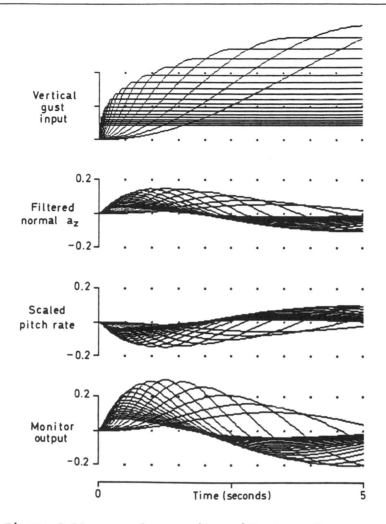

Figure 6-29a: q and a_z monitor with changed aerodynamics

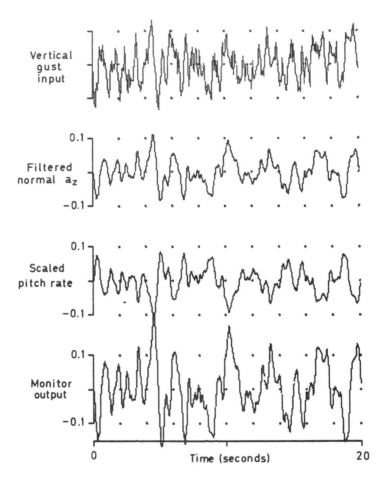

Figure 6-29b: q and a_z monitor with changed aerodynamics and random
vertical gust input

6.7 SIGNALS WITH OVERSWINGS

The monitors illustrated in Section 6.6 all used signals whose bandlimited response to a ramp input took the form of a predominant single pulse. Many circumstances exist, however, in which, in the absence of further signal shaping, the band-limited response has two or more sequential overswings. In order to minimise response times it is advantageous to pass such signals through an additional filter which converts them to a single pulse before use (Figure 6-16).

Responses with overswings most commonly occur in two types of system:

(a) Poorly damped systems
(b) Systems with numerator zeros in their transfer functions.

Two different types of filter were found to be best at removing the multiple responses in these cases. As the multiple responses arise as a result of the convolution of a single ramp input with the system transfer function, such filters are referred to as *deconvolution filters* (see Section 6.5).

A. Poorly damped systems

A poorly damped system typically has a resonant response with a sequence of overswings, each decaying in amplitude by roughly a constant factor from the last. This type of response is most readily converted into a single pulse by a notch filter. The frequency of the notch is tuned to that of the resonance, and its depth depends on the damping of the system, increasing as the damping reduces.

Figure 6-30 shows the responses of an underdamped system and the same system followed by a notch filter. The latter has completely removed the overswings.

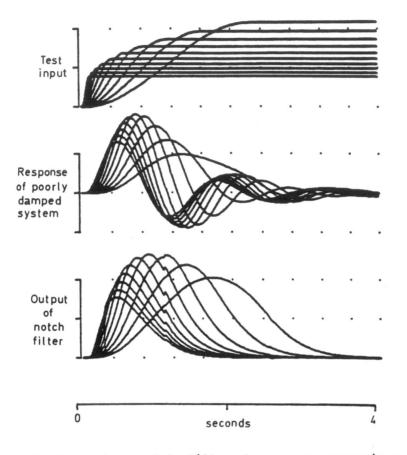

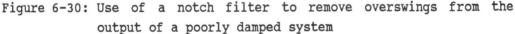

Figure 6-30: Use of a notch filter to remove overswings from the
          output of a poorly damped system

B. System with single overswing

Responses with only one overswing are often more difficult to reduce
to a single pulse than ones with a sequence of overswings. The main
reason is that the latter type of response is mainly confined to a
narrow frequency band, which can be suppressed by a simple notch
filter leaving the initial octave-band pulse largely unaffected. The
single overswing, on the other hand, frequently results from a transfer function which has a wide bandwidth, but which also has numerator
zeros in the right-hand plane. Any simple filtering operation applied
to such a response, to suppress the overswing, frequently has the
result of elongating or suppressing the initial pulse at the same

time. A deconvolution filter which operates effectively without changing the shape of the initial pulse is shown in Figure 6-31. In this filter a time delayed version of the output is reshaped by a lag and added to the input, thereby cancelling out the overswing. The responses of this filter are shown in Figure 6-32.

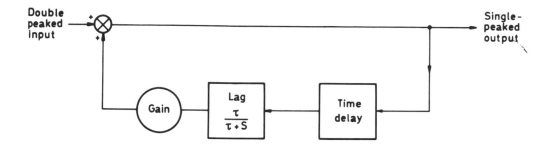

Figure 6-31: Recursive filter converting double-peak into single-peak signal

Care must be taken when using this filter, since it may become unstable if the feedback gain is of magnitude one or greater. For this reason it cannot completely cancel overswings when their amplitude approaches that of the initial pulse.

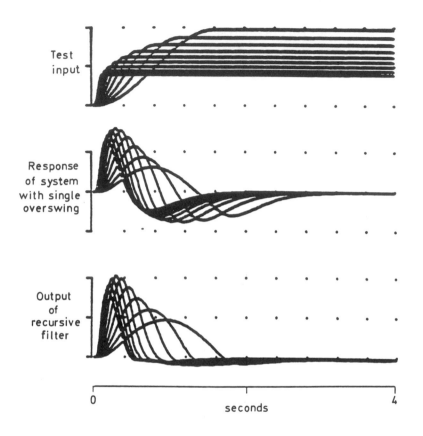

Figure 6-32: Use of the recursive filter to remove one overswing from a response

6.8 MONITOR FOR FLIGHT-CONTROL COMPUTER

In the previous examples the transfer functions whose outputs were monitored generally included the dynamic response of the aircraft in addition to effects of the flight-control system. In the present section it is proposed that band-limiting filters may be used in a more specific manner to monitor functions of the flight-control computer alone.

The structure of the monitor is illustrated in Figure 6-33. We consider a simple linear control law in which the only inputs to the flight-control computer are pilot demand and the response of a single sensor, such as a pitch-rate gyro. (Control laws with multiple sensor inputs may be treated in an analogous manner.) The output X is an actuator-position or actuator-rate demand signal. X may, in principle, be expressed as the sum of the outputs of two independent transfer functions S_s and S_p acting on the sensor signal and pilot input respectively. F_1 is a band-limiting filter whose output is the band-limited signal A_1. F_2 and I_p (Figure 6-33) are band-limiting filters operating directly on the sensor and pilot-input signals respectively.

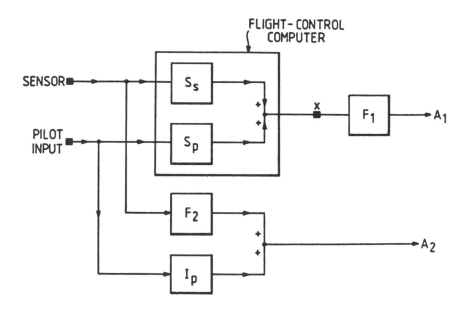

Figure 6-33: Monitor for flight control computer

The design process is as follows. First the components illustrated in Figure 6-33 are embedded in a closed-loop simulation model, including effects of aircraft dynamic response, in which the actuator-demand signal X results in associated aircraft motion and the sensor signal responds to both pilot and gust inputs. In the following step the filters F_1 and F_2 are designed, as described in previous sections, such that the outputs A_1 and A_2 respond as matched channels for increments in gust input. In general it will also be possible to constrain F_1 so that, in conjunction with the combined outputs of S_s and S_p, A_1 responds as an increment-detector output for pilot inputs.

Finally, the band-limiting filter I_p is designed such that A_1 and A_2 respond as matched detectors for increments in pilot input. A_1 and A_2 thus have matched responses to both gust and pilot inputs and the difference signal $(A_1 - A_2)$ may be employed as an indicator of the correct functioning of the flight-control computer.

Although a model of the complete aircraft dynamics is employed in the design process, its role is primarily to ensure that optimum pass bands (in the sense of high signal energy and associated high information content) are chosen for the band-limited signals A_1 and A_2. The resulting monitor is relatively insensitive to the aircraft dynamic model and, for instance, will usually not require scheduling, through the flight envelope, for changes in aircraft dynamics. On the other hand, it will require scheduling to match associated schedules in the flight-control laws as embodied in S_s and S_p.

A similar comment applies to cross-coupling between lateral and longitudinal aircraft motions. The monitor outlined in previous sections, when applied to signals such as normal acceleration that respond to aircraft rolling manoeuvres, will in general require augmentation involving band-limited cross-feeds from sensors of lateral motion, such as roll rate, to take account of dynamic interaction lying within the pass band of interest. However, with the configuration shown in Figure 6-33, any sensor response to lateral motions with significant energy lying within the common pass-band will have essentially identical effects on the filtered signals A_1 and A_2. Even so, the possibility of cross-coupling effects on the difference signal $(A_1 - A_2)$ should be taken into account in threshold specification and overall monitor assessment.

The monitor shown in Figure 6-33 may be generalised to include the case where the control laws incorporate multiple sensor signals. An example with two sensor signals, namely aircraft pitch rate (q) and aircraft normal acceleration (a_z) is illustrated in Figure 6-34. The output (X) from the flight-control computer, used as a further input to the monitor, is the elevator-demand signal (η_D). As there are two sensor signals, the band-limiting filter F_2 (Figure 6-33) is replaced by two filters F_2', F_2'' (Figure 6-34). By analogous means, an arbitrary number of sensor signals can be catered for. As in the examples discussed earlier, the band-limiting is achieved by filters of low order.

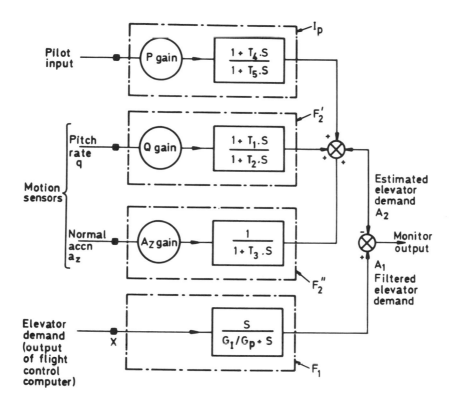

Figure 6-34: Example of monitor for flight control computer

Figures 6-35 and 6-36 show the band limited signals A_1 and A_2, and the resulting error signal, for pilot and gust inputs comprising families of ramps. For the gust input the error signal is zero to within the graph-plotting accuracy.

In contrast, the error signals corresponding to control-computer faults are illustrated, in conjunction with pilot inputs, in Figures 6-37 and 6-38. The former shows the situation of a completely failed a_z channel within the control computer and in the latter the q channel has had its gain reduced by a factor of two. The magnitudes of the error signals confirm the general result obtained from assessments of this type, namely that by monitoring over a limited, but appropriately chosen, passband adequate indication of a system fault can be provided using filters of quite low order.

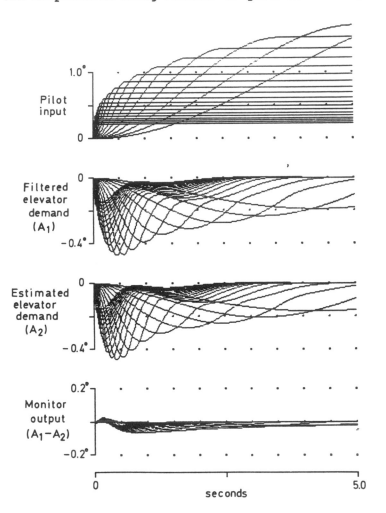

Figure 6-35: Flight control computer monitor with pilot input ramps

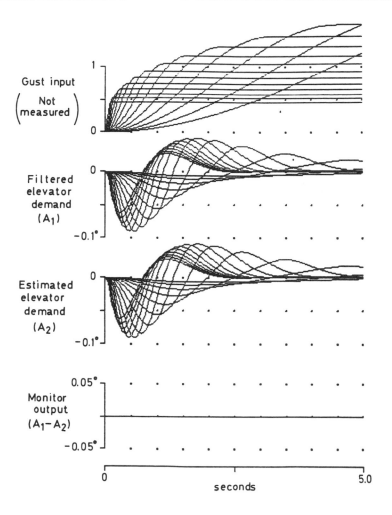

Figure 6-36: Flight control computer monitor with gust input ramp

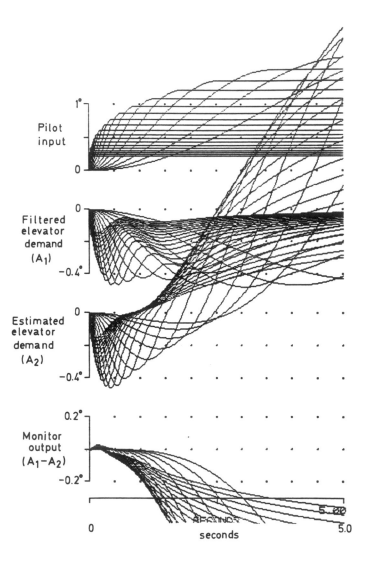

Figure 6-37: Flight control computer monitor with fault in a_z channel
           of computer

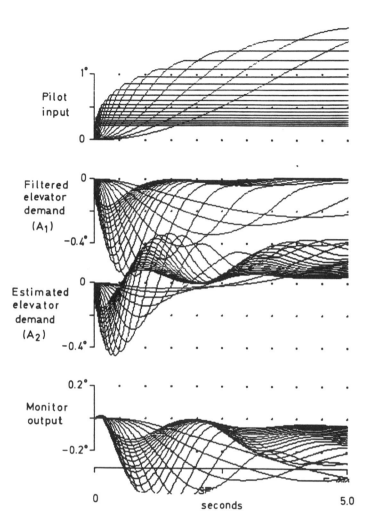

Figure 6-38: Flight control computer monitor with pitch rate feedback
halved

APPENDIX 6.1: USE OF STATISTICAL MODELS

Statistical models of external inputs play a role both in the monitor design process and in performance assessment, particularly the prediction of false alarms.

As outlined in Section 6.2, the model used for monitor design comprises families of smooth increments (Figure 6-4(b)), satisfying the equation:

$$w = \Delta t^k \tag{6-A.1}$$

(Figures 6-6, 6-19(a), 6-27(a)).

This model forms the basis (see Jones (1985) and (1986(b))) of the *method of equivalent deterministic inputs*, which may also be used for performance assessment. The method may be summarised schematically as follows:

(a) **Stochastic process** → (b) **equivalent deterministic inputs (family of equiprobable patterns or profiles)**
→ (c) **maximum peak in filter response, $|Y|_m$, (worst-case analysis)**
→ (d) **threshold exceedance rate, n_y, for peaks in stochastic response.**

When the stochastic input is a self-similar process, for which the power spectrum takes the form $1/f^B$, the smooth increments, or ramps, satisfying equation (6-A.1) form an equi-probable family, provided that:

$$k = \frac{B - 1}{2} \tag{6-A.2}$$

This model is well validated (Jones and Haynes (1984), Foster, (1983), Jones (1980) and Jones (1986(b))), when the input process is atmospheric turbulence (and the ramp profiles correspond to discrete gusts). It has also been found to provide an adequate representation for pilot inputs (unpublished). In the turbulence case it is a standard result that $B = 5/3$ (and thus $k = 1/3$). Pilot inputs are more variable but generally lie in the range $1/4 < k < 3/4$. The same value, $k = 1/3$, as is used for turbulence is thus typical.

The application of the method of equivalent deterministic inputs to the prediction of filter response takes the form of a worst-case analysis, in which, for a prescribed family of equi-probable profiles (as outlined above) the overall maximum peak in filter response, $|Y|_m$, is found. In effect, this analysis takes the form of a search for a condition of *transient resonance*, the resonance condition associating the filter being evaluated with a particular member (the *tuned input*) of the prescribed family of deterministic inputs (Figure 6-6).

The maximum peak response $|Y|_m$, obtained as a result of a deterministic worstcase analysis, may be used to derive the threshold exceedance rate n_y in response to the associated stochastic input. Details of this step have been reported in RAE reports by Jones (1986(a)), (1986(b)). Here we simply note that the quantity $|Y|_m$, although derived by a deterministic procedure, provides the basis for a statistical analysis of filter response.

Furthermore, to first order, the deterministic *tuned input* (Figure 6-6) and its associated filter response may be shown to provide optimum predictions of the stochastic input and response waveforms, given the occurrence of a peak in stochastic response. The prediction is optimum in the sense of providing the most probable input and response wave-forms in the vicinity of the prescribed peak.

In the case of filters in the form of band-pass channels with bandwidth roughly of order one octave, the statistical relationship between the occurrence of a peak in response and the shape of the neighbouring response waveform is particularly strongly constrained, Marr et al. (1979). In fact, if we exclude fluctuations below a relatively low threshold level, such a waveform can be reconstructed to first order simply from the information provided by the location and amplitude of the peaks.

If follows that if two octave-width channels are designed such that they have a common tuned input, and also have matched peak responses to this tuned input, they will be matched to first order for the associated stochastic input. This result is the basis of the deterministic filter design procedure described in Section 6.2 of the main text.

The existence of a statistical model for external inputs to the system (pilot-demands and gusts) allows the setting of threshold levels in a rational manner which takes account of the costs or penalties associated with an excessive false-alarm rate (too low a threshold) and with delayed detection of real system failures (too high a threshold). In principle, by attaching numerical values (loss functions) to such costs, an *expected risk* may be evaluated and the threshold set to minimise this risk (Bayes' solution to the decision problem). In practice, confidence in the performance of monitor designs would depend to a considerable degree upon extensive simulation studies using measured turbulence records and a piloted closed-loop simulator.

APPENDIX 6.2: EQUATIONS OF MOTION OF THE AIRCRAFT MODEL

A general aircraft model would have the form:

$$I \, \underline{a} = f \tag{6-A.3}$$

where $\underline{a}$ is a vector of linear and rotational accelerations, $\underline{I}$ is the aircraft inertia and $\underline{f}$ is a vector of aerodynamic forces and moments, which are non-linear functions of the aircraft state vector $\underline{x}$.

For the majority of the work reported here, a small-perturbation linearised version of this full model has been used, which has the form:

$$\underline{\dot{x}} = \underline{A} \, \underline{x} + \underline{B} \, \underline{u} \tag{6-A.4}$$

where $\underline{u}$ is the vector of inputs. The matrices of coefficients, $\underline{A}$ and $\underline{B}$, are derived from the full force equations for the particular speed and height investigated. These linearised equations separate naturally into decoupled symmetric and anti-symmetric motions, of which the symmetric (longitudinal - pitch and heave) have been selected for study.

The individual equations and coefficients are:

$$\dot{w} = -Z_w (w + w_g) + (1 - Z_q) q - Z_\eta \, \eta$$

$$\dot{q} = -M_w (w + w_g) - M_{\dot{w}} \, \dot{w} - M_q \, q - M_\eta \, \eta$$

$$a_z = q - \dot{w}$$

where:

q is pitch rate

w is angle of incidence away from trim

η is the elevator angle away from trim

w_g is the change in incidence due to a gust

a_z is normal acceleration

For the simulator results in Figures (6-2) and (6-9) to (6-13) the full force equations were used, including anti-symmetric (lateral) motions.

Chapter Seven

PROCESS FAULT DIAGNOSIS BASED ON DYNAMIC MODELS AND PARAMETER ESTIMATION METHODS

Rolf Isermann

Fault detection and fault diagnosis of technical processes are rapidly gaining in importance due to the advancing developments in applied control. Computer-based fault supervision methods have been developed which enable process faults to be detected and localised during normal plant operation. The use of process models enables the estimation of process state-variables and parameters which are influenced by faults. The chapter concentrates on fault diagnosis based on process parameters. The general procedure comprising parameter estimation, feature extraction, fault decision and classification is outlined. Experimental results are given for the detection of several typical faults in an electrical drive/centrifugal pump set and a steam heated exchanger.

7.1 INTRODUCTION

The main interest in the automation of technical processes in recent years can be observed in instrumentation, feedforward and feedback control, alarm monitoring and protection, documentation, manual operation and optimisation. Good progress can also be seen in the technology and performance of modern measurement and control systems. The improvements in process control are, on one-hand based on the development of the components for sensors, transducers, control systems and actuators and, on the other hand, on the understanding and modelling of process dynamics together with applied control theory.

There have also been developments in components for *process supervision* including alarm monitoring and protection. However, the

implemented methods are still rather simple and consist mainly of
limit value checking of some easily available single signals. In
contrast to the field of control, methods based on modern dynamic
systems theory are hardly applied. But there is now an increasing
interest in the field of process supervision, which includes fault
detection and diagnosis, see for example Willsky (1976), Himmelblau
(1978), Pau (1981), Isermann (1980, 1984a, 1984b), Gertler and
Singer (1985). Some reasons for this increasing interest are as
follows:

(a) increasing demands on reliability and safety of technical
    processes in general,
(b) high investment capital for processes,
(c) increasing interlinkage of processes,
(d) increasing complexity of processes,
(e) unmanned automation, and
(f) errors due to manual operations.

The goal of supervision is to detect process changes and faults
during normal operation and to take actions to avoid damage to the
process or injury to human operators. For this reason a supervision
loop is implemented which contains the following tasks:

(a) fault detection and diagnosis,
(b) fault evaluation,
(c) decision on operating state (stop or change), and
(d) fault elimination.

An essential prerequisite for improving the supervision of a process
control system is an early *process fault detection* and *localisation* such that
sufficient time is allowed for the following actions:

(a) fault elimination,
(b) prevention of further fault development,
(c) operation in an alternative mode, and
(d) scheduling of next maintenance and repair procedures.

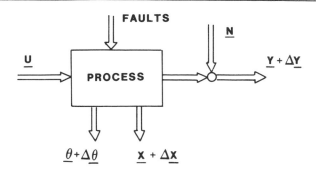

Figure 7-1: Representation of a process which is influenced by a fault

    $\underline{U}$, $\underline{Y}$: measurable input and output variables

    $\underline{\Theta}$: process parameters

    $\underline{X}$: process state variables

    $\underline{N}$: disturbance variables (noise)

Figure 7-1 shows a process which can be described by a mathematical process model of the following form:

$$\underline{Y} = f[\underline{U},\underline{N},\underline{X},\underline{\Theta}] \qquad\qquad (7\text{-}1.1)$$

where $\underline{U}(t)$ and $\underline{Y}(t)$ are measurable input and output variables, $\underline{\Theta}$ constant or slowly time-varying process parameters, $\underline{X}(t)$ process state variables, $\underline{N}(t)$ disturbance variables. A process fault generally causes changes $\Delta\underline{\Theta}(t)$ of process parameters and changes $\Delta\underline{X}(t)$ of process state variables, depending on the type of fault. These changes then lead to changes $\Delta\underline{Y}(t)$ of the output signals, according to the process dynamic and static characteristics. Fault detection methods may be classified according to the use of the following quantities, cf. Isermann (1981, 1984a):

(a) measurable signals $\underline{U}$, $\underline{Y}$,

(b) state variables $\underline{X}(t)$, mostly unmeasurable,

(c) process parameters $\underline{\Theta}(t)$, mostly unmeasurable, and

(d) characteristic quantities $\eta = f[\underline{U},\underline{Y},\underline{\Theta}]$.

The usual way is to check the range of the measurable signals $\underline{Y}(t)$. Measurements are normal if:

$$\underline{Y}_{min} < \underline{Y}(t) < \underline{Y}_{max} \qquad\qquad (7\text{-}1.2)$$

This is called *range or limit checking*. It can also be applied to the trend $\underline{Y}(t)$ of the signals.

Another approach is to *analyse the signals* $\underline{Y}(t)$, which often consist of lower frequency components $\underline{Y}_{LF}(t)$ and higher frequency components $\underline{Y}_{HF}(t)$. In particular, the analysis of higher frequency components (vibrations etc.) by autocorrelation or spectral analysis can give additional information concerning the inner state of a process.

If both output and input variables are available *characteristic quantities* η (like efficiencies, consumption rate, wear rate, etc.) can be calculated. Changes $\Delta\eta$ may give overall information on internal changes. However, in most cases a detailed fault diagnosis is not possible.

If a *process model* is known, one can try to estimate the usually unmeasurable *process state variables* $\underline{X}(t)$ or *process parameters* $\underline{\Theta}(t)$ based on the measurable inputs $\underline{U}(t)$ and outputs $\underline{Y}(t)$ by using state variable or process parameter estimation methods and to detect changes $\Delta\underline{X}(t)$ or $\Delta\underline{\Theta}(t)$. This approach in other contexts is known as *Analytical Redundancy* (see earlier chapters of this book for definition and additional discussion) and has the following advantages:

(a) In terms of signal flow the state variable $\underline{X}(t)$ or process parameters $\underline{\Theta}(t)$ are, in many cases, closer to the process faults. Hence the process faults may be detected earlier and localised more precisely than by range checking of $\underline{Y}(t)$.

(b) A process fault usually causes changes of several output variables $\Delta Y_i(t)$ with different signs and dynamics. The model-based fault detection now takes into account all these detailed changes, provides a data reduction and determines (theoretically) the state variable or the process parameter which has been changed directly by the fault. Hence, it can be expected that a significant change $\Delta X_j(t)$ or $\Delta\Theta_j(t)$ can be extracted and that the fault detection selectivity will be improved.

(c) Closed-loops generally compensate for changes $\Delta\underline{Y}(t)$ of the outputs by changing the inputs $\underline{U}(t)$. Therefore deviations caused by faults cannot be recognised by range checking alone. Model-based fault detection methods automatically consider the relations between inputs and outputs and are therefore also applicable to closed-loop systems.

(d) If one wants to detect more process faults one can try to implement more sensors which measure, as directly as possible, the required information. However, in general one must use many more sensors including some non-robust sensors. This may become expensive and more sensitive to sensor-based faults. Model-based fault detection methods need, in principle, only a few robust sensors.

The price one has to pay for these advantages is the higher effort expended in modelling the process, the development of the fault monitor itself (software package) and the additional computer capacity required.

The application of these model-based fault detection methods is not restricted to continuous process supervision in normal operation, but can also be used on request to enable the following procedures to be achieved:

(a) preventive maintenance,
(b) maintenance on request (instead of fixed schedules),
(c) remote diagnosis of processes, and/or
(d) automatic inspection of products without disassembling in manufacturing.

This contribution concentrates on model-based fault detection methods based on parameter estimation.

A survey of various process fault detection methods has been given by Isermann (1984a). The present survey includes therefore only a small reference list. In Section 7.2 the principle of the method is described. This is followed by methods for fault diagnosis. Section 7.3 shows results which were obtained for an electrically driven centrifugal pump system. Section 7.4 provides a discussion of experiments with a heat exchanger.

7.2 FAULT DIAGNOSIS BASED ON PROCESS PARAMETERS

7.2.1 General procedure

A generalised scheme of technical fault diagnosis is shown in Figure 7-2. Three phases can be distinguished as follows:
(a) *Data processing*

The measured signals are processed by methods of filtering and estimation such that the information reduction becomes suitable for fault detection and diagnosis. The reduced information, for example, exists in filtered signal components, correlation functions, or in parameter or state-variable estimates (if process models are applied).

(b) *Fault detection*

Using only the reduced process information, features are extracted which allow the detection of faults in the process. Changes of these features are then determined with reference to the normal process. These changes are subsequently used to recognise the event of a fault and the time of its occurrence. For this task statistical decision methods may be used.

(c) *Fault diagnosis*

After a fault event has been detected, the features and their changes are submitted to a classification procedure with the aim to determine the fault type, fault location, fault size and the cause of the fault. Therefore classification pattern recognition methods, respectively can be used. (It should be noted that sometimes fault decision and classification are so combined that their clear separation is not possible.)

This general procedure holds as well for fault diagnosis methods based on *signal models* and *process models* in non-parametric or parametric form. In the case of parametric process models the reduced information may consist of *process model parameters* or *process model state variables*.

In the subsequent discussion only *fault diagnosis methods* with *process model parameters* are considered. A more specific scheme for this class of fault diagnosis is shown in Figure 7-3. This shows agreement with Figure 7-2. The fundamental idea is that many process faults appear as changes of process coefficients p like resistances, capacitances, inductances, mass, stiffness, etc. These process coefficients are contained in the parameters $\underline{\Theta}$ of a process model. Process model parameters are understood as constants or time-dependent coefficients in the process which appear in the mathematical description of the relationship between the input and output signals, the process model. A distinction can be made between *static process models*, e.g. in the form of a polynomial equation:

$$Y(U) = \beta_0 + \beta_1 U + \beta_2 U^2 + \dots \qquad (7\text{-}2.1)$$

and *dynamic process models* which, for processes with lumped parameters, are often expressed in differential equation form as follows:

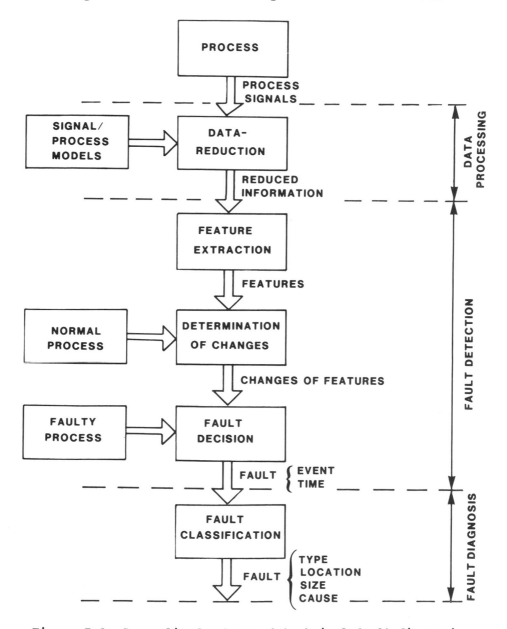

Figure 7-2: Generalised scheme of technical fault diagnosis

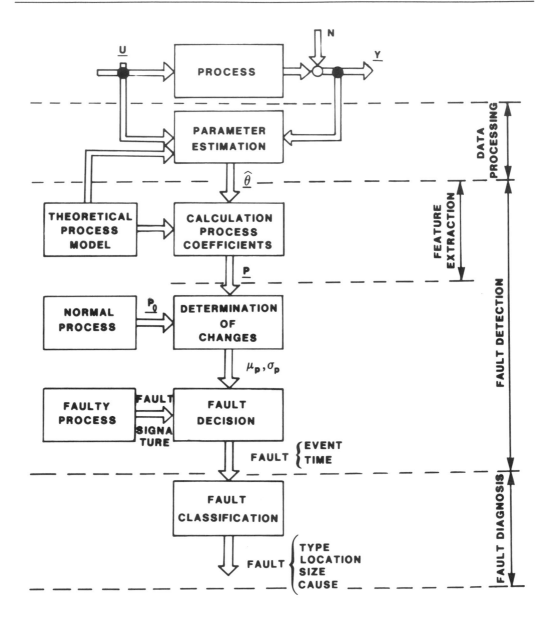

Figure 7-3: Fault diagnosis based on parameter estimation and theoretical modelling

$$a_0 y(t) + a_1 \dot{y}(t) + a_2 \ddot{y}(t) + \ldots + y^{(n)}(t)$$

$$= b_0 u(t) + b_1 \dot{u}(t) + b_2 \ddot{u}(t) + \ldots + b_m u^{(m)}(t) \qquad (7\text{-}2.2)$$

These are, in the simplest case, linearised about one operating point (of course, non-linear differential equations can also be used). The process model parameters are given by:

$$\underline{\theta}^T = [\beta_0 \ \beta_1 \ \beta_2 \ \ldots],$$

or

$$\underline{\theta}^T = [a_0 \ \ldots \ a_{n-1} : b_0 \ \ldots \ b_m] \qquad (7\text{-}2.3)$$

These are now intricate relationship of several physical process coefficients $\underline{p}$. To determine these process coefficients and their changes the following procedure has been developed (compare with the single blocks in Figure 7-3):

(a) Establishment of the process model for the measurable input and output signals:

$$Y(t) = f[\ U(t), \underline{\theta}\ ] \qquad (7\text{-}2.4)$$

(b) Determination of the relationship between the model parameters θ_i and the process coefficients p_j:

$$\underline{\theta} = f(\underline{p})$$

(c) Estimation of the model parameters θ_i based on the measured signals $Y(t)$ and $U(t)$.

(d) Calculation of the process coefficients:

$$\underline{p} = f^{-1}(\underline{\theta}) \qquad (7\text{-}2.5)$$

(e) Determination of changes Δp_j of the process coefficients with reference to normal values.

(f) Fault detection based on the process coefficient changes and fault signatures of a model for the faulty process in which the relationships between process faults and changes of the process coefficients Δp_j has been established a priori.

(g) Fault classification to determine the fault type, localisation and size.

This fault detection is therefore based on:

(a) theoretical process modelling
(b) parameter estimation for continuous-time models
(c) statistical decision and classification.

Prerequisites for the application of this method are:

(a) The process models have to describe the process behaviour pre-
    cisely
(b) powerful parameter estimation methods are available which yield
    precise parameter estimates
(c) sufficient excitation of process dynamics by appropriate input
    signals is possible
(d) unique determination of the required process coefficients
    (certain exceptions are possible, see for example Isermann
    (1984a) and (Section 7.4).

7.2.2 Parameter estimation

The basic process model is developed using theoretical process
modelling. The physical laws which govern the process behaviour must
therefore be known. By stating the balance equations, the
physical-chemical state equations and the phenomenological laws,
continuous-time models for the dynamics result in ordinary or partial
differential equations. Hence, the structure of the models is known
and the parameters θ_i have to be estimated in order to detect changes
during normal operation. A survey of parameter estimation methods for
linear continuous-time models based on sampled signals was given by
Young (1981), see also the summary by Isermann (1984a). At that time
mainly the least-squares (LS) and instrumental variables methods had
been developed. The least-squares method is still the most robust and
recommended if the superimposed noise is small. The numerical
properties can be improved by using a square-root filtering approach.
The differential equation (7-2.2) of a stable linear process with
lumped parameters and transfer function:

$$G_p(s) = \frac{y(s)}{u(s)} = \frac{B(s)}{A(s)} = \frac{b_0 + b_1 s + \ldots + b_m s^m}{a_0 + a_1 s + \ldots + s^n} \qquad (7-2.6)$$

can be written in the form:

$$y^n(t) = \psi^T(t)\underline{\theta} + e(t) \qquad (7\text{-}2.7)$$

with the vector of the signals and their derivatives:

$$\psi^T(t) = [-y^{(n-1)}(t) \; -y^{(n-2)}(t) \; \ldots \; -y^{(1)}(t)$$

$$- y(t) \vdots u^{(m)}(t) \; u^{(m-1)}(t) \; \ldots \; u^{(1)}(t) \; u(t)] \qquad (7\text{-}2.8)$$

and the parameter vector:

$$\underline{\theta}^T = [a_{n-1} \; a_{n-2} \; \ldots \; a_1 \; a_0 \vdots b_m \; b_{m-1} \; \ldots \; b_1 \; b_0] \qquad (7\text{-}2.9)$$

Measurements of the input and output signals are made at discrete times $t = kT_0$, $k = 0,1,2,\ldots,N$ with T_0 the sampling time. The deviations:

$$y(t) = Y(t) - Y_\infty \; ; \; u(t) = U(t) - U_\infty \qquad (7\text{-}2.10)$$

are determined and also the corresponding derivatives. Then $N + 1$ equations are obtained:

$$y^n(k) = \psi^T(k) \; \hat{\underline{\theta}} + e(k) \qquad (7\text{-}2.11)$$

with $e(k)$ as the equation error, resulting in a vector equation:

$$\underline{y}^n = \underline{\Psi} \, \hat{\underline{\theta}} + \underline{e}. \qquad (7\text{-}2.12)$$

Minimisation of the loss function:

$$V = \sum_{k=0}^{N} e^2(k) = \underline{e}^T\underline{e} \qquad (7\text{-}2.13)$$

yields the well-known nonrecursive LS-estimation (normal) equations:

$$\hat{\underline{\theta}} = [\underline{\Psi}^T\underline{\Psi}]^{-1} \underline{\Psi}^T \underline{y}^n. \qquad (7\text{-}2.14)$$

A certain problem is the determination of the signal derivatives. State variable filters with transfer function given as:

$$F(s) = \frac{1}{1 + f_1 s + \ldots + f_n s^n} \qquad (7\text{-}2.15)$$

have shown to be well suitable, Geiger (1984). They provide

simultaneously the time derivation without differentiation and noise filtering, Young (1981). For the filter parameters the following approximation is recommended:

$$f_i \approx \hat{a}_i \qquad\qquad (7\text{-}2.16)$$

The LS-method then results in only consistent estimates, if the noise is zero. In the stochastic case biased estimates are obtained. The numerical properties of the basic LS-method can be improved by bringing it into a square-root filter form. Various non-recursive and recursive algorithms are possible, see Goedecke (1985), Geiger (1985).

7.2.3 Determination of process coefficients

It has already been shown by Isermann (1984(a)) that a unique determination of process coefficients:

$$\underline{p} = f^{-1}(\underline{\Theta})$$

is not always possible and that the use of dynamic models instead of static models facilitates the monitoring of more process coefficients. To get some more insight *two* simple examples of lumped parameter processes are considered:

Example 7-1: First order electrical circuit.

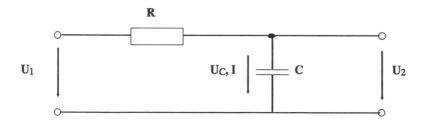

$$U_1(t) - U_R(t) - U_c(t) = 0; \quad U_R = RI; \quad I = C \, \dot{U}_c(t)$$

(a) input: voltage U_1; output: voltage U_2

$$a_1 \dot{U}_2(t) + U_2(t) = b_o U_1(t); \quad a_1 = RC; \quad b_o = 1.$$

The dynamic model has only one parameter a_1 which contains *two* process coefficients. Hence the process coefficients cannot be determined uniquely. One of them, either R or C has to be assumed as known. The static model (d/dt=0) does not give any information on R and C (however, if $b_0 \neq 1$ a fault of the capacitor (in the form of a leakage current) is indicated).

(b) input: voltage U_1; output: current I

$$a_1 \dot{I}(t) + I(t) = b_1 \dot{U}_1(t); \quad a_1 = RC; \quad b_1 = C.$$

In this case the dynamic model has *two* parameters a_1 and b_1 from which the process coefficients R and C can be determined uniquely.

Example 7-2: Second order mechanical system.

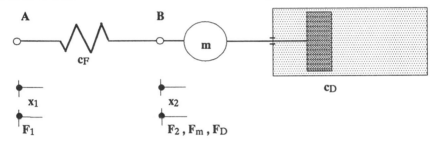

Forces at point B: $F_1 + F_2 + F_m + F_D = 0$

spring: $F_1 = c_F x_1; \quad F_2 = -c_F x_2$

mass: $F_m = -m \ddot{x}_2$

damper: $F_D = -c_D \dot{x}_2$

(a) input: *displacement x_1*; output: *displacement x_2*

$$a_2 \ddot{x}_2(t) + a_1 \dot{x}_2(t) + x_2(t) = b_0 x_1(t)$$

$$a_2 = m/c_F; \quad a_1 = c_D/c_F; \quad b_0 = 1.$$

From *two* parameters a_1, a_2 the process coefficients m, c_F, c_D and c_F cannot be determined uniquely. (However, a fault of the spring or the damper is indicated if $b_0 \neq 1$.)

(b) input: *force F_1*; output: *displacement x_2*

$$a_2 x_2(t) + a_1 \dot{x}_2(t) + x_2(t) = b_0 F_1(t)$$

$$a_2 = m/c_F; \quad a_1 = c_D/c_F; \quad b_0 = 1/c_F.$$

The *three* parameters a_1, a_2, b_0 allow the unique determination of the *three* process coefficients m, c_F, c_D. From the static model only $c_F = \hat{b}_0$ can be estimated.

These examples show:

(a) The unique determination of process coefficients depends on the selected input and output variables. There exists an *identifiability condition* for the process coefficients, which is valid for linear single-input, single-output passive lumped processes.
In order to determine the process coefficients uniquely the measured variables have to be selected as follows with regard to the signal flow diagram:
   (i) either as input variables which influence the state variables between the process coefficients,
   (ii) or as directly measured state variables between the process coefficients.
(b) If not all process coefficients can be determined some of them have to be assumed as known.
(c) Dynamic models enable the determination of more process coefficients than static models.

For higher order and more complex processes identifiable process coefficients depend on the individual model structure and choice of measurement signals, Nold and Isermann (1986).

7.2.4 Fault decision and classification

After the process coefficients $\underline{p}$ are determined, significant changes $\Delta\underline{p}(j)$ have to be detected with reference to normal values as a basis for the fault decision and the fault classification. As the process coefficients usually appear as stochastic variables a *Bayes decision* can be applied. The observed coefficients at time j are described by the vector $\hat{\underline{p}}(j)$ and for time j = 1,2,...,N by the process coefficient matrix:

$$\hat{\underline{p}} = [\ \hat{p}(1)\ \ \hat{p}(2)\ \ldots\hat{p}(N)\ ] \qquad (7\text{-}2.17)$$

Now it is assumed that the $\hat{p}(j)$ have a normal distribution and are statistically independent. The Bayes decision then leads to the following algorithms, see Geiger (1985):

Phase I: Determination of the normal state (training):

$$\hat{\mu}_{pio} = \frac{1}{N'}\ \sum_{j=1}^{N'}\ \hat{p}_i(j)\quad 1 \le i \le \ell \qquad (7\text{-}2.18)$$

$$\hat{\sigma}^2_{pio} = \frac{1}{N'}\ \sum_{j=1}^{N'}\ [\ \hat{p}_i(j)\ -\ \hat{\mu}_{pio}\ ]^2 \qquad (7\text{-}2.19)$$

Phase II: Detection of changes:

$$\hat{\mu}_{pi}(k) = \frac{1}{N}\ \sum_{j=1}^{N}\ p_i(k\text{-}j) \qquad (7\text{-}2.20)$$

$$\hat{\sigma}^2_{piI}(k) = \frac{1}{N}\ \sum_{j=1}^{N}\ [\ p_i(k\text{-}j)\ -\ \hat{\mu}_{pio}\ ]^2 \qquad (7\text{-}2.21)$$

$$\sigma^2_{piII}(k) = \frac{1}{N}\ \sum_{j=1}^{N}\ [\ p_i(k\text{-}j)\ -\ \hat{\mu}_{pi}(k)\ ]^2 \qquad (7\text{-}2.22)$$

$$d_i(k) = \frac{\sigma^2_{piI}(k)}{\sigma^2_{pio}}\ -\ \ln \frac{\sigma^2_{piII}(k)}{\sigma^2_{pio}}\ -1 \qquad (7\text{-}2.23)$$

$$d_2(k) \le 2\ \ln \frac{NP_o}{1-P_o}\quad \text{no change} \qquad (7\text{-}2.24)$$

$$d_i(k) > \ln \frac{NP_o}{1-P_o}\quad \text{change of } p_i. \qquad (7\text{-}2.25)$$

d_i denotes the *test quantity* and P_o is the *a-priori* probability of the no-fault case ($0 \le P_o < 1$). As a result, significant changes of single process coefficients p_i or combinations of several process coefficients p_i are indicated. By a comparison with the fault signatures the process coefficient changes can be classified and the fault diagnosis can be performed, see Section 7.2.1.

7.3 FAULT DIAGNOSIS OF A D.C MOTOR-CENTRIFUGAL PUMP

The described fault diagnosis approach based on *parameter estimation* was investigated for a centrifugal pump with a water circulation system, driven by a speed-controlled direct current motor, see Figure 7-4. A detailed description of this motor has been given by Isermann (1984a) and Geiger (1985). Therefore only a brief outline is given here. The emphasis of this section lies in new results on the detection of implemented faults.

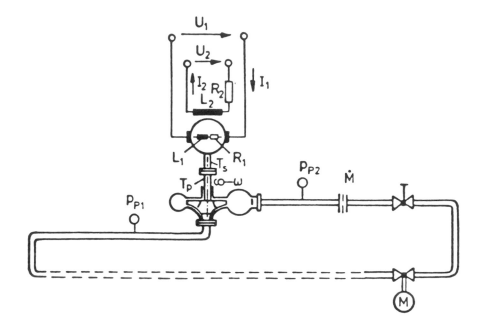

Figure 7-4: Scheme of the speed controlled d.c. motor and centrifugal pump

Motor: P_{max} = 4kW; n_{max} = 3000 rpm

Pump: H = 39m; $\dot{V}_{max}$ = 160 m^3/h; n_{max} = 2600 rpm

7.3.1 Mathematical model of the process dynamics

The following variables are measured:
U_2 armature voltage,
I_2 armature current,

ω angular velocity,

H pump total head, and

$\overset{\bullet}{V}$ volume flow rate.

The basic equations after some simplifying assumptions are, see Geiger (1985):

(a) armature circuit:

$$L_2 \frac{dI_2(t)}{dt} = -R_2 I_2(t) - \Psi\omega(t) + U_2(t) \tag{7-3.1}$$

(b) mechanics of motor and pump:

$$J\frac{d\omega}{dt}(t) = \Psi I_2(t) - c_{RO} - \rho g h_{TH1}\omega(t) \; \overset{\bullet}{V}(t) \tag{7-3.2}$$

(c) hydraulics of the pump:

$$H(t) = h_{NN}\omega^2(t) \tag{7-3.3}$$

(d) hydraulics of the pipe system:

$$a_B \frac{d\overset{\bullet}{V}(t)}{dt} = -h_{RR}\overset{\bullet}{V}^2(t) + H(t). \tag{7-3.4}$$

The model is non-linear and contains *nine* process coefficients:

$$p^T = [\; L_2, \; R_2, \; \Psi, \; J, \; c_{RO}, \; h_{TH1}, \; h_{NN}, \; a_B, \; h_{RR} \;]. \tag{7-3.5}$$

For the parameter estimation the equations are brought into the form:

$$y_j(t) = \psi_j^T(t) \; \hat{\underline{\Theta}}_j, \quad j = 1,2,3,4 \tag{7-3.6}$$

where:

$$\left. \begin{array}{ll} y_1(t) = dI_2(t)/dt & y_2(t) = d\omega(t)/dt \\[2mm] y_3(t) = H(t) & y_4(t) = dV(t)/dt \end{array} \right\} \tag{7-3.7}$$

$$\left. \begin{array}{l} \psi_1^T(t) = [\; I_2(t), \; \omega(t), \; U_2(t) \;] \\[4mm] \psi_2^T(t) = [\; I_2(t), \; 1, \; \omega(t)V(t) \;] \\[4mm] \psi_3^T(t) = [\; \omega^2(t) \;] \\[4mm] \psi_4^T(t) = [\; H(t), \; V^2(t) \;] \end{array} \right\} \tag{7-3.8}$$

$$\underline{\Theta}_1^T = [\ a_{11}\ a_{12}\ b_1\ ]$$

$$\underline{\Theta}_2^T = [\ a_{21}\ a_{22}\ a_{23}\ ]$$

$$\underline{\Theta}_3^T = [\ a_{31}\ ]$$ (7-3.9)

$$\underline{\Theta}_2^T = [\ a_{41}\ a_{42}\ ]$$

Based on the following parameter estimates:

$$\hat{\underline{\Theta}} = [\ \hat{\underline{\Theta}}_1^T\ \hat{\underline{\Theta}}_2^T\ \hat{\underline{\Theta}}_3^T\ \hat{\underline{\Theta}}_4^T\ ]^T$$ (7-3.10)

it should be noted that all process coefficients p can be determined uniquely as:

$$\hat{L}_2 = \frac{1}{\hat{b}_1}\ ;\ \hat{R}_2 = -\hat{a}_{11}\ \hat{L}_2$$

$$\hat{\Psi} = -\hat{a}_{12}\ \hat{L}_2;\ \hat{J} = \hat{\Psi}/\hat{a}_{12}$$

$$\hat{c}_{RO} = -\hat{a}_{22}\ \hat{J};\ \hat{h}_{TH1} = -\hat{a}_{23}\ \hat{J}/\rho g$$ (7-3.11)

$$\hat{h}_{NN} = \hat{a}_{31};\ \hat{a}_B = 1/\hat{a}_{41}$$

$$\hat{h}_{RR} = -\hat{a}_{42}\ \hat{a}_B$$

7.3.2 Experimental results

The d.c. motor is controlled by an a.c./d.c. converter with cascade control of the speed and using the armature current as an auxiliary control variable. The manipulated variable is the armature voltage U_2. A microcomputer DEC-LSI 11/23 was connected on-line to the process. For the experiments the reference value W(t) of the speed control has been changed in a stepwise manner with a magnitude of 750 rpm every 2 min. The operating point was n = 1000 rpm, H = 5.4 m and

$V = 6.48$ m^3/h. The signals were sampled with sampling time $T_0 = 5$ msec and 20 msec over a period of 2.5 and 10 sec, so that 500 samplings were obtained. These measurements were stored in the core memory and subsequently the parameters were estimated. The available computation time was about 100 sec. Hence, one set of parameters and process coefficients was obtained every 120 sec. For the training phase 50 coefficient sets were used.

Table 7-1 gives an overview of significant changes of process coefficients for *nineteen* different artificially generated faults. A selection of experiments will now be considered in more detail.

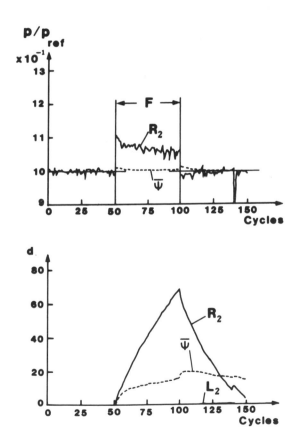

Figure 7-5: Change of process for a reduction of the brush contact
surface of about 50 %

A. Faults in the d.c. motor

Experiment A3: *Change of carbon-brushes,* **Figure 7-5**

A reduction of the brush contact surface of 50 % increases mainly the armature resistance R_2. With increasing operating time R_2 decreases due to the grinding effect.

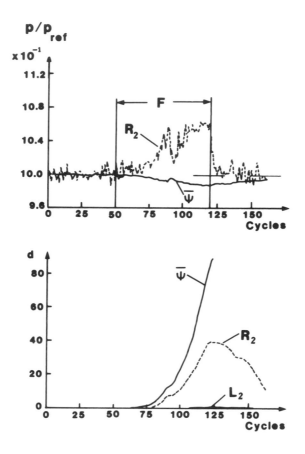

Figure 7-6: Change of process coefficients for a reduction of the cooling air flow

Experiment A5: *Disturbance of the air cooling,* **Figure 7-6**

A stepwise reduction (25%, 50%, 75%, 100%) of the air flow (e.g. due to plugging by dirt) leads to a temperature change of the whole motor and therefore to an increase of the resistance in the armature circuit and excitation circuit. Therefore R_2 increases and the magnetic flux linkage Ψ decreases. This is an example where the coefficients move in opposite directions.

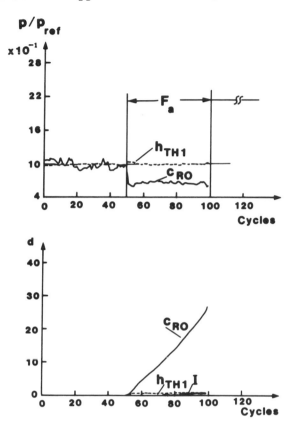

Figure 7-7: Change of process coefficients after washing out the lubrication from the ball bearings

B. Faults in the centrifugal pump

Experiment P1a: *Ball bearing without lubrication*, **Figure 7-7.**

If the grease of the ball bearing is washed out, the adhesive friction coefficient c_{RO} first decreases by about 30 %.

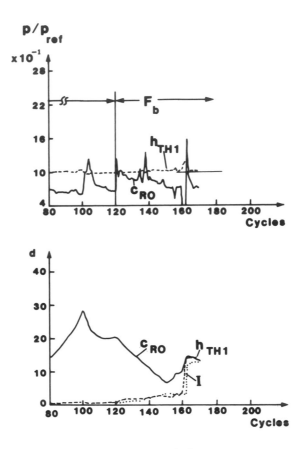

Figure 7-8: Change of process coefficients after the ball bearings
            get dirty by iron cutting

Experiment P1b: *Ball bearing with dirt*, Figure 7-8.

By adding fine iron cuttings to the ball bearing the coefficient c_{RO}
increases its mean as well as its variance.

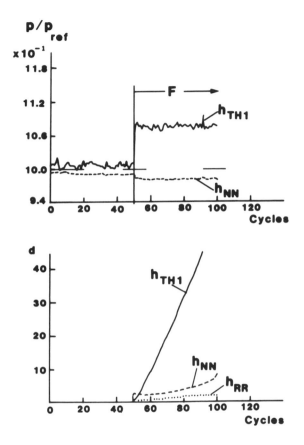

Figure 7-9: Change of process coefficients after increasing the slot clearance of the pump

Experiment P3: *Increase of slot clearance*, Figure 7-9.

An increase of the clearance of the slot between the pump wheel and the pump case increases the internal losses. Therefore h_{TH1} increases and h_{NN} decreases.

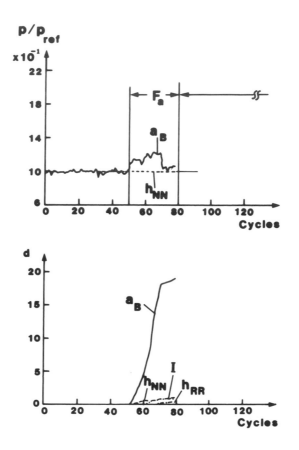

Figure 7-10: Change of process coefficients in the case of small
cavitation and gas bubble generation in the pump

C. Faults in the fluid part

Experiment F1a: *Small cavitation*, Figure 7-10.

A small cavitation and gas bubble generation in the pump by lowering
the entrance pressure is indicated by an increasing coefficient a_B,
which is the time constant of the pipe system. The system was
considered to be noise-free.

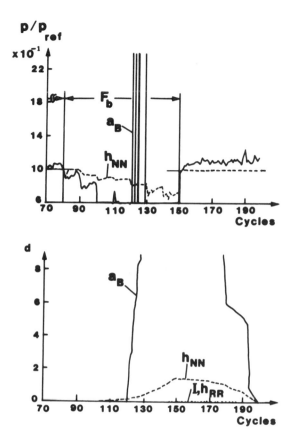

Figure 7-11: Change of process coefficients in the case of medium cavitation and gas bubble generation in the pump

Experiment F1b: *Medium cavitation*, Figure 7-11.

The effects of F1a are now much stronger.

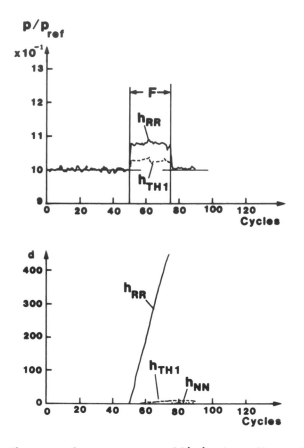

Figure 7-12: Change of process coefficients after closing the valve
position in the pipe from 100 % to 80 %

Experiment F4: *Increased resistance in the pipesystem*, Figure 7-12.

The resistance in the pipe could be increased by closing the valve
from 100 % to 80 % as a result the hydraulic friction coefficient h_{RR}
increases and also the internal loss coefficient h_{TH1} of the pump
(due to the new operating point). For a more detailed description and
discussion of all experiments see Geiger (1985).

These experiments have demonstrated that in all cases where a
significant change of process coefficients could be expected the
fault could be detected. In these cases a fault diagnosis could also
be performed, based on the pattern given in Table 7-1.

7.3.3 Mathematical model of the process statics

It is now considered which process coefficients can be determined if only the static behaviour is taken into account and the same signals are measured as in the dynamic case. With $d/dt = 0$ it follows from equation (7-3.1) through (7-3.4):

$$R_2 I_2 = \Psi\omega + U_2 \tag{7-3.12}$$

$$\Psi I_2 = c_{Ro} + \rho g h_{TH} \omega \hat{V} \tag{7-3.13}$$

$$H = h_{nn}\omega^2 \tag{7-3.14}$$

$$H = h_{RR}\hat{V}^2 \tag{7-3.15}$$

Hence the coefficients L_2, J and a_B can no longer be determined. equation (7-3.14) and (7-3.15) allow the calculation of h_{NN} and h_{RR} directly. By *changing the speed* ω signals for different operating points can be measured. Theoretically equation (7-3.12) then yields R_2 and Ψ. However, as $R_2 I_2 \ll \Psi\omega$, R_2 can hardly be obtained with sufficient accuracy. Ψ can be determined approximately. Based on Equation (7-3.13) then c_{Ro} and h_{TH1} can be estimated.

In comparison to the dynamic model the static model allows to determine *five* process coefficients instead of *nine*. Therefore the following faults of Table 7-1 cannot de detected: A2, A3, A5, K1, F1a and b, F2. A second type of experiment can be performed by varying the pipeline resistance h_{RR} for the case of $\omega = $ const., e.g. by changing the *valve position*. Then R_2 can be estimated and also the coefficients Ψ, c_{Ro}, h_{TH1}, h_{NN}, h_{RR}, i.e. *six* process coefficients. The following faults cannot be detected: K1, F1a and b, F2. However, the accuracy of these estimates has to be examined by experiments.

These considerations show that the use of *dynamic* models possible with *static* models. In this way acquisition of more faults can be detected.

Some other examples where fault detection based on process parameters was investigated are:

(a) Hydraulic drive of a machine tool, Hohmann (1977)
(b) Jet engines, Baskiotis, Raymond and Rault (1979, see also chapter 8 of this book)
(c) Electric motor for machine tools, Filbert and Metzger (1982).

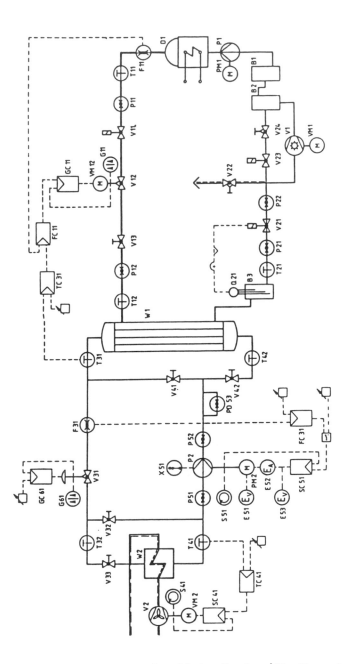

Figure 7-13: Scheme of the thermal pilot plant with the steam heated heat exhanger

7.4 FAULT DIAGNOSIS OF A TUBULAR HEAT EXCHANGER

7.4.1 Introduction

The fault diagnosis via parameter estimation was also investigated for a process which cannot be modelled with high accuracy. An industrial size steam heated heat exchanger was implemented in a pilot plant, consisting of an electrically powered steam generator, a steam-condensate circulation, a water circulation and a cross-flow heat exchanger to transport the heat from water to air, see Figure 7-13.

7.4.2 Mathematical model of the process dynamics

Following variables are measured: (see Figure 7-14)

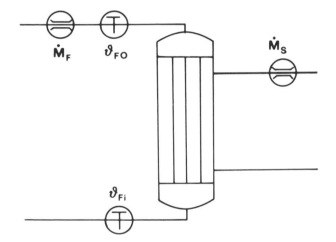

Figure 7-14: Tubular heat exchanger and measured variables

$\dot{M}_S$ mass flow of the steam

$\dot{M}_F$ mass flow of the fluid (water)

ϑ_{Fi} inlet temperature of the fluid

ϑ_{Fo} outlet temperature of the fluid

The fluid outlet temperature ϑ_{Fo} is considered as the output variable, the other three measured variables as input variables. To model the dynamic behaviour, the heat exchanger is subdivided into the tubular section, the water head, a transport delay and the temperature sensor. The basic balance equations for a heated tube are obtained by considering an infinitesimally small element of length dz and assuming distributed parameters for the fluid and lumped parameters for the tube wall, see Isermann (1969) and Goedecke (1986):

$$\frac{\partial h_F(z,t)}{\partial t} + w_F(t) \frac{\partial h_F(z,t)}{\partial z} = \frac{1}{A_F \rho_F} \frac{\partial \dot{Q}(z,t)}{\partial z} \qquad (7\text{-}4.1)$$

$$\dot{Q}_{SW}(z,t) - \dot{Q}_{WF}(z,t) = \frac{d}{dt} \dot{Q}_W(z,t) \qquad (7\text{-}4.2)$$

and after introducing the phenomenological equations for the heat transfer and $h_F = c_F \vartheta_F$, it follows that:

$$\frac{\partial \vartheta_F(z,t)}{\partial t} + w_F(t) \frac{\partial \vartheta_F(z,t)}{\partial z} = \frac{U_F \alpha_{WF}}{A_F \rho_F c_F} [\vartheta_W(z,t) - \vartheta_F(z,t)] \qquad (7\text{-}4.3)$$

$$\alpha_{SW} U_W [\vartheta_S(z,t) - \vartheta_W(z,t)] - \alpha_{WF} U_F [\vartheta_W(z,t) - \vartheta_F(z,t)]$$

$$= A_W \rho_W c_W \frac{d\vartheta_W(z,t)}{dt} \qquad (7\text{-}4.4)$$

The various symbols used have the following meaning:

A	cross-sectional area	w	velocity
c	specific heat capacity	z	location
h	enthalpy	α	heat transfer coefficient
l	length of tubes	ϑ	temperature
M,$\dot{M}$	mass, mass flow	ρ	density
Q,$\dot{Q}$	heat, heat flow		
r	evaporation heat		
U	periphery		

together with the subscripts:

F	Fluid (water)	W	Wall
K	Condensate	i	inlet
S	Steam	o	outlet

In addition, balance equations are stated for the steam space and the shell tube. The equation system is then linearised around the operating point, several simplifying assumptions are made and by applying the Laplace transform with respect to time t and spatial variable z one obtains three transfer functions, e.g.:

$$G_{\vartheta F}(s) = \frac{\Delta \vartheta_{Fo}(s)}{\Delta \vartheta_{Fi}(s)} = e^{-T_t s} \cdot e^{-\kappa_F \frac{T_{WF}s}{1+T_{WF}s}} \qquad (7\text{-}4.5)$$

with:

$$T_t = \frac{\ell}{w_F} \; ; \; \kappa_F = \frac{T_t}{T_F} = \frac{\alpha_{WF}\ell \; U_F}{w_F A_F \rho_F c_F} \; ; \; T_{WF} = \frac{A_W \rho_W c_W}{\alpha_{WF}U_F} \qquad (7\text{-}4.6)$$

This transcendental transfer function can be approximated by a lumped parameter model with dead time as follows:

$$G_{\vartheta\vartheta}(s) \approx \tilde{G}_{\vartheta F}(s) = K_\vartheta \frac{1 + T_{D\vartheta}s}{1 + T_{1\vartheta}s} e^{-T_t s} \qquad (7\text{-}4.7)$$

$$T_{1\vartheta} = \frac{\kappa_F}{1-e^{-k_F}} T_{WF} \; ; \; T_{D\vartheta} = T_{1\vartheta}e^{-\kappa_F} \; ; \; K_\vartheta \approx 1 \qquad (7\text{-}4.8)$$

Similarly one obtains the approximate transfer functions:

$$\tilde{G}_{s\vartheta}(s) = \frac{\Delta \vartheta_{Fo}(s)}{\Delta \dot{M}_S(s)} = \frac{K_D}{(1+T_{1D}s) \; (1+T_{2D}s)} e^{-T_{tD}s} \qquad (7\text{-}4.9)$$

with:

$$\left. \begin{array}{l} K_D = \dfrac{r}{\dot{M}_S c_F} \; ; \; T_{1D} = \dfrac{\ell}{w_F} \left[1 + \dfrac{A_W \rho_W c_W}{A_F \rho_F c_F} \right. \\[3em] T_{2D} = \dfrac{A_W \rho_W c_W}{\alpha_{WF}U_F} \cdot \dfrac{1}{\left[1 + \dfrac{A_W \rho_W c_W}{A_F \rho_F c_F}\right]} \end{array} \right\} \qquad (7\text{-}4.10)$$

and:

$$\tilde{G}_{F\vartheta}(S) = \frac{\Delta\vartheta_{Fo}(s)}{\Delta\dot{M}_F(s)} = \frac{K_F}{[1+T_{1F}s]\,[1+T_{2F}s]}\, e^{-T_{tF}s} \tag{7-4.11}$$

with:

$$\left.\begin{aligned}
K_F &= \frac{\vartheta_W - \vartheta_F}{\dot{M}_F}\frac{\ell}{w_F}\frac{\alpha_{WF}U_F}{A_F\rho_F c_F} \\[2mm]
T_{1F} &= T_{1D}\;;\;T_{2F} = T_{2D}
\end{aligned}\right\} \tag{7-4.12}$$

These equations show that it is not possible to determine all process coefficients uniquely. In the case of $\tilde{G}_{S\vartheta}$ for Example 7-2 parameter estimates correspond to *nine* process coefficients, and in the case of $\tilde{G}_{S\vartheta}$ three parameter estimates correspond to *ten* process coefficients. By assuming some of the process coefficients to be known e.g. the following process coefficients can be determined:

$$\tilde{G}_{\vartheta\vartheta}:\;\left.\begin{aligned}
\alpha_{WF} &= \frac{\dot{M}_F c_F}{\ell\,U_F}\left[-\ln\frac{T_D}{T_{1\vartheta}}\right] \\[3mm]
A_W\rho_W c_W &= \frac{\dot{M}_F c_F}{\ell}\,[T_{1\vartheta} - T_{D\vartheta}]
\end{aligned}\right\} \tag{7-4.13}$$

$$\tilde{G}_{S\vartheta}:\;\left.\begin{aligned}
\alpha_{WF} &= \frac{A_F\rho_F c_F}{T_{2D}\,U_F}\left[1 - \frac{1}{T_{1D}w_F}\right] \\[3mm]
A_W\rho_W c_W &= \frac{T_{1D}\dot{M}_F c_F}{\ell} - A_F\rho_F c_F \\[3mm]
r &= K_D \dot{M}_F c_F
\end{aligned}\right\} \tag{7-4.14}$$

$$\tilde{G}_{F\vartheta}:\;\left.\begin{aligned}
\alpha_{WF} &= \frac{A_F\rho_F c_F}{T_{2F}U_F}\left[1 - \frac{1}{T_{1F}w_F}\right] \\[3mm]
A_W\rho_W c_W &= \frac{T_{1F}\dot{M}_F c_F}{\ell} - A_F\rho_F c_F \\[3mm]
\vartheta_W - \vartheta_F &= K_F T_{2F}\dot{M}_F - \frac{w_F}{\ell}\frac{T_{1F}w_F}{T_{1F}w_F - 1}
\end{aligned}\right\} \tag{7-4.15}$$

7.4.3 Experimental results

The parameter estimation experiments are based on transient functions of the fluid outlet temperature ϑ_{FO} after steps in the input variables ϑ_{Fi}, M_S and M_F in the direction of decreasing temperature ϑ_{FO}. The operating point was:

$$\dot{M}_F = 3000 \text{ kg/h}; \quad \dot{M}_S = 50 \text{ kg/h}; \quad \vartheta_{Fi} = 60^\circ \text{ C}; \quad \vartheta_{FO} \approx 70^\circ \text{ C}$$

The sampling time was $T_O = 50$ msec, the time period of one experiment 360 sec, so that 720 samplings were used. For the parameter estimation the method of total recursive least-squares was applied by using *a digital state variable filter* approach. The determination of the normal state (training phase) was based on *sixty* transient responses for each transfer function. Then *thirty* transient responses were carried out for each transfer function and for each of *four* artificially generated faults:

F1: air (inert gas) in the steam space,
F2: opened condensate valve,
F3: closed condensate valve, and
F4: plugged tube.

(In total 540 experiments were carried out, lasting about 150 operating hours.)

Figure 7-15 shows (a) one measured transient function and (b) the corresponding time history of the parameter estimates. Good convergence of the parameter estimates was obtained in all cases. A verification of the measured and the calculated transient function shows very good agreement Figure 7-15c. In Table 7-2 the parameter estimates are given for $G_{S\vartheta}$. Table 7-3 indicates that for each of the *four* faults different changes of the parameter estimates are obtained. Corresponding results were obtained for changes of the fluid inlet temperature, $G_{\vartheta\vartheta}$ and changes of fluid flow, $\tilde{G}_{F\vartheta}$. Based on the parameter estimates $\hat{K}_D$, $\hat{T}_{1D}$, $\hat{T}_{2D}$ the obtainable process coefficients due to equation (7-4.14) are shown in Table 7-4.

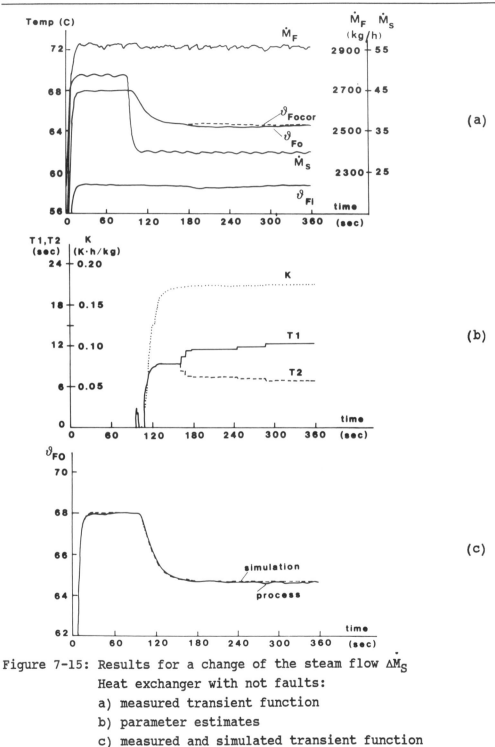

Figure 7-15: Results for a change of the steam flow $\Delta\dot{M}_S$
            Heat exchanger with not faults:
            a) measured transient function
            b) parameter estimates
            c) measured and simulated transient function

The differences between the coefficients used for theoretical modelling and the estimated ones for the normal state can be at least partially explained by the usual tolerances of the tabulated values and by the assumption and neglection in the modelling. The differences between the faulty state and the normal state are quite significant. However, they can only be explained partially by the physical effects of the faults. One reason is also that the investigated faults are not directly mapped by the obtainable process coefficients. In addition the calculated process coefficients are obviously rather sensitive to changes of the process parameter estimates and to the values of the coefficients which have to be assumed as known. This is well known for processes with heat transfer. Therefore it is recommended that the fault detection for this type of process, which is a higher order distributed parameter system and is approximated by lower order lumped models, is based on model parameter estimates like gain and time constants. This means that a detailed theoretical modelling is not necessary. The case study has shown that the considered four faults could be detected by using patterns of changes according to Table 7-3. This holds for all three transfer functions. Hence, it is sufficient to use only one transfer function. The most significant changes were obtained for steam flow changes, G_S . The use of static models allows only to recognise changes from the normal state. Hence a fault detection and diagnosis in this case is only possible by applying dynamic models.

Table 7-1: Detected significant changes of process coefficients after appearance of a fault. P_O = 0.5. In brackets the sequence of appearance is shown.

EXPERIMENT	L_2	R_2	Ψ	J	c_{RO}	h_{TH1}	h_{NN}	a_B	h_{RR}
A1: EXC. RESISTANCE INCREASE	-	-	-(1)	-	-	-	-	-	-
A2: ARM. RESISTANCE INCREASE	-	+(1)	-(2)	-	-	-	-	-	-
A3: AFFECTED BRUSHES	-	+(1)	-(2)	-	-	-	-	-	-
A4: NEW BRUSHES	-	-	-	-	-	-	-	-	-
A5: INSUFFICIENT COOLING	-	+(2)	-(1)	-	-	-	-	-	-
A6: COLD DRIVE	-	-	+(1)	-	-	-	-	-	-
K1: SHAFT DISPLACEMENT	-	-	-	-(1)	-	-	-	-	-
P1A: BEARING WITHOUT GREASE	-	-	-	-	-(1)	-	-	-	-
P1B: BEARING WITH DIRT	-	-	-	+(3)	+(2)	+(1)	-	-	-
P2: SIDE THRUST COMPEN. DEF.	-	-	-	-	-	-	-	-	-
P3: SPLITERING CLEARANCE	-	-	-	-	-	+(1)	-(2)	-	-
P4: AFFECTED IMPELLER	-	-	-	-(3)	-	+(2)	-(1)	-	-
P5A: PUMP CASING DEFECTIVE I	-	-	-	-	+(1)	-	-	-	-
P5A: PUMP CASING DEFECTIVE II	-	-	-	-	-(3)	+(2)	-	+(1)	-
F1A: LITTLE CAVITATION	-	-	-	-	-	-	-	+(1)	-
F1B: MEDIUM CAVITATION	-	-	-	-	-	+(3)	-(2)	-(1)	-(4)
F2: INSUFFICIENT VENTING	-	-	-	-	-	+(4)	-(2)	-(1)	-(3)
F3: FLUID TEMPERATURE INCREASE	-	-	-	-	-	-	-(1)	-	-(2)
F4: VALVE POSIT. INCREASE	-	-	-	-	-	+(2)	-	-	+(1)

Table 7-2   Parameter estimates for steam flow changes $G_{S\vartheta}$.

Fault	Mean and Standard Deviation	$\hat{k}_D$ [KH/kg]	$\hat{T}_{1D}$ [sec]	$\hat{T}_{2D}$ [sec]
None	μ	0.1708	12.38	7.21
	σ	0.0032	1.63	1.07
F1	μ	0.1896	7.26	7.26
	σ	0.00716	0.73	0.73
F2	μ	0.1268	7.62	7.62
	σ	0.00369	0.35	0.35
F3	μ	0.1899	13.89	3.81
	σ	0.0042	0.82	0.44
F4	μ	0.1689	13.65	6.01
	σ	0.00322	1.50	0.81

Table 7-3   Changes of the parameter estimates of Table 7-2.

Fault	$\hat{k}_D$	$\hat{T}_{1D}$	$\hat{T}_{2D}$
F1	>	<<	=
F2	<<	<<	>
F3	>	>	<<
F4	=	>	<

<u>Table 7-4</u>   Some process coefficients for steam flow changes.

Process coefficients	Theoretical modelling	Normal state	Faulty state			
			F1	F2	F3	F4
$r[kJ/kg]$	2309	2145	2382	1593	2385	2122
$A_W\rho_W c_W[kJ/Km]$	3.806	7.343	0.200	0.703	9.455	3.715
$\alpha_{WF}[kJ/Km^2 s]$	2.692	1.267	0.058	9.186	2.752	1.712

7.5 CONCLUSION

The use of mathematical models and measurable input and output
signals allows to estimate usually unmeasurable internal quantities
like parameters which can then be used for process fault detection.
These quantities express changes generated by process faults
directly. The model-based fault detection considers all included
changes of input and output signals, provides a data reduction and
tries to determine the process parameter which has been affected by a
fault. Therefore process faults can be detected earlier and localised
more precisely.

A generalised procedure for fault diagnosis based on process
parameters has been described. It has also been shown how the
process coefficients can be determined from the estimated system
model parameters and how this can be followed by a fault decision
and a fault diagnosis.

The methods were first investigated and tested experimentally for
a d.c. motor-centrifugal pump system which can be modelled as a nonlinear lumped system. Based on *five* measured signals *nine* process
coefficients could be monitored. Then *nineteen* process faults were
generated artificially. In all cases where a significant change of
process coefficient could be expected the fault could be detected and
localised. As a further application, an industrial size heat
exchanger was used as a representation of a process with distributed
parameters which can only be modelled approximately by lower order

lumped models. It has been shown that *two* measured signals are sufficient to monitor *three* process model parameters and to detect *four* artificially generated faults.

The case studies have demonstrated that the calculated process coefficients $\underline{p} = f^{-1}(\hat{\underline{\Theta}})$ allow a detailed fault detection and diagnosis, as long as the process can be modelled precisely with lumped parameters (so called well-defined processes) and if enough measured signals are available so that the identifiability condition is satisfied. If the process can only be described by approximate models (so called not well-defined processes) and if only a few signals are measurable the process parameter estimates $\hat{\underline{\Theta}}$ still contain valuable information for fault detection. However, the fault diagnosis can then not be as detailed as with reliable process coefficient estimates. Both examples have shown that the dynamic process behaviour yields considerably more information on process faults than can be achieved in the static case.

Chapter Eight

MODEL-BASED MODULAR DIAGNOSIS METHOD
WITH APPLICATION TO JET ENGINE FAULTS

Andre Rault and Chrysostome Baskiotis

After a general presentation of the hierarchy of problems covered by
fault detection, diagnosis and predictive maintenance of complex
systems, the basic methodology used is described as well as the
necessary mathematical tools. An example of jet engine fault diagnosis gives further insight into the problem.

8.1 INTRODUCTION

The problem of system availability has always been of main concern to
military personnel who want to be able to perform their mission when
ordered. In the nuclear industry as well as in high risk chemical
processes security is the prevailing factor. In the growing automated
industries, it has only recently become apparent that process
availability is a first objective to be satisfied before any optimisation. It is much easier to increase process availability from 60%
up to 95% than to gain 2 or 3 points in quality thanks to optimisation; this economic argument has therefore stimulated large
interest towards fault detection and diagnosis problems.

It is an area where the research effort has only been emphasised
comparatively recently (Willsky (1976)) and in which an effort of
integration of multi-disciplinary tools has to be made.

After discussing a formulation of the problems associated with
system availability this chapter will present a general methodology
for system diagnosis and go through its implementation to a jet
engine case study.

8.2 PROBLEM STATEMENT

System availability is the result of the mastery of faults or failures as it is defined as the ratio of the actual running time of the process over the desired running time.

In order to improve system availability, it is important to achieve the following:

(a) - fault detection before getting to the failure stage,
(b) - fault localisation and diagnosis,
(c) - forecast of fault evolution,
(d) - eventual system reconfiguration,
(e) - maintenance and/or rescheduling to satisfy mission or manufacturing requirements.

Actually, a hierarchy of problems can be set up according to the nature of occurring faults, typically a process is characterised by a certain number of state variables (whose knowledge permits the prediction of the process behaviour in the immediate future) of observations or outputs and of inputs. These variables signify the state of operation of the system and are the basic elements given to the operator who supervises the process evolution. So, basically a process is said to be in a *normal state of operation* if its observed variables are in the neighbourhood of a predefined set point.

Actually, the significant variables of a process are linked by causal relationships and we will define a *state of fault* and/or *failure* as a change in these relationships.

The relationships between the observed variables can be obtained by a mathematical model which will exhibit a certain number of parameters often referred to as structural parameters. The actual values of these parameters can be obtained through system identification. A change of value of these parameters signifies a structural change i.e. a fault.

Various levels of deterioration can be defined:

(a) - unsteady faults (random structural changes);
(b) - steady faults or failures (permanent structural changes);
(c) - catastrophic faults or failures (structural changes creating catastrophes).

Faults appear often in that order and progressive deterioration
may lead to catastrophy especially when an automatic control
compensates for it. Faults can also be separated in two classes:
evolving faults mainly due to ageing and *cataleptic* ones whose occurrence
is purely random.

The term *failure* is used when there is actual physical breakage
preventing normal operation. From this classification, a time scale
hierarchy immediately appears and it is schematically presented here
after.

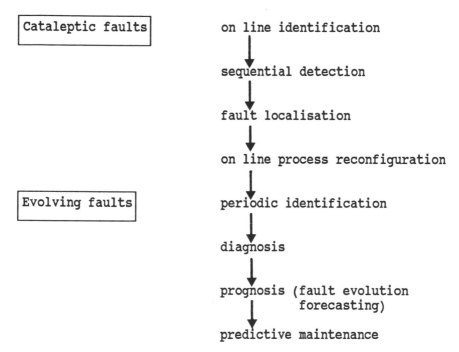

The schematic analysis shows that cataleptic fault detection should
be carried as a real time task while the evolving faults will be
looked for and diagnosed periodically; the learning of their
evolution will permit modelling of the wear process and thus fore-
casting the risk of failure or unavailability.

Thus, these two tasks are hierarchically embedded and it is
interesting when possible to adopt the same structural representation
for identification.

A typical implementation of fault detection for industrial
processes has been presented in (Rault et al. (1984)) and therefore
the rest of this chapter will concentrate on the diagnosis problem

through its implementation on jet engines.

8.3 METHODOLOGY

The basis of any external diagnosis method resides on the hypothesis
that any internal structural change of the system induces a
behavioural change. Typically, in rotating machines, vibration level
and frequency changes indicate the possibility of some fault.
Similarly, performance variations, static or dynamic, are deterioration. Actually, a good diagnosis system should integrate all the information sources. In the application study presented here, only the
performance analysis is covered because, at the start of the project,
the necessary expertise and competence were not available; however
the general methodology is applicable to any type of information.

8.3.1 Main steps

Behavioural information is generally weakened through the course of
time (performance variations are weak) since the purpose is to detect
faults at their early stage. Therefore it is interesting to find a
form of data reduction which yields parameters independent of the
experimental environment and enhance faults at their early stage.
Static and dynamic modelling constitute an appropriate data reduction
of performances which yield parameters independent of environmental
disturbances provided that the appropriate identification procedure
has been chosen (Rault (1984)). Dynamic parameter variations also
appear to be quicker indicators of faults and this will be shown in
the application to jet engine fault diagnosis.

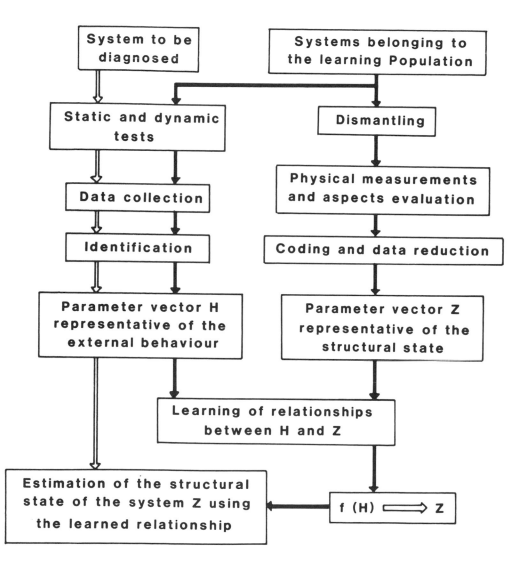

Figure 8-1: General methodology of diagnosis

The problem can be separated into four phases:

(a) - Measurement of the external behaviour through static and dynamic performances.
(b) - Determination and measurement of the mechanical state on a test population (learning phase).
(c) - Establishment of relationships between the external behaviour and the structural state.
(d) - Implementation of the methodology: diagnosis of the structural state directly from the external behaviour measurement.

The general methodology is described in the synopsis given in Figure 8-1.

The two main phases, *learning* and *implementation* are differentiated as follows. During the learning procedure, systems belonging to the test population follow the procedures indicated by the paths with simple arrows. In the normal implementation mode, diagnosis is performed by following the double trace arrows. This methodology has been applied and implemented on jet engines Marbore II and Larzac of the French Air Force, but it is basically applicable to any system whose wear and tear has an effect on its external behaviour.

8.3.2 Mathematical tools used

Basically, the static and dynamic data reduction is performed using system identification which could be implemented on-line. This type of technique is well-known (Richalet et al (1971), Rault et al. (1971)) and not subject to particular difficulty provided the data acquisition phase has been properly solved.

A more difficult problem to solve is the data reduction of the physical parameters of the learning population because it makes use of statistical tools less familiar to the system engineer. Indeed, the problem is as follows: given the physical measurements representative of the observation of the structure of the systems belonging to the learning population it is necessary to code and reduce these data. The behavioural and the structural data once collected have to be related. Such degradation of a module induces a change in the static and dynamic parameters. This type of analysis is performed firstly in a qualitative manner using correspondence

analysis (Benzecri (1977)); a quantitative measure is more difficult
to be given because of the statistical nature of the data significant
of the structure. The authors (Baskiotis et al (1979)) have derived
a fuzzy discriminant analysis which gives as answer the degree of
membership to a class. This method will be presented with the
application.

8.3.3 Necessary characteristics of a diagnosis method for practical implementation

The use of a diagnosis method should not be a burden for the user and
therefore should have the following properties:

(a) - *robustness*: the final diagnostic measure should be insensitive
      to small parameter variations,
(b) - *flexibility*: any change in the basic fault classification or
      definition has to be taken into account easily,
(c) - *adaptivity*: structural modifications (mechanical redesign of a
      part) or operational changes should be included without having
      to reinitiate the learning phase,
(d) - *transparency*: necessary tests for the diagnosis should be carried
      out without having to interrupt the normal operating mode
      (during flight for an aircraft or during production for a
      process).

These qualities should be considered at an early stage of the design
so that its implementation does not introduce supplementary
constraints to the normal system exploitation. This remark is not
valid for the jet engine application as this is still at the research
stage.

8.4 JET ENGINE APPLICATION

The methodology has been applied to two types of jet engines: a
single spool engine called *Marbore* and a twin spool of recent design
called *Larzac*. In this section the various steps of the methodology
will be scanned with highlight points taken from both of these two
examples.

8.4.1 External behaviour

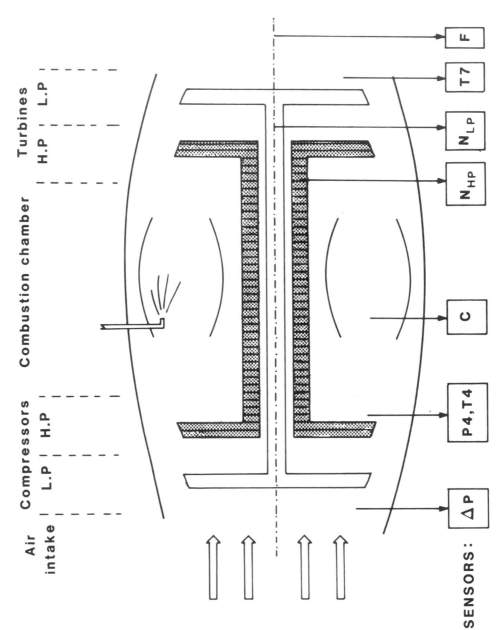

Figure 8-2: Larzac jet engine: schematic diagram

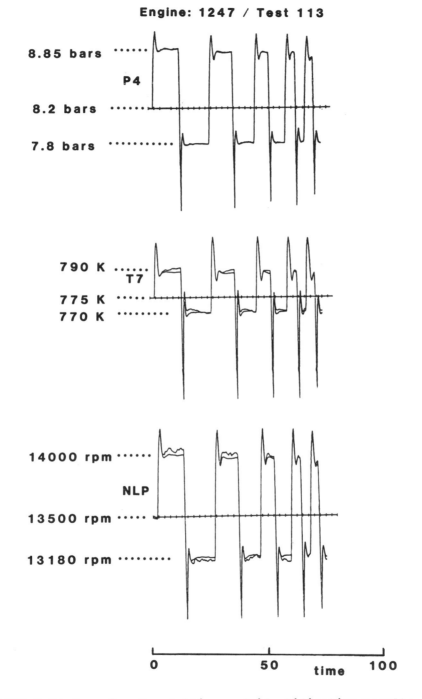

Engine: 1247 / Test 113

Figure 8-3: Example of modelling and identification results

The Larzac engine is a twin spool type which has been modelled for linear domain operation by a discrete-time state variable model given by equation (8-4.1). The two state variables are the low pressure (NLP) and high pressure (NHP) spool regimes, the input is the fuel flow C.

$$
\begin{bmatrix} NHP(k+1) \\ NHP(k+1) \end{bmatrix} = \begin{bmatrix} a_{11} & 0 \\ 0 & a_{22} \end{bmatrix} \begin{bmatrix} NHP(k) \\ NLP(k) \end{bmatrix} + \begin{bmatrix} b_1 \\ b_2 \end{bmatrix} C(k)
$$

$$
\begin{bmatrix} TT7(k) \\ TT4(k) \\ PT3(k) \\ P4(k) \end{bmatrix} = \begin{bmatrix} 0 & c_{12} \\ c_{21} & 0 \\ 0 & c_{32} \\ c_{41} & 0 \end{bmatrix} \begin{bmatrix} NHP(k) \\ NLP(k) \end{bmatrix} + \begin{bmatrix} d_1 \\ d_2 \\ d_3 \\ d_4 \end{bmatrix} C(k) \qquad (8\text{-}4.1)
$$

A schematic diagram of the engine with the sensors is given in Figure 8-2.

Temperatures and pressures at various levels in the vein are measured. This model is of the representation type and has little thermodynamic interpretation.

In this application, the external or ambient behaviour was monitored through a specific series of tests performed either on the aircraft at the fixed point area or on the test bench at the periodic visits every 300 hours.

Tests consist of recording the static regime at four operating points and the dynamic regime in the neighbourhood of these points. An example of the identification results around one operating point is given in Figure 8-3.

8.4.2 External behaviour and structural state

The objective of this phase is to establish relationships between the external behaviour and the structural state of the engine (basically its internal mechanical state).

The structural state is analysed on a learning population which should scan at the best the possible faults, as the engines have to be dismantled and inspected. From inspection, information is coded to constitute a vector Z to indicate the structural state. A certain number of components and clearances are measured as well as graded on their aspect.

The number of degrees of degradation retained depends on the number of engines belonging to the learning population. The larger the population is, the more precise the diagnosis will be. In the Larzac application we have retained only two levels of degradation:

(a) healthy or slightly degraded,
(b) degraded.

The number of elements diagnosed is *five* corresponding to the number of physical modules:

(a) LP (low pressure) compressor,
(b) HP (high pressure) compressor,
(c) combustion chamber,
(d) HP turbine,
(e) LP turbine.

The vector Z has thus 5 components taking discrete values 0 or 1, thus indicating $2^5 = 32$ possible classes of structural state.

Given the learning population, the next step is to verify that the spatial classification of the Z vectors is in agreement with that of the H vectors.

Consider the space on R^n where n is the number of coordinates of H. In R^n the population will constitute a *cloud* of points whose distribution must be analysed. The mathematical tool used to perform this analysis is the principal components analysis which projects the cloud of points into a subspace of lower dimension with minimal loss of information (Anderson (1985)).

Analysis of this projection will show whether the partitioning of the cloud into classes is possible or not. This procedure may be iterative as some coordinates have to be eliminated since they do not carry any information and otherwise serve to increase the entropy of the cloud.

Figure (8-4) gives the projection onto the first two axes of a certain number of engines using uniquely static parameters. It shows that static information enables the analyst to discriminate the level of degradation of the engine as a whole.

STATIC PARAMETERS

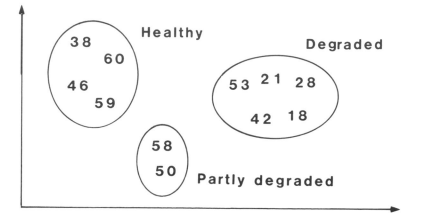

Figure 8-4: Corresponding analysis, static parameters

DYNAMIC PARAMETERS

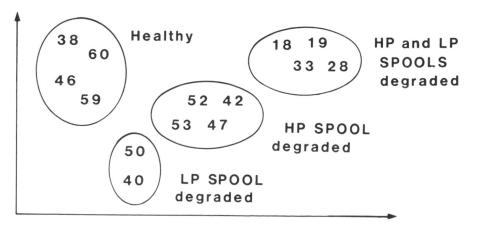

Figure 8-5: Corresponding analysis, dynamic parameters

Figure (8-5) shows that the use of dynamic parameter variations permits the discrimination not only of the degradation level but also of the degraded modules.

Test of an engine before HP Spool change (103) and after (108).

STATIC PARAMETERS

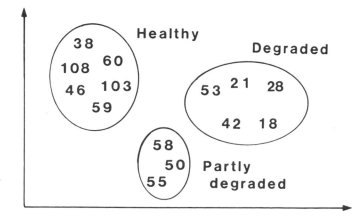

Figure 8-6: Correspondance analysis, <u>static parameters</u>

DYNAMIC PARAMETERS

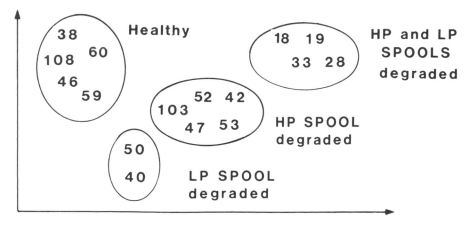

Figure 8-7: Correspondance analysis, <u>dynamic parameters</u>

By using an iterative procedure on the learning population a basis for diagnosis is thus established.

Based on the results shown previously, a demonstrative example of diagnosis has been carried out on an engine before a change of HP spool module (labelled as <u>103</u> in Figure 8-6) and after a change has occurred (labelled as <u>108</u>).

Figure (8-6) shows that the static parameters information is able to give the proper diagnosis whilst Figure (8-7) shows that the information corresponding to dynamic parameters yields the correct diagnosis.

This analysis is purely a *qualitative* one and constitutes the step of classification on the basis of the learning population. Of course, the more completely the learning population scans the space of possible faults the simpler the classification will be. In the present case, as the Larzac is a new engine, a difficulty arises due to the lack of faulty engines; indeed the actual learning population is limited to 20 engines.

The next step is to establish a classification to determine a *quantitative* diagnosis and to ensure the adaptivity of the diagnosis method for new information. For this, we have developed the fuzzy discriminant analysis whose effectiveness has been shown on the older Marbore engines.

8.4.5 Diagnosis

Consider an engine, whose external behaviour vector H is known and for which the structural vector Z (i.e. diagnosis) is looked for.

Presently several methods exist to solve this classification problem; however, before discussing these the following remarks should be made:

(a) - classes are not separated; for example *HP compressor degradation* is not independent of *HP turbine degradation*. It means that engines may belong to both classes; each engine may belong to several classes.
(b) - clearly not every class defined by the Z vector can be present in the learning population. This is typically the case for the Larzac engine, the learning population is built up as faults appear. Thus certain classes of faults corresponding to critical components or modules are present whilst others are empty.

Therefore, instead of using a typical discriminant analysis which associates the candidate systematically to one class and only one, it is more appropriate to attribute a degree of membership to each of the determined classes. Such an approach can be developed using fuzzy set theory (Baskiotis et al. (1979)).

Given a learning population $X = x_1, \ldots, x_n$.

Each element is characterised by $m + p$ items:

$$x_i = (x_{i1}, \ldots, x_{im}, \ldots, x_{im+p})$$

According to the p last components, the ensemble X is partitioned into q classes $C_1, \ldots, C_q$

The classification problem is thus as follows:

Let $y \in R^m$ be an element for which the m first items are known; given these values, the degree of membership of y to any class C_k is determined.

C_k is divided in two sub-classes:

$$C_k^S(y) = \{x \in C_k \mid d(x,y) \leqslant d(x,C_k)\}$$

$$C_k^D(y) = \{x \in C_k \mid d(x,y) > d(x,C_k)\}$$

$$(8\text{-}4.2)$$

$d(x,y)$ is the distance between two elements x, and y and:

$$d(x,C_K) = \frac{1}{|C_k - x|} \sum_{z \in C_k} d(x,z)$$

$$(8\text{-}4.3)$$

where $|x|$ denotes the modulus of x.

Using these two sub classes one can compute:

(a) – *a similarity index*:

$$S(y,C_k) = \begin{cases} \dfrac{1}{|C_k^S(y)|} \sum \dfrac{d(y,x)}{d(x,C_k)} & \mid x \in C_k^S(y) \text{ if } C_k^S(y) \neq 0 \\[4mm] 1 & \text{if } C_k^S(y) = 0 \end{cases}$$

$$(8\text{-}4.4)$$

(b) – *a dissimilarity index:*

$$D(y,C_k) = \begin{cases} \dfrac{1}{|C_k^D(y)|} \sum \dfrac{d(x,c_k)}{d(y,x)} & \mid x \in C_k^D(y) \text{ if } C_k^D(y) \neq 0 \\[4mm] 1 & \text{if } C_k^D(y) = 0 \end{cases}$$

$$(8\text{-}4.5)$$

$$S(y,C_k), \ D(y,C_k) \ \epsilon \ [0,1]$$

The degree of membership $\mu(y,C_k)$ of y to a class C_k should represent the degree of neighbourhood of y to all the elements. Therefore, we can define the degree of membership as the average between the membership to the similarity class and to the dissimilarity class with the following properties:

$$\mu S(y,C_k) \geqslant 1 - S(y,C_k)$$

$$\mu D(y,C_k) \leqslant D(y,C_k) \qquad\qquad\qquad (8\text{-}4.6)$$

$$\mu(y,C_k) \ = \ \frac{\mu S(y,C_k) + \mu D(y,C_k)}{2}$$

These yield $\mu(y,C_k) = 0.5[1 - \alpha S(y,C_k) + \beta D(y,C_k)]$ \qquad (8-4.7)

where α , $\beta \ \epsilon \ [0,1]$

At the present time the Larzac learning population is considered too small, and consequently, the diagnosis we have made is qualitative (correspondence analysis). However, fuzzy discriminant analysis has been used to diagnose the Marbore engines with a degree of success of over 80 % on a validation population of 52 engines.

8.5 CONCLUSION

This Chapter has presented the general principles of fault detection, diagnosis within the context of predictive maintenance. The methodology of diagnosis has been analysed through a jet engine fault monitoring application. It has been restricted to performance analysis measurements but is applicable to any source of behavioural information as long as a learning population has been formed.

Through the Larzac Jet engine example, it has been shown that amongst the performance analysis parameters the dynamic ones are better indicators of faults. It has also been shown that the methodology has to be adaptive and progressive as the learning population cannot be complete; a specific measure of the degree of membership of a newcomer to incomplete classes has been developed and has proved its efficiency on the Marbore engine application.

The main drawback of this approach is that it needs a learning population which may be a lengthy operation when one is considering equipment whose average lifetime is quite large (at least 1500 hours for jet engines); of course the adaptive nature of the methodology permits the addition of new information but at the start of such a set up one should be aware that the data acquisition of the learning population is a heavy investment before any possible diagnosis.

This drawback is overcome for the case of a large population of interest, thus permitting a fast-running population to be seen.

A current application on pneumatic door jacks of the Paris Subway system does not have this drawback. Thus the predictive maintenance problem for the subway will be tackled in the near future.

The authors believe that such a systematic and rational approach is a solution to the fault detection and diagnosis problem and can integrate all sources of information; however all the environmental constraints (sensors, tests, etc.) have to be overcome at the start in order not to create a supplementary burden to the normal operational conditions.

Chapter Nine

FAULT DETECTION IN NUCLEAR REACTORS WITH THE AID OF PARAMETRIC MODELLING METHODS

M. Kitamura

Some recent developments in the methods of fault detection and identification (FDI) in nuclear power plants (NPPs) are presented. An
emphasis is placed on FDI-information extraction from fluctuating
components inherently observed in the process signals rather than
from gross behaviour of the signals themselves. Parametric modelling
techniques have been utilised as a unified approach to alleviate the
difficulty in determining physical principle models of the plant for
wide operating conditions. The Least-Squares (LS) modelling principle
has been adopted extensively to enhance the applicability of the
method to short-time observation data. Different models have been
employed to extract FDI-relevant information through time and
frequency domain analysis, and direct parameter estimation. The
proposed FDI methods are grouped into two categories; serial and
parallel redundancy. The applicability of the FDI is demonstrated
through simulation and actual experiments at several nuclear power
stations.

9.1 INTRODUCTION

Several representative methods developed for fault detection and
identification (FDI) in nuclear power plants (NPPs) are presented in
this paper. The scope is focused on a specific approach to FDI,
namely the use of the parametric modelling method for extracting
FDI-related information. This approach has been chosen in view of its
high potential in dealing with various subprocesses of a large-scale
complex plant in an unified manner. Emphasis of descriptions will be

placed on derivation of redundant information rather than on
organisation of the actual FDI-scheme, since we believe that the
former is the most essential issue for successful FDI.

9.1.2 Background to problem

Safe and reliable operation has always been recognised as an issue of
prime importance in NPPs. Efforts have been paid for constructing and
operating the NPPs to satisfy very strict safety standards, such as
fail-safe and fool-proof design, ample design margin, inherent safety
(i.e. self termination of power excursion), automated shutdown
mechanism, emergency core cooling system, instrumentation with
redundancy and multiplicity, etc. A wide variety of FDI techniques
has also been introduced and practised, with the aim for minimising
the potential damage of plant constituents by conducting early
detection and warning.

FDI techniques currently in use by the industrial sector are
conservative and traditional. Typical examples include calibration
against standards, limit checking, mutual consistency checking
(majority voting), surveillance during periodic shutdown and perturbation tests before restart (see also discussion by Isermann in
Section 7.1). Although these traditional techniques have contributed significantly to attain high level safety of NPPs, they are by no
means perfect nor free from errors. The limitations of the
traditional FDI techniques were dramatically disclosed by the
accident at *Three Mile Island-2* (TMI-2). Through intensive studies of
the accident by many independent investigations, improvements of NPP
instrumentation to provide higher FDI-capability has been identified
as one of the most crucial issues of common interest.

Actually, a considerable amount of research activity has built
up towards the development of FDI techniques. Most of this work has
been in the academic sector, over the last two decades. One major
approach to FDI in NPPs is to obtain information about component
integrity via analysis of fluctuating components in process signals,
often referred to as reactor noise. The conventional as well as
modern techniques of random signal analysis have been employed, leading to many results evidently demonstrating the possibility of incipient fault detection in commercial NPPs (Williams (1977; 1982)).

Other approaches to FDI in NPPs, based on the concept of analytic redundancy, have also been attempted (Kitamura (1980), Clark and Campbell (1982), Tylee (1982)) for a few subsystems of NPPs. The techniques of Kalman filtering and the Luenburger observer have been utilised to derive redundancy (i.e. extra signal) relationships for fault detection by cross-checking. The importance of FDI with incipient fault detection capability has been made clear through these early attempts based on simple processes. In spite of these positive advances, however, the nuclear industry had been somewhat reluctant to accept the advanced FDI techniques as the standard procedure for routine use. Some reasons for the reluctance are summarised below:

(a) *Lack of familiarity*: Such advanced techniques are less well known than the traditional manual methods based on operational or main-tenance personnel. It is inevitable for personnel to tend to avoid the possibly powerful but apparently unfamiliar techniques, unless they are provided in a compact, easily comprehensible manner.

(b) *Fear of economic loss*: The advanced FDI methods can possibly be so sensitive that they require too many diagnostic and/or remedi-cal actions involving considerable expenditure in time and money, together with possible loss of power generation. Also, the advanced FDI methods can be a cause of additional false alarms/ shutdowns which lead to costly consequences.

From the author's viewpoint, the FDI techniques should be modified to alleviate the above-mentioned causes of reluctance to industrial acceptance. Firstly, the fundamental principle of FDI should be clearly articulated in conjunction with each method under consideration. The superficial concept of the method is, of course, explained in technical documents, but the concept is apparently quite diverse. The potential misunderstanding and confusion caused by the diversity can be most effectively removed by explaining the method from a more fundamental perspective of FDI, as will be shown in a subsequent section. Secondly, more emphasis should be placed on FDI in its incipient, rather than evolved, phase. Then the FDI method can be used not as a source of unwanted extra alarms disturbing the scheduled operation, but as the source of valuable information

supporting the maintenance personnel in identifying the potential
source of malfunction to be fixed. Last but not least, the technical
content should be arranged with simplicity to avoid unnecessary
confusion. If similar FDI performance is expected from two different
technical candidates, the one with simpler content should always be
adopted.

9.1.3 Present approach

We have made several attempts to develop the FDI techniques in NPPs
based on the consideration described above. First, we focused on the
concept of redundancy as the most fundamental element of FDI. Every
FDI technique requires a certain set of redundant information for
conducting mutual comparison, detecting inconsistency, and for identifying the failed or faulty component/subsystem if possible. The
diverse concepts of various FDI techniques can be better understood
by introducing this aspect.

The *conventional redundancy*, whether hardware-made or analytic, is
categorised as *parallel redundancy* in our terminology, because the
comparison is made between the reference quantities estimated in
parallel from multiple (either similar or dissimilar) information
sources.

The usage of reactor noise analysis for FDI might appear to be
quite different at first, having nothing to do with redundancy. A
typical practice is to estimate statistical characteristics (often
called noise signatures) such as auto-power spectral density (APSD)
function or the transfer function, collect an ample amount of these
signatures during normal operation, prepare the so-called baseline
signatures to be used as the standard, and to compare the everyday
signatures with the baseline ones to detect abnormal changes. One
should notice that the concept of redundancy is again the most
essential in this approach; the redundancy is, in fact, provided
serially and not in a parallel way in time. The necessity of
baseline signatures indicates the crucial importance of redundancy
for FDI.

The proposed FDI methods are grouped into two categories; serial
and parallel redundancy. The former implies FDI-methods based on the
comparison of signal characteristics with the pre-established

standard characteristics. The latter implies methods to derive a redundant estimate of a signal by analysing dissimilar signals.

The majority of FDI techniques based on reactor noise analysis have been designed by utilising the serial redundancy; whilst the FDI by analytic redundancy applications have been based on the parallel approach. It should be noticed, however, that other pairings are also possible and promising. The reactor noise analysis can be used for parallel redundancy, and the analytic redundancy used for serial redundancy.

As we have two different approaches (i.e. noise analysis and analytic redundancy) and two types of redundancy, we could pursue four paths toward the FDI in NPPs. Available FDI-related information can most efficiently be covered by systematic utilisation of all the four paths. However, we decided to pursue only the two paths, namely, parallel and serial redundancy using noise analysis. Because of the second principle mentioned in the preceding section, the FDI techniques should be designed to have the capability of dealing with incipient faults, which are characterised by subtle or small changes in signal fluctuations rather than by changes in the gross behaviour of signals. To fulfil this requirement, the use of noise analysis is more promising than the analytic redundancy based on physical-principle models which are suitable for characterising the gross behaviour.

One major cause of difficulty in utilising physical principle models in NPPs is the strict regulation on the operation. Injection of test signals is prohibited except during the scheduled restart period. Also, the NPPs are usually base-loaded to operate in the constant power mode because of the relatively low cost in power generation. The physical-principle models, valid for transient analysis and control system design, are not suitable for characterising the reactor noise. The determination of specific physical models representing the fluctuating components is quite difficult, if not impossible, because the statistical properties of the reactor noise are dependent on the operating condition and history of the plant. The use of the noise analysis approach to extract redundant information for FDI is imperative in this context.

Among various available methods of handling reactor noise, we have chosen the parametric modelling approach in view of the third principle, i.e. *technical simplicity*. Derivation of FDI-information is carried out by the single methodology of parametric modelling rather than by introducing a wide variety of methods of random signal analysis.

Univariate and multivariate (UAR and MAR) models, single-input, single-output (SISO) models and multiple-paths delay (MPD) models are employed in the present study. The LS modelling principle has been adopted as a unified method of determining the model parameters from the data, since the method coupled with the Householder transformation allows us to determine the optimal model parameters with high numerical stability and efficiency (Golub (1965), Kitagawa and Akaike (1978)).

The additional benefit of using the LS modelling approach is the applicability of the method to non-steady state data. Because of the high numerical stability, the modelling algorithm works successfully even when the number of available samples is too small for other modelling algorithms. The parametric models estimated for a set of consecutive data blocks, each consisting of a small number of samples, can be used to characterise the time-dependent variation of the stochastic nature of plant variables. In short, the guidelines adopted in this chapter for developing the FDI techniques in NPPs can be summarised as follows.

(a) The concepts of parallel and serial redundancy have been introduced to provide a systematic perspective for FDI.

(b) The fluctuating or stochastic components of signals, rather than bulk behaviour, were treated as the information carrier in order to ensure applicability of the method for incipient phase FDI.

(c) The parametric modelling approach has been adopted as the unified method for processing the signals, allowing a simple organisation of the FDI-techniques.

(d) The LS modelling algorithm has been most commonly employed to determine model parameters from signals, with a view to maintaining methodological simplicity and enhancing the applicability to small sample measurements.

A short review of structure and dynamics of typical NPPs is given in Section 9.2. The parametric models and the modelling algorithms are given in Section 9.3. Examples of FDI by serial and parallel redundancy are described in Sections 9.4 and 9.5 respectively. In most of the examples, signals from actual NPPs have been processed. The faults considered are, however, simulated ones, as it has been impossible to introduce artificial faults or failures in operating NPPs. Some concluding remarks are given in Section 9.6.

9.2 DESCRIPTION OF NUCLEAR POWER PLANTS

9.2.1 Introduction

A short description of typical NPPs are given in this section to provide a rough scope of the objective process. More detailed descriptions can be found elsewhere (Lahey and Moody (1977), Kerlin (1978)). The system organisations of two representative NPPs, namely, the pressurised water reactor (PWR) and the boiling water reactor (BWR) system, are described together with the fundamental dynamic relationships between plant variables.

9.2.2 Pressurised water reactor

A simplified schematic diagram of a modern PWR is shown in Figure 9-1. The main subsystems are reactor core, primary water loop, pressuriser, steam generator, turbine, condensor, and control systems. The primary water loop implies the coolant circuit consisting of the reactor vessel, piping from reactor vessel to the steam generator, primary-side water channels in the steam generator, the piping from steam generator to reactor vessel, and the primary water pumps. The cooling water driven by the pumps removes heat from the reactor core, delivers the heat to the secondary side of the steam generator, and returns to the core.

The piping in which water flows from the reactor vessel to the steam generator is called the *hot leg*, while the piping from the steam generator to the reactor vessel is called the *cold leg*. The pressure in the primary loop is maintained high enough to prevent boiling of the water by the pressuriser.

The pressuriser is basically a tank connected to the hot leg as shown in Figure 9-1. The tank is partially filled with water; the rest of the tank is filled with steam at saturation temperature corresponding to the pressure. The steam volume expands or contracts to accommodate water volume changes accompanying temperature changes in the loop. It is also used to maintain the primary loop pressure by adding or removing heat. The addition of heat is performed with electric heaters in the tank. The removal is performed by reducing the heater power and by spraying the cold water obtained from the cold leg into the tank.

The typical PWR steam generator is of the *recirculation type*. The primary water enters the steam generator at the bottom, passes through a plenum, then goes to steam generator tubes. As the tubes are in the inverted U configuration, the water flows upward and then downward through the steam generator. The water leaving the tubes flows into an outlet plenum then leaves the steam generator. The heat of the primary water is transferred to the secondary water, which enters the steam generator at about the same elevation as the top of the U-tube bundle. The feedwater mixes with water separated from two-phase (i.e. steam-water) mixture leaving the tube region. The mixed flow enters the section called the riser, where it receives heat from the primary water. The heat is adequate to evaporate a certain fraction of the secondary water. The steam leaves the steam generator, flows to the turbine, condensed to water, and then returns to the steam generator. This flow passage is called the secondary loop.

The secondary loop is similar to that of the fossil-fuel power plant. In this Chapter, we present the results of studies on the primary side which is more specific because of the nuclear reactions in the reactor core. For quantitative information, some important design parameters of a typical PWR are listed in Table 9-1.

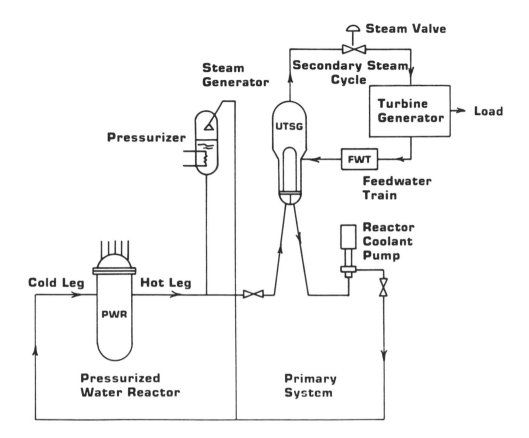

Figure 9-1: Schematic diagram of a typical PWR

TABLE 9-1 Typical PWR design parameters

Thermal power	2440 (MW)
Electric power	826 (MWe)
Primary Pressure	157 (Kg/cm^2)
Coolant flow rate	45000 (Tonne/hr)
Coolant temperature (inlet)	287 (^oC)
Coolant temperature (outlet)	322 (^oC)
Core diameter (effective)	3.04 (m)
Core height (effective)	3.66 (m)
Number of fuel assemblies	157
Fuel rods per assembly	15x15
Steam pressure (turbine inlet)	52.1 (kg/cm^2)

The PWR plant has three major controllers; (a) *reactor control system*, (b) *pressuriser pressure control system*, and (c) *steam generator control system*. The reactor control system adjusts the reactivity of the core using control rods and by dissolved boronic acid in the primary water. In essence, they are used to adjust the amount of parasitic absorption of neutrons, and thus to adjust the amount of heat generation. The mechanism of pressuriser pressure control has been described in preceding paragraphs. A small change in the primary loop pressure is usually observed because the pressuriser is designed to compensate for relatively slow pressure changes. The steam generator control system, called three-element water level controller, is widely known as the level controller of boilers and requires no further description in this Chapter.

The reactor control systems use control rod motions to achieve the desired power generation and desired average coolant temperature at each load level. Usually, the control rods are manipulated in response to an error in average coolant temperature relative to the set point, which is a predetermined function of load level. The control rods also respond to the mismatch of turbine power and reactor power. The second control policy is adopted to cause the reactor power adjustment to take place sooner after the load change than would occur if the controller were to wait for the change in power caused by the change in the heat removal from the steam generator. The second policy will be widely utilised when the PWRs are operated in the load-following mode. At present, however, most PWRs are operated by employing the first control policy for scheduled constant-power generation, because of the economical reason mentioned earlier.

9.2.3 Boiling water reactor

The BWR is characterised by the schematic diagram shown in Figure 9-2 and by the design parameters given in Table 9-2. The topology of the scheme is simpler than that of the PWR in the sense that the steam is generated directly in the reactor core. The heat generated in the reactor core by nuclear fission is removed by the steam at saturation temperature to be transferred to the turbine. The decrease in coolant inventory by the evaporation is compensated by the feedwater, which is obtained by the condensation of exhaust steam from the turbine. The amount of feedwater is controlled according to the same principle as the PWR steam generator, i.e. three-element control.

TABLE 9-2   Typical BWR design parameters

Thermal power	2380 (MW)
Electric power	780 (MWe)
Primary Pressure	70.7 (Kg/cm^2)
Coolant flow rate	33800 (Tonne/hr)
Coolant temperature (inlet)	278 ($^\circ C$)
Coolant temperature (outlet)	286 ($^\circ C$)
Core diameter (effective)	4.03 (m)
Core height (effective)	3.66 (m)
Number of fuel assemblies	548
Fuel rods per assembly	7x7
Steam pressure (turbine inlet)	66.8 (kg/cm^2)

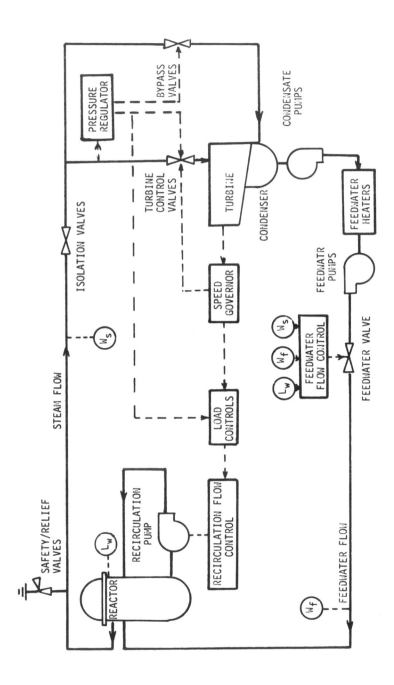

Figure 9-2: Schematic diagram of a typical BWR

In a BWR, power generation can be controlled in four different ways; by moving control rods, by changing recirculation water flow rate, by changing operating pressure of the reactor core, and by changing the subcooling of the feedwater. Effect of withdrawal or insertion of a control rod is the same as in the case of a PWR. An increase in the recirculation rate causes a decrease in the amount of steam inventory in the core, which causes an increase in reactivity and thus in power generation. These two control actions are usually employed during power operation.

An increase in reactor pressure causes a collapse of steam bubbles in the core, together with an accompanying increase in the subcooling of inlet water. The subcooling increase is also observed independently when the subcooling of feedwater is increased by decreasing the amount of heat supplied by the feedwater heater. In both cases, the consequence is the increase in reactivity and thus in reactor power.

In the four processes mentioned above, the increase in reactivity forces the reactor power to rise until the neutron balance is established again by the effect of the negative reactivity, which is due to the increase of steam inventory in the core induced by the power rise. When the control rods are used to change the power, the axial distribution of power generation is modified in magnitude and in shape. In other processes, the power shape remains almost the same whilst the magnitude changes.

The existence of the tightly-coupled dynamic relationships as described in this section suggests the possibility of establishing various new schemes of analytical redundancy for efficient FDI in PWRs and BWRs.

9.3 PARAMETRIC MODELLING METHOD

9.3.1 Modelling procedures

The LS modelling procedure is described in this section by taking the univariate autoregressive (UAR) model as an example. The UAR model is defined as:

$$y(k) = \sum_{m=1}^{p} a(m)y(k-m) + v(k) \tag{9-3.1}$$

for time series data [$y(k)$; $k=1,2,\ldots,N$], where $v(k)$, a white noise source driving the process, is assumed to satisfy the following conditions:

$$E[v(k)] = 0, \tag{9-3.2a}$$

$$\begin{aligned} E[v(k)v(j)] &= \sigma^2 \quad &&\text{for } k=j; \\ &= 0 \quad &&\text{for } k \neq j. \end{aligned} \tag{9-3.2b}$$

The parameter p is the autoregressive (AR) model order, and $E[\ ]$ is the expectation operator. To determine the AR model coefficients $[a(m);\ m = 1,2,\ldots,p]$, we solve the overdetermined system of simultaneous equations:

$$\underline{Y} = B\underline{a}\ ; \tag{9-3.3}$$

where:

$$\underline{Y} = [\ y(p+1),y(p+2),\ldots,y(N)\ ]'\ , \tag{9-3.4a}$$

$$\underline{a} = [\ a(1),a(2),\ldots,a(p)\ ]'\ , \tag{9-3.4b}$$

$$B_{i,j} = y(p+i-j);\quad i=1,2,\ldots,N-p;$$

$$j=1,2,\ldots,p. \tag{9-3.4c}$$

The solution of the overdetermined equations in the sense of least-squares is equivalent to minimisation of the Euclidean norm of the vector $\underline{Y} - B\underline{a}$. Thus, the problem is redefined as determination of the AR coefficients so that:

$$\|\ \underline{Y} - B\underline{a}\ \| = \text{minimum}, \tag{9-3.5}$$

where $\|\ldots\|$ represents the Euclidean norm.

An orthogonal transformation Q is applied to solve equation (9-3.3). The matrix Q is chosen so that the resultant matrix takes the form:

$$QB = R = \left| \begin{array}{c} \hat{R} \\ \hline 0 \end{array} \right| \tag{9-3.6}$$

where the matrix $\hat{R}$ is an upper-triangular square matrix. The upper-triangularisation is actually performed by applying a sequence of Householder transformation (Golub (1965)). The transformed equation is easily solved by the backward substitution.

An estimate of the variance of $v(k)$, to be used in the model order determination and spectral estimation, is obtained from the minimised norm of equation (9-3.5). The model order p is determined by the *Akaike Information Criterion* (AIC) minimisation (Akaike (1974)), where the model corresponding to the minimum of the quantity defined as:

$$AIC = (-2) \log (\text{maximum likelihood})$$
$$+2 \quad (\text{number of adjustable parameters}) \qquad (9-3.7)$$
$$= N\log a^2 + 2p \text{ , for UAR model.}$$

is adopted as the best model for the data sample. The AIC minimisation principle might be replaced by other, theoretically more sophisticated criteria (Akaike (1978), Hannan and Quinn (1979)). For the present purpose, however, the AIC minimisation is as good as others.

The above modelling procedure is extented to other models without difficulty. Generalisation can be made to the following multivariate autoregressive (MAR) model:

$$\underline{Y}(k) = \sum_{m=1}^{p} A(m)\underline{Y}(k-m) + \underline{V}(k) \qquad (9-3.8)$$

where $\underline{Y}(k) = [\ y_1(k),y_2(k),\ldots,y_n(k)\ ]$ is an n-dimensional vector of measured signals, $A(m)$ is the AR coefficient matrix, and $\underline{V}(k)$ is a vector of driving noise sources. The other class of parametric representation called the single-input single-output (SISO) model is defined as:

$$y(k) = \sum_{m=1}^{p} a(m)y(k-m) + \sum_{m=1}^{q} b(m)x(k-m) + v(k). \qquad (9-3.9)$$

Another parametric representation, called the multiple paths delay (MPD) model, is defined as:

$$y(k) = a(1)x(k-p) + a(2)x(k-q) + v(k). \qquad (9-3.10)$$

The parameters of all these models are obtained by the same procedure with modified definitions of $\underline{a}$ and B.

It should be noted that different modelling algorithms may be used for the parametric models. However, the LS modelling procedure tends to be advantageous over the other algorithms along with the decrease in number of samples used for the modelling (Kitamura, Washio, Kotajima and Sugiyama (1985)), and thus is conveniently applied when the observation time span is limited by operational and other reasons.

9.3.2 Extraction of FDI-related information

(a) *Time domain applications*
A direct usage of the parametric model is to use it as an estimator of one process variable from the other variable(s). The SISO model is applied for this purpose in this study. The so-called *one-step-ahead predictor* is derived from the model parameters as:

$$y_o(k) = \sum_{m=1}^{p} a(m)y(k-m) + \sum_{m=1}^{q} b(m)x(k-m), \qquad (9-3.11)$$

where the output signal $y(k)$ is estimated based on the previous observations of both input and output. Another estimation scheme called *the input-output predictor* is also derived as:

$$y_i(k) = \sum_{m=1}^{p} a(m)y_i(k-m) + \sum_{m=1}^{q} b(m)x(k-m). \qquad (9-3.12)$$

In this scheme, the output is estimated based on a knowledge of the previous input values only.

The predictors (9-3.11) and (9-3.12) are both usable as an estimator of $y(k)$ to supply a redundant signal for organising a parallel redundancy scheme. For detecting a change in statistical characteristics of $y(k)$ and/or $x(k)$, the differences $[y_o(k)-y(k)]$ and $[y_i(k)-y(k)]$ are more sensitive to the changes than the original signals.

The other time domain application is to estimate response functions between the variables. When the MAR model is obtained, the impulse response of signal y_i to signal y_j is derived from the MAR coefficients by simple recursive computation. The step and ramp response functions are easily derived by integrating the impulse response. The step response is used to estimate the response-time. The ramp response is used to estimate the delay time between the two signals. These estimates are utilised to organise serial redundancy schemes.

(b) *Spectral domain applications*

The auto-power spectral density (APSD) of signal y is obtained from the UAR model by the formula:

$$APSD_y(f) = 2 \sigma^2 \Delta t / |H(f)|^2 \qquad (9\text{-}3.13a)$$

$$H(f) = 1 - \sum_{m=1}^{p} a(m)\exp(-j2\pi fm\Delta t)^2 . \qquad (9\text{-}3.13b)$$

The above follow by assuming the whiteness of the noise source v(k). It is sometimes possible to identify resonances in the APSD reflecting the value of other process variable(s). For instance, the APSD of neutron signal measured at a PWR is known to contain several peaks related to mechanical vibration of the components (Türkcan (1982), Bastl and Wach (1982)). The resonance magnitude reflects the amplitude of the mechanical vibration, while the resonance frequency reflects integrity of the vibrating component. This type of information extraction is further pursued in this study, mostly for synthesising parallel redundancy schemes.

The MAR model is also transformed to spectral representation as

$$PS(f) = 2\Delta t \ [\ A(f) \ ]^{-1} \ Q_V \ [\ A(f)^* \ ]^{-1}; \qquad (9\text{-}3.14a)$$

$$A(f) = I - \sum_{m=1}^{p} A(m)\exp(-j2\pi fm\Delta t), \qquad (9\text{-}3.14b)$$

where $*$ denotes complex-conjugation. The matrix Q_V is the covariance of V(k). The (i,i) element of the matrix PS(f) represents the APSD of signal y_i; the (i,j) element the cross-power spectral density (CPSD) between signal y_i and y_j. The spectral signatures such as phase angle, coherence function, and transfer function are easily derived from the matrix PS(f) (Upadhyaya, Kitamura, Kerlin (1980)). Serial as well as parallel redundancy can be synthesised from these estimates.

(c) *Direct estimation of physical parameters*

The estimates of process variables and parameters described so far are regarded as *indirect* estimates since they are obtained via two stages (i.e. first determine a model then derive estimates). Attempts

have been made to develop methods for direct parameter estimation for
simplicity and convenience of the FDI scheme. The estimation of delay
time, for instance, is carried out by fitting the MPD model of
equation (9-3.10) to bivariate time series data [x(k),v(k);
k=1,2,...,N]. The coefficients [a(1),a(2)] are easily determined for
fixed (p,q). The optimal (p,q) pair is determined by two-dimensional
grid search of the minimum of AIC in the (p,q) space. The delay times
corresponding to the multiple (i.e. two in this formulation) paths
are simply given as p and q times the sampling interval.

It should be noted that the delay time can also be estimated by
other techniques, e.g. by cross-correlation, phase angle, cepstrum,
etc. The performances of these techniques are, however, strongly
dependent on signal properties (periodic, one-shot, sinusoidal, white
noise, etc.) and measurement qualities (signal to noise ratio,
duration of measurement, etc.). The MPD modelling approach is in
principle suited to deal with the signals observed with multiple
delay channels.

9.4 FAULT DETECTION BY SERIAL REDUNDANCY

9.4.1 Introduction

The methods described above have been applied to signals measured at
commercial NPPs and at other experimental facilities. The capability
of the methods is evaluated in terms of numerical experiments to
monitor sensor integrity, controller performance, and plant stability. This section summarises the results based on the serial redundancy approach, viz. comparison of the current characteristics with the
baseline ones.

9.4.2 Sensor fault detection

(a) *In-situ monitoring of temperature sensors*
The integrity of temperature sensors in the reactor cooling system is
of critical importance for ensuring the operational safety of an NPP.
The standard approach to examine the integrity has been a static
calibration intermittently conducted during scheduled shutdown
periods. The time-honoured method needs to be upgraded to have the

capability of monitoring dynamic performance of the temperature sensors. Furthermore, the method should be applicable to the sensors during normal operation of the plant. In fact, the US Nuclear Regulatory Commission required that utilities operating NPPs make in-situ response time measurements of sensors installed in the plant safety systems (U.S. Nuclear Regulatory Commission, 1978). The dynamic performance is represented by the response of the sensor output signal to a step increase in the measured quantity, i.e. temperature in this case.

The impulse response is derived from the UAR model, equation (9-3.1), simply by computing the response of the model to the unit impulse as;

$$y_I(k) = \sum_{m=1}^{p} a(m)y_I(k-m) + v(k); \qquad (9-4.1)$$
$$v(0)=1, \ v(k)=0 \text{ for } k>0.$$

Once the impulse response is determined by equation (9-4.1), the step response is derived by integrating the impulse response as:

$$y_S(t) = \int_0^t y_I(t')dt' \qquad (9-4.2)$$

A simple trapezoidal rule integration scheme is employed to evaluate the step response since the integrand is available only at discrete sample points. The response time of the sensor is estimated from the $y_S(k)$ as the time duration for the response to reach 63.2 % of the final value.

This method has been applied to temperature sensors at core-exit and at the primary coolant loop (i.e. the hot and cold legs) of a PWR. Typical step responses are depicted in Figure 9-3, showing that the estimated responses are reasonably reflecting the physical characteristics of temperature sensors. Estimates of the response time are shown in Figure 9-4 for several sensors, indicating that the ones at core-exit have significantly faster responses than the ones at the coolant loop. Also, a systematic difference is found between the responses of sensors at cold legs and at hot legs. These observations have been validated by examining the design data; the thermocouples at core-exit have a diameter of 1.6 mm, whilst the thermocouples in the coolant loop have the identical diameter of 3.2 mm. This difference in diameter is consistent with the difference in response time: the larger diameter implies the slower response and

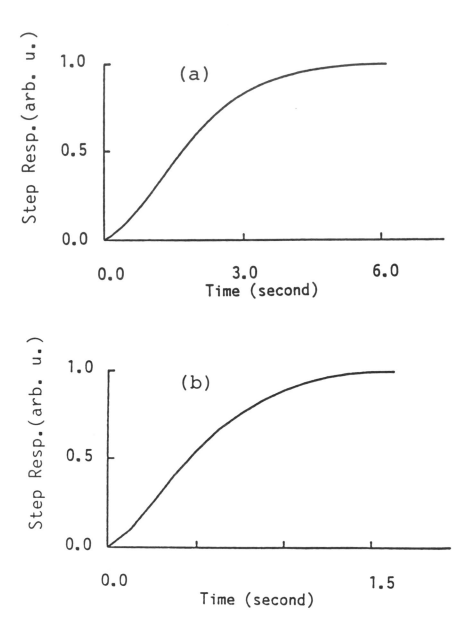

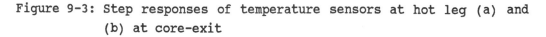

Figure 9-3: Step responses of temperature sensors at hot leg (a) and (b) at core-exit

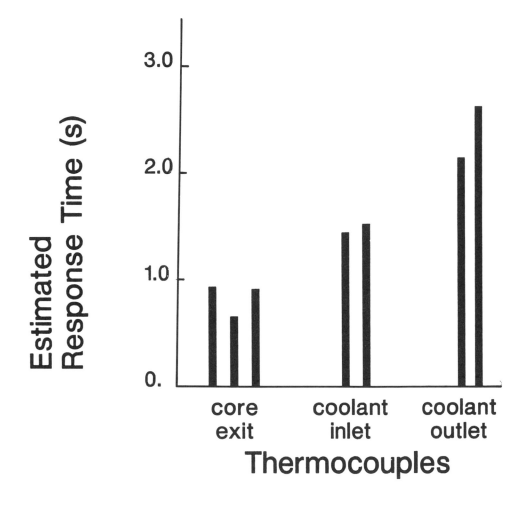

Figure 9-4: Summary of estimated response times.

vice-versa. Furthermore, the systematic difference in response time among the sensors at the cooling loop is attributed to another design feature: the thermocouples at the hot leg have an extra protection sheath for the purpose of withstanding higher temperature. The sheath is apparently responsible for the slower response due to its extra heat capacity. It should be noted that the validity of the present method has also been confirmed through comparisons with active tests in which sensor response was estimated by electrical heating of the sensors (Upadhyaya and Kerlin (1978)).

The results clearly demonstrate the possibility of on-line FDI of temperature sensors since a substantial change in the response time is an indication of change in the sensor structure or material properties.

(b) *Estimation by time-domain prediction*
The prediction schemes of equations (9-3.11) and (9-3.12) have been tested as potential tools for deriving redundant temperature signals by applying them to data distorted with simulated faults. One of the in-core neutron signals located at the uppermost elevation in the instrumentation string and a signal from a temperature sensor located at the core-exit were selected as test data. The axial separation of these sensors are about 58 cm.

Numerical experiments indicated that the core exit temperature can be estimated by the SISO scheme of equation (9-3.11) with the in-core neutron signal as input $x(k)$. The scheme is useful for detecting and identifying sensor faults as exemplified by the simulation tests described below.

The types of sensor faults simulated in this study are; *bias, drift and parasitic noise* (i.e. impulsive and Gaussian). Firstly, a SISO model was determined and innovation sequences (i.e. values of the instantaneous error in the model) were estimated by processing the normal data. Next, a fault was introduced to one channel of $[x(k), y(k)]$, and innovation sequences were again estimated. Detectability of the fault was evaluated in terms of changes in statistical characteristics of the innovation sequence before and after the onset of the fault.

Four feature parameters; (i.e. moments of order 2, 3, 4 and zero-crossing rate) were estimated from the innovation. The feature parameters obtained for a block of data define a sample point in a multi-dimensional parameter space. The fault was judged to be detectable when the sample points obtained from distorted and normal data blocks formed isolatable clusters in the feature space.

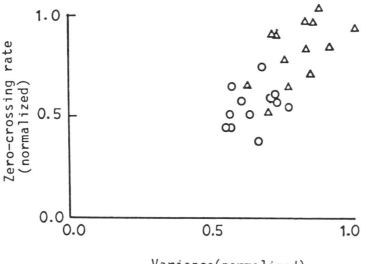

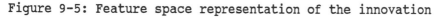

Figure 9-5: Feature space representation of the innovation

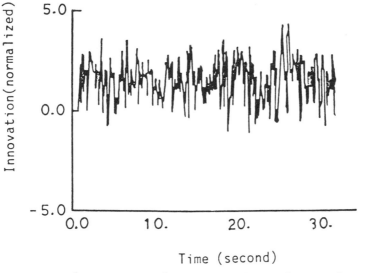

Figure 9-6: Time trace of the innovation

A typical example of feature space representation of the innovation is given in Figure 9-5. Gaussian noise was superimposed onto the temperature signal. The magnitude of the noise is considered sufficiently small that the onset of the fault is hardly detectable by visual inspection of the innovation illustrated in Figure 9-6, where the malfunction starts at the middle of the time span. The sample points in two-dimensional feature space (variance, zero-crossing rate) are clustering into two groups, suggesting that the fault is at least detectable, in principle. Although the two clusters overlap, it is possible to define a discriminant to classify the two clusters by the standard method of statistical pattern recognition. Of course, the separation becomes clearer for larger disturbances.

(c) *Pressure sensor monitoring*

Pressure sensors are also implemented in safety protection systems of NPPs. Attempts have been made to establish a reliable method for process-noise-excited in-situ monitoring of pressure sensors (Jacquot, Poujol, Beaubatie and Ciaramitaro (1983)). The step response method is not valid in this case, since the form of the step response is not so simple as that of a temperature sensor. The output signal from a pressure sensor system (sensor itself and the sensing line) is usually characterised by several resonance components. Thus it was judged more reasonable to monitor the overall pattern of APSD than to monitor the step response. In applying this method (i.e. APSD monitoring), the results are influenced by a number of physical elements; the process noise character, features of the transmitter, and the physical status of the sensing line, which could be disturbed by anomalies such as blockage, bubble formation, leakage, etc.

The APSD of the pressure signal is computed by equation (9-3.14) after UAR modelling of the signal. Several feature parameters are derived from the APSD in order to characterise the APSD pattern in a quantitative manner. These parameters are:

(a) peak frequency;
(b) peak intensity; and
(c) stationarity parameter.

The peak frequency and intensity are estimated for a set of dominating peaks in the APSD. The set is specified by the analyst, based on either physical or empirical *a-priori* knowledge. The stationarity parameter is defined as the ratio of APSD-value at the lowest frequency to the average APSD. The ratio increases when the mean value of the signal is disturbed by superposition of trend or drift, since the ratio reflects the resultant increase in low frequency components in the APSD. In the most extreme case, the APSD is known to take the pattern of 1/f spectra. On the other hand, the parameter becomes smaller when the macroscopic pattern of the APSD is of low-pass filter type; it approaches to zero when the APSD is of band-pass or high-pass type. In this regard, the stationarity parameter is useful in characterising not only the stationarity but also the global pattern of the APSD.

A set of experiments was carried out with a small-scale water loop to evaluate the applicability of the method. Three nuclear-grade pressure sensors were connected to the loop with copper sensing tubes of about 0.5 cm inner diameter. A small centrifugal pump was used to circulate the water in the loop and to provide the needed noise source to cause pressure fluctuation. Each of the sensing lines was designed to have a valve and a rubber stopper to simulate various anomalies. The valve is used to simulate a blockage; the rubber stopper is used as a port to introduce an air bubble with an injector.

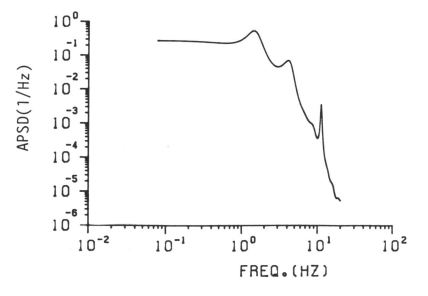

Figure 9-7: APSD of pressure signal (normal)

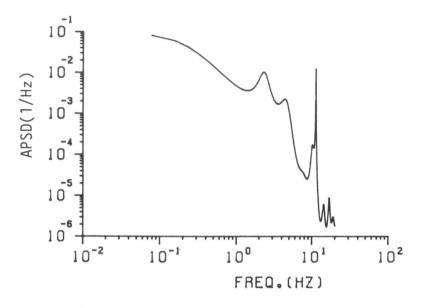

Figure 9-8: ASPD of pressure signal (blockage)

An example of the APSD during normal operation is depicted in Figure
9-7, showing that the macroscopic pattern is of low-pass type, with
several small but noticeable peaks at intermediate and high frequency
regions. In this experiment, the valve in the sensing line was
gradually closed to simulate the blockage. After the onset of
blockage, significant changes were observed in the APSD pattern as
illustrated in Figure 9-8. The peaks became more and more prominent
along with the development of the blockage, whilst the overall
magnitude of signal fluctuation became smaller. The macroscopic APSD
tended to be the 1/f noise pattern.

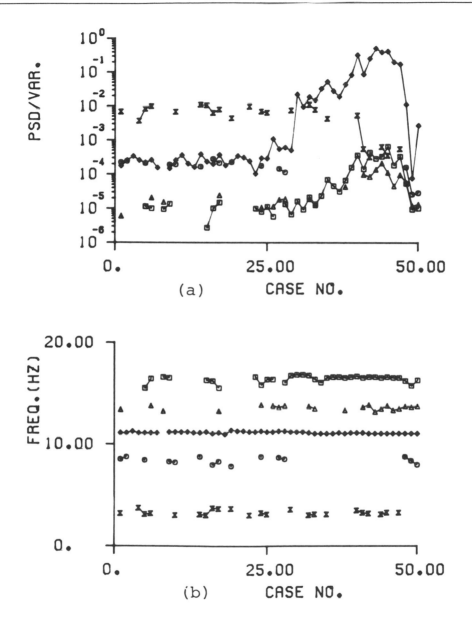

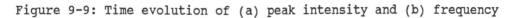

Figure 9-9: Time evolution of (a) peak intensity and (b) frequency

The time-evolution of the blockage is better represented by changes in the feature parameters. The occurrence and the evolution are reflected well by the peak intensity as shown in Figure 9-9. Peak frequencies of the prominent peaks are also depicted in Figure 9-9 for the sake of comparison. Each feature parameter in these plots was estimated from the data corresponding to a period of 12.8 seconds. Note that the peak intensity is normalised by the variance of the signal, namely, the total power of the fluctuation. The increase in the normalised intensities along with the passage of time, i.e. along with the development of blockage, indicates the relative dominance of the peaky components over the continuous ones in the APSD.

In contrast to the behaviour of peak intensity, no systematic evolution was observed in the peak frequencies. This observation suggests that the peaks are attributable to the generation process independent of the opening of the valve in the sensing line. One of the candidates is the beam-mode mechanical vibration of the sensing line. Another possible cause is the standing wave phenomenon in the sensing line. Further studies are needed for exact identification of the cause of each peak.

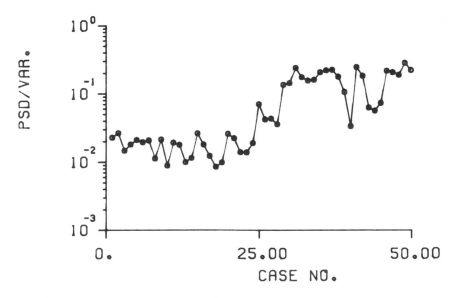

Figure 9-10: Time evolution of stationarity parameter

Usefulness of the stationarity parameter in detecting and quantifying the blockage was also confirmed as depicted in Figure 9-10. The increasing dominance of the lowest frequency component reflects the change in the APSD pattern and thus the evolution of the blockage. The stationarity parameter is convenient for quick monitoring, while the peak frequency and intensity are suitable for detailed fault diagnosis. The capability of the feature parameters was confirmed for other fault modes also.

The essence of the techniques described in this section is to monitor the quantities representing statistical characteristics of the signal. One natural criticism against this approach is that the proposed techniques have no capability of distinguishing between sensor faults and inherent changes in noise properties. This criticism is theoretically valid. In actual applications, however, the difficulty can be alleviated by examining whether corresponding changes are observed in other signals also. If the apparent change in the step response or in feature parameters is caused by an inherent change in noise properties, the effect of the change should be observed in dynamically related signals of the same or even dissimilar kind.

The same technique can be applied to performance monitoring of other important sensors as well. In fact, the feasibility has been confirmed for differential pressure sensors (Jacquot, Poujol, Beaubatie and Ciaramitaro (1983)) and for neutron detectors (Türkcan (1982)). The demonstrated results also seem to encourage further attempts to apply the technique to a wider range of problems.

9.4.3 Controller fault detection

(a) *Specific difficulty in controller monitoring*
The detection and identification of controller faults is also highly important for safety operation of NPPs. For this purpose, it is needed to measure transfer functions of the components consisting of the control system. This problem is generally more difficult than straightforward monitoring of the signal characteristics, since the components are operated in a closed-loop system; the difficulty in measuring transfer functions of components in a closed-loop is widely recognised (Astrom and Eykhoff (1971)).

One promising approach to this identification problem is the use of MAR modelling (Akaike (1968)). Once an MAR model is determined for multivariate time series, transfer functions between the modelled signals are known to be derived from the model parameters. The possibility of identifying closed-loop dynamics is an attractive feature of the MAR modelling, leading to a number of technical challenges with fruitful results (Akaike (1978)). However, difficulties are still experienced in many applications mostly owing to lack of identifiability under certain conditions.

Consider a feedback system represented by signals $y_1(k)$ and $y_2(k)$. Parameters of the bivariate AR model written as equation (9-8) are determined by the LS fitting. The transfer functions from y_1 to y_2 denoted G_{21}, and from y_2 to y_1, denoted G_{12}, are easily derived from the MAR coefficients. The physical credibility of the derived transfer functions is spoiled, however, unless both of the noise sources v_1 and v_2 have sufficiently large enough magnitude. In other words, the derived transfer functions become erroneous when one of the noise sources has too small a variance to perturb the system and is thus regarded as negligible. The same discussion holds for systems represented by more than two variables. The transfer functions G_{ij} ($i \neq j$) from the signal y_j to y_i derived from the MAR model are physically unreliable unless all noise sources have enough power. Physically, this is because the influence of transfer functions are not conveyed correctly to the time series data under such a condition. Mathematically, the full rankness of the covariance matrix of noise sources defined as;

$$Q_V = E[\ \underline{V}(k)\underline{V}'(k)\ ] \hspace{3cm} (9\text{-}4.3)$$

is needed for the observed time series data to reflect the dynamics of all the transfer functions in the system. Note that the matrix is estimated as a by-product of the MAR modelling procedure.

Unfortunately, such a situation, i.e. the loss of rank, is often encountered in many practical processes including control systems. One practical means to ensure the full-rank conditions is to add artificial perturbations to make up for negligible noise sources v_i. This modification of experimental condition is known to be effective in many industrial processes, but has seldom been accepted in the nuclear industry due to strict regulations against modification of operational conditions.

To overcome this difficulty, we have pursued another possibility named *partial identifiability* (Kitamura, Matsubara, Oguma (1979)). Through theoretical and experimental studies, it has been found that some of the transfer functions are derived from the MAR model even for a non-full-rank Q_V matrix. One of the important conclusions of this study is summarised below:

Suppose that one noise source v_j is negligible whilst the others have substantial magnitude. Then most of the transfer functions cannot be estimated correctly. Even in this case, the transfer functions from y_i to y_j (i.e. G_{ji}) are correctly reflected in the time series data and are thus identifiable by the MAR modelling. In other words, the transfer functions from other signals to the specific one without noise source are potentially identifiable whilst the remaining transfer functions are not.

We have deliberately used the word *possibility* in the above statement, since the above discussion has been derived by considering information content in the measured signals rather than in the MAR model. It is thus important that any condusions drawn take account of this restriction. The credibility of the model is strongly dependent on other technical details of the modelling, e.g. sampling interval, signal combination, model order, etc. Practical identifiability is attained only when these decisions are also made in a suitable manner. In the following example of partial identification, the explanation is simplified by avoiding statements related to the practical aspects. It must be stressed that these decisions have been made with due caution based on several guidelines established through other studies which are beyond the scope of this paper. Brief comments on this issue will be addressed at the end of this section for further information.

(b) *Identification and monitoring of controller dynamics*
As stated above, the MAR modelling approach itself can be performed without any problems. The difficulty arises at the stage of deriving the transfer functions. The full-rankness of the Q_V matrix should be examined in order to obtain the knowledge about credibility of the derived transfer functions. Instead of conducting the rank test in a standard manner, a simple but efficient procedure was introduced for computational load reduction (Kitamura, Matsubara and Oguma (1979)).

An indicator of the significance of each noise source v_i is defined as:

$$VR_i = q_{ii}/(\text{variance of } y_i). \hspace{3cm} (9\text{-}4.4)$$

Here, q_{ii} corresponds to the i-th diagonal element of Q_V; VR stands for *variance reduction*. The Q_V matrix is judged to violate the requirement of full-rankness when one or some of the VR_i is extremely smaller than the rest. The validity of this simplified procedure is based on the assumption that the off-diagonal elements of the Q_V are negligibly small, which is usually the case provided that the modelling is carried out successfully.

We have applied the method to identify the pressure controller dynamics of a BWR. The reactor was operated at a low power and the pressure was controlled by the bypass-mode. Three signals, (a) reactor pressure (P), (b) main steam flow (W_s), and (c) bypass valve opening (BPR), were measured and MAR model parameters were determined. The resultant values of VRs are listed in Table 9-3. Apparently, the VR for the BPR signal is markedly smaller than other VRs, suggesting that the transfer functions from P to BPR and from W_s to BPR are identifiable despite the loss of rank.

TABLE 9-3   VR values obtained by the MAR model.

Signal	VR-value
Pressure (P)	7.6×10^{-2}
Steam flow (W_s)	2.3×10^{-2}
Bypass valve openting (BPR)	1.3×10^{-3}

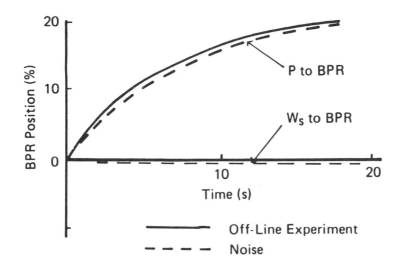

Figure 9-11: Step responses of BPR to P and W_S obtained from noise analysis (dashed line) and off-line test (solid line)

The corresponding step responses derived from the MAR model are depicted in Figure 9-11. The time-domain representation was adopted for convenience of comparison with the results obtained by an off-line transient experiment in which step responses of a bypass-valve controller were measured under open-loop operation mode. The fairly good agreement between the corresponding curves seems to support the present method of partial identification. The results are also consistent with our physical knowledge: the valve opening follows the step increase in the pressure to extract more steam flow from the reactor dome, and thus to moderate the pressure increase.

The negligible response of BPR to W_S is also reasonable: the valve opening is not designed to respond to the steam flow variation but to the pressure. The valve opening influences the steam flow but the reverse is not true. The experimental observations indicate that identification and monitoring of dynamic performance of controllers are feasible even when the full-rank condition is violated. By routine application of the present method during commercial operation, the degradation in the controller performance can be detected with a higher possibility than by the conventional practice (i.e. inspection during shutdown).

The MAR modelling has been extended to identification of

feedwater control system (F.C.S.) and pressure control system (P.C.S.) of a BWR by other research groups (Kanemoto, Ando, Yamamoto, Kitamoto and Nunome (1982), Kanemoto, Ando, Yamamoto, Idesawa and Itoh (1983)). Other difficulties were experienced in the attempts due to the effect of:

(a) direct coupling between some of the signals, and
(b) pure-integral transfer functions.

   More specifically, the condition (a) takes place when some of the transfer functions are characterised by far smaller time constants than the others. The apparent direct coupling can be avoided by reducing the sampling interval. But then the MAR model tends to lose its capability of representing other transfer functions having larger time constants. The condition (b) is related to the intrinsic characteristics of the physical system. For instance, the incremental change in reactor water level is exactly determined by integrating the mismatch between the feedwater flow and the steam flow.

   The MAR model obtained by the standard modelling procedure is known to provide poor estimates of the transfer functions under the condition (a) and/or (b) simply because the model structure is not suited to represent such processes. Examples of modifications introduced to solve the difficulties are:

(a) introduction of $A(0)$ matrix in the model,
(b) elimination of physically nonexisting $a_{ij}(m)$s in $A(m)$, and
(c) difference operation of the *integrated* signals.

The elements of the $A(m)$ matrix to be eliminated are determind based on *a priori* knowledge about the process. These elements are forced to be zero prior to the modelling, then the nonzero elements are determined by the LS algorithm. The meaning of operation (c) would be obvious as this is the standard procedure for eliminating the pure integration effect.

   The details and effectiveness of these modifications are not described here as they can be found elsewhere (Kanemoto, Ando, Yamamoto, Kitamoto and Nunome (1982), Kanemoto, Ando, Yamamoto, Idesawa and Itoh (1983)). Such modifications are unavoidable in most practical applications. However, the examples cited in this section are, in the author's view, sufficient to demonstrate the possibility of *in-situ* monitoring of controller performance and the possibility of fault detection by means of parametric modelling.

9.4.4 Plant Stability Monitoring

In addition to the integrity of plant components such as sensors and controllers, attention must be paid to the dynamic characteristics of the plant itself due to its own ultimate importance. Experimental and computational evaluations of stability margin have been carried out periodically in most NPPs to ensure the self-regulating characteristics. The experimental technique is, however, applicable only before restart of commercial operation following the scheduled shutdown. The possibility of on-line monitoring of the stability margin has been suggested from early stages of NPP development; but the actual implementation of the on-line technique became feasible after recent development in computer technology. The problem of instability has been considered mostly for BWR plants because of its intrinsic property; i.e. tight coupling between nuclear reaction and coolant boiling in the reactor core. Practical experience as well as theoretical studies have indicated that the stability margin is reduced when the BWR is operated at a low flow, high power mode.

Two types of instabilities are considered possible in a BWR; i.e. *reactivity* instability and *channel* (or *local*) instability. The former arises due to the interaction of the neutronic and the thermalhydraulic processes, and is related to the stability of neutron flux response to a perturbation in reactivity, such as the control rod action and coolant flow rate change. The latter is caused by the two-phase flow dynamics of a heated coolant channel, or of multiple channels. Early detection of these instabilities is highly desirable because the technique will contribute to reduce the number of scrams (i.e. forced shutdowns).

As a natural extension of the techniques described in the preceding sections, we have developed an efficient method of stability estimation by measuring only the neutronic power of the reactor. The stability of industrial processes is usually determined by estimating a quantity called *decay ratio (DR)* defined as;

$$DR = A_2/A_1 \tag{9-4.5}$$

which corresponds to the ratio between the second and first peaks of the impulse response. The smaller (larger) DR implies the higher (lower) stability. For an unstable system, the relation DR > 1 should hold. Instead of measuring the response to reactivity perturbation, the UAR model of the neutronic signal was utilised to derive the impulse response by the same procedure as given in 4-A. The physical assumptions underlying this procedure are that the noise source v(k) in equation (9-3.1) has the physical dimension of the reactivity, and that the noise source satisfies the statistical property of equation (9-3.2).

The method has been applied to several BWRs in the United States and in Japan (Upadhyaya and Kitamura (1981), Mitsutake, Tsunoyama and Nanba (1982), Kanemoto, Tsunoyama, Ando, Yamamoto and Sandoz (1984)). Two impulse responses obtained for different BWRs are illustrated in Figure 9-12. The decay ratios estimated are 0.024 and 0.37 for the responses in Figures 9-12(a) and (b), respectively. Interestingly, the data of Figure 9-12(a) were measured at 100 % power, 100 % flow condition, while the data of Figure 9-12(b) at 62 % power, 42 % flow condition. The flow to power ratio is significantly smaller in the latter case, which is known to be less stable according to the theoretical prediction. The results suggest that the monitoring of BWR stability is possible at least in the *qualitative* sense by the present method.

Actually, the stability can be monitored even on the *quantitative* basis. Other studies proved that the DRs estimated by the present method are in good agreement with the DRs obtained by perturbation test, in which the artificial disturbance was introduced by modulating the opening of a valve in the steam line. The disagreements in the DRs were mostly less than 5 % (Mitsutake, Tsunoyama and Nanba (1982)), which is good enough to validate the use of the present method of parametric modelling. It should be stressed again that the perturbation experiment is quite costly and time-consuming. On the other hand, the present method requires a small number of computations by using the normal operation record.

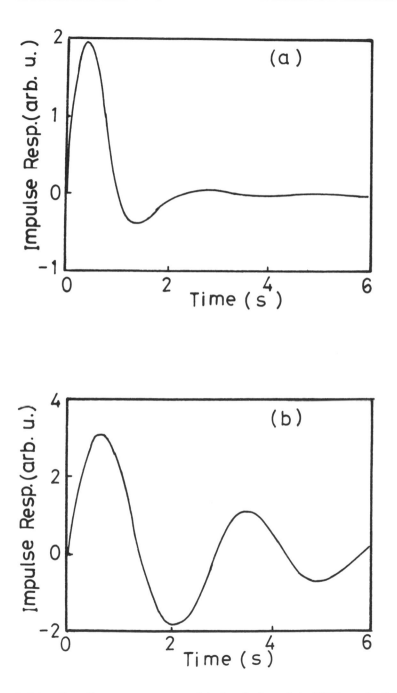

Figure 9-12: Impulse responses obtained for (a) BWR1 and (b) BWR2

The same procedure was applied to monitor the local stability of a BWR. The only difference lies in the processed signal. The signal obtained from a neutron sensor called *local power range monitor (LPRM)* was analysed in this case, while the signal from *average power range monitor (APRM)* was used in the estimation of the reactivity stability. A typical number of instrumentation tubes equipped in the BWR core is 40. Since four LPRMs are installed in each instrumentation tube corresponding to four axial elevations with an equal spacing, the total number of LPRM is 160 in this BWR. The APRM signal represents bulk power in the core by averaging signals from as many as 40 LPRMs distributed throughout the core, resulting in loss of information about local behaviour of neutronic power. The local information can only be obtained by processing the LPRM signal.

The results are summarised in Figure 9-13, where the estimated APSDs are illustrated instead of the impulse response for the purpose of a better visual comprehension. The APSD of Figure 9-13(a) obtained for the LPRM near the core periphery represents a more stable characteristic than the APSD of Figure 9-13(b) for the LPRM near the core centre. The difference can be attributed to the relatively low power density in the peripheral region. Though further studies need to be conducted for clarifying the exact cause of the space-dependency of stability, the result is regarded as positive evidence of the usefulness of stability monitoring by the proposed method. The need for the stability monitoring will become more apparent in the near future when the BWR is required to operate in a load-following mode, i.e. in wider flow to power ratios.

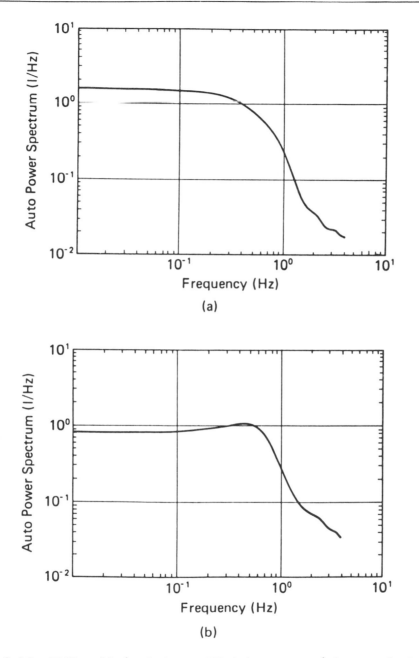

Figure 9-13: APSDs obtained for LPRM (a) near periphery and (b) near
centre of the core

9.5 SIGNAL ESTIMATION FOR PARALLEL REDUNDANCY

9.5.1 Introduction

Another type of FDI scheme can be synthesised by introducing noise-dependent signal estimates to provide extra redundancy. The FDI scheme is viewed as a natural extension of standard analytic redundancy in the sense that the redundant estimates are provided in time-parallel manner. The approach to realise the parallelism is, however, quite different from the standard analytic redundancy. In what follows, only the methods of deriving redundant estimates will be described, since the FDI scheme can be synthesised without any difference from the case of standard analytic redundancy.

The temperature and flow rate of the coolant are among the most important quantities to be monitored carefully, since it provides us with the information about the integrity of the coolant loop. These quantities are usually measured by multiple sensors to attain a high integrity; i.e. the hardware redundancy is already implemented. The redundancy is, however, not perfectly fault-free, since the probability of common-mode faults is not negligible. The terminology, common-mode fault, is used to specify faults which cause simultaneous faults of redundant components. Repeated fabrication errors, a faulty calibration instruction, and a power supply fault are typical examples of the common-mode fault. The temperature sensor overscale experienced at the TMI-2 accident is another well-known example of the common-mode fault. The possibility of common-mode fault can be effectively reduced by introducing redundant signals measured by totally different principles. The methods described in this section were developed with the aim of attaining the diversity in measurement principles.

9.5.2 Temperature estimation

(a) *Estimation by spectral analysis*
The coolant temperature in the primary loop of a PWR must be estimated from the measurements of quantities other than temperature. One of the effects of temperature on physical properties of materials in the reactor core is the variation of density. The

density effect was expected to appear in neutron and pressure
signals. Through detailed analyses of these signals in a PWR during
various operation conditions, it was recognised that the resonance
frequency of a specific peak in the estimated APSDs showed a common
systematic variation.

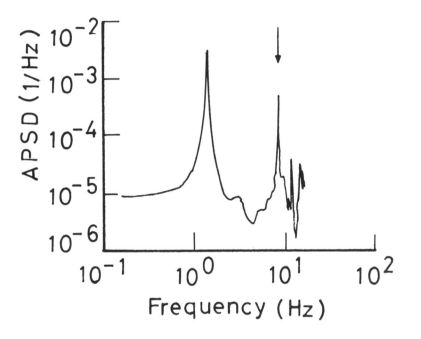

Figure 9-14: Typical APSD of primary pressure

Typical APSD of pressure signal measured at the primary cooling
circuit are illustrated in Figure 9-14. It has been confirmed that
the resonance at 6.5 Hz represents the standing wave (Türkcan
(1982)). The standing wave was found to exist over the whole primary
circuit, with nodes inside the reactor vessel. An attempt was made to
estimate coolant temperature from the resonance frequency. Since the
relation:

$$F_r = V_c/L_c \qquad\qquad\qquad (9\text{-}5.1)$$

holds for the resonance frequency F_r, sound velocity within the
coolant V_c and effective circuit length L_c, it is expected that F_r
changes depending on the V_c and thus on the temperature of coolant.

The values of F_r were estimated during cool-down periods of the reactor to evaluate this effect. Typical results are plotted in Figure 9-15 together with the variance of the signal. Here, the APSDs were obtained via UAR modelling for each case corresponding to a time duration of 8 seconds separated by a gap of 2400 seconds. The resonance frequency shifted around 2HZ during the overall measurement corresponding to temperature decrease of about 120°C, whilst the variance remained roughly constant. This result suggests that the temperature can be estimated with a modest accuracy of 6°C, since spectral resolution of 0.1HZ or less is attained without difficulty.

Similar results could be obtained by using other methods of spectral analysis, as far as the specific example cited is concerned. For faster transients, however, the LS-UAR modelling approach becomes advantageous over the others. The measurement time of less than 10 seconds is too short for the conventional methods to monitor the resonance frequency around 6 to 9HZ.

The time duration needed to estimate the resonance frequency is short enough to justify the claim that a redundant temperature signal can be derived from the pressure signal. The temperature resolution of 6°C is too crude for a detailed measurement, but is sufficient for a redundant signal to back up the common cause failure of ordinary temperature sensors. A better resolution can be attained easily by increasing the time duration at the cost of degraded on-line performance.

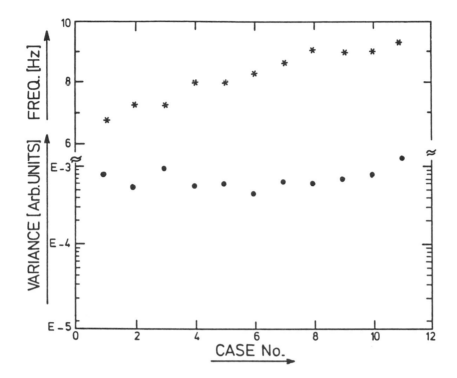

Figure 9-15: Drift of resonance frequency during cool-down operation

9.5.3 Flow velocity estimation

The estimation of coolant velocity by analysing neutron signals is a standardised technique in boiling water reactors (BWRs) where the neutron field is sufficiently perturbed by random fluctuation of coolant density caused by boiling. The axial propagation of random fluctuation driven by the coolant flow is detected successively by the four LPRMs implemented in one instrumentation tube. The conventional methods such as cross-correlation analysis and phase angle analysis of the LPRM signals have been utilised to estimate the time delay of the random components commonly observed in the signals.

In PWRs, the random fluctuation component propagating through the coolant channel is far weaker than in BWRs because no boiling is present. Accordingly, it is quite difficult to obtain an accurate estimate of the coolant velocity by the method successfully applied to the BWR. Several attempts have been made to estimate the PWR coolant velocity despite the difficulty. The performance of two methods, i.e. ramp response method and MPD modelling method, are evaluated.

As described briefly in Section 9.3, the impulse, step and ramp responses between the signals are obtained once the MAR model is determined to characterise the multivariate signals. More precisely, the impulse response of signal $y_i(k)$ to signal $y_j(k)$ is estimated recursively from the MAR model by setting:

$$y_j(k) = 1 \quad \text{for} \quad k=0, \quad \text{for } k>0. \tag{9-5.2}$$

Note that the procedure is slightly different from the one employed in Section 9.4 to estimate the sensor response time; the response was derived by setting $v(0)=1$ in equation (9-4.1) since the model is determined for a single variable. In equation (9-5.2), the perturbation is imposed directly on the $y_j(k)$ itself rather than on $v_j(k)$ to approximate the actual situation more closely. The step and ramp responses are easily derived from the impulse response by successive numerical integration. In principle, the delay time can be estimated from either the step and the ramp responses. Practically, the latter is more convenient. The zero-crossing point of the ramp function is usually determined with less ambiguity than the intersection of the step function with a specified threshold, owing to the smoothing effect of the integration.

The second method, MPD modelling, has also been utilised to estimate the delay time by executing the procedure explained in 9.3. As the procedure is essentially a LS fitting with only two adjustable parameters, the method is expected to be suitable for microprocessor-based implementation.

The delay time between two in-core neutron detectors of a PWR was estimated by applying the ramp response method and the MPD modelling. The geometrical spacing between the two detectors is known to be 0.776 m. The typical step and ramp responses are depicted in Figure 9-16. The sampling interval is 0.125 second; the number of samples in one run is 256. The delay time, and the coolant velocity as well, is thus obtained for every 32 seconds.

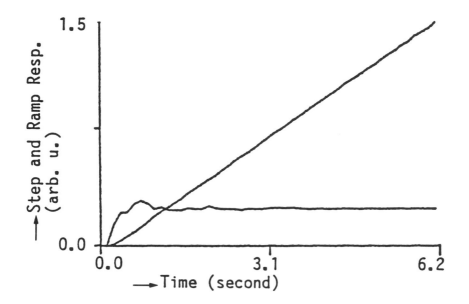

Figure 9-16: Step and ramp responses estimated for two in-core
neutron detectors

For the ten cases analysed, an average value of the delay was 0.22
second, with casewise variation as large as 0.12 second. The
corresponding estimate of coolant velocity is 3.53 m/s for the
average estimate of delay time. The average estimate is somewhat
lower than the rated coolant velocity, 4.28 m/s. Though the rated
value is covered by the dispersed estimates, the accuracy is
insufficient for detecting the sensor faults in the early (incipient)
phase. A better accuracy can be attained by increasing the number of
samples, if the resultant degradation in time resolution is
acceptable.

The result obtained by the MPD modelling of the same data tended
to indicate shorter delay time with larger uncertainty. In some
cases, the model indicated that the delay time is less than the
sampling interval. The observation suggests that the performance of
the MPD modelling method is strongly degraded by the weakness of the
propagating component in the two signals.

The method was successful in another numerical experiment with
the pressure signals from the small water loop described in Section
9.4. In this experiment, one of the pressure signals was used as the

input x(k) and the output signal y(k) was generated by feeding the
x(k) to a multi-paths delay system then adding white noise. The
purpose of this simulation experiment is to examine the capability of
the MPD modelling method to estimate the known delay times. The
magnitude of the delay was modified in a stepwise manner to simulate
a process with time-varying delay parameters. Recollect that the
pressure signal is characterised by several resonances; the methods
based on cross-correlation or phase angle analysis are difficult to
apply for such signals.

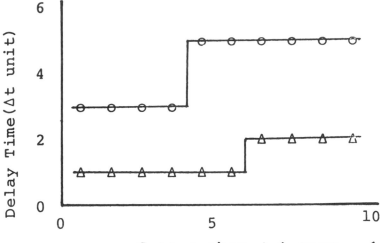

Figure 9-17: Delay times estimated by the MPD modelling. Solid lines
            represent the true values

As shown in Figure 9-17, each of the multiple components was
correctly estimated without difficulty. This result seems to justify
our interpretation that poor performance in the in-core neutron
signal analysis can be attributed to the weakness of the propagating
signal component. The preprocessing of the signals to enhance the
propagating component would be one of the possible remedies.

9.5.4 Other possibilities

In addition to the methods cited in this section, several attempts have been made to derive a redundant estimate of signals by analysing fluctuation of other signal(s). Examples of promising estimation schemes are:

(a) coolant velocity from a pair of (in-core neutron, core-exit temperature) signals,
(b) pressuriser water level from a pressure signal, and
(c) coolant pump rotation from a pressure signal.

In (a), the phase angle estimated by the MAR modelling was analysed to derive the delay time. The statistical uncertainty in the result was found to be less than the method described in B mainly owing to the smaller distance between the sensors. At the same time, however, the method (a) was found to underestimate the flow velocity. To solve the difficulty, a correction might be needed to compensate the difference in response times between the neutron and temperature sensors (Sweeney, Upadhyaya and Shieh (1985)).

The pressuriser water level was found to be related to a resonance frequency in the APSD of the pressure signal fluctuation. The water volume in the pressuriser shows a time-dependent periodic variation since the water can flow in and out through the surge line connected to the hot leg piping. The frequency of the periodic variation, reflected in the variation of the primary pressure, is dependent on the total mass of the water in the pressuriser. This fact allows the estimation of level by analysing the pressure APSD in method (b) above.

The rotating performance of the coolant pump is known to be reflected in a specific resonance in the APSD of pressure signal. The driving force of the pump has a periodic component characterised by a frequency defined by the pump rotation times the number of impellers. A change in the pump rotation can be clearly detected by monitoring the resonance frequency in the pressure APSD.

These examples are by no means exhaustive but would be sufficient to validate our claim that the fluctuation components in the process signals are rich sources of redundant information crucial for synthesising FDI-schemes. Further investigations based on this claim and guideline are currently in progress.

9.5 CONCLUDING REMARKS

The methods of parametric modelling have been developed and applied to extract FDI-related redundant information from the neutronic and process signals of NPPs. Emphasis has been placed on information extraction from random components in the signals rather than from their gross behaviour. The concepts of serial and parallel redundancies have been introduced to classify and clarify the apparently diverse spectrum of techniques employed. The usefulness as well as the potential of the methods has been demonstrated through numerical experiments of:

(a) performance monitoring of temperature and pressure sensors,
(b) performance monitoring of a pressure controller,
(c) stability monitoring of a BWR,
(d) coolant temperature estimation from pressure signal, and
(e) coolant velocity estimation by neutronic signal analysis.

The idea to derive diagnostic information from fluctuating components of process signals is not totally new in the field of nuclear engineering. Actual realisation of the idea was, however, made feasible by the development of the parametric modelling methods described in this paper. The use of LS modelling is helpful in attaining higher numerical efficiency and stability, which are crucial factors for application of the proposed methods to automated FDI. Further efforts to improve the amount and quality of the extracted redundant information are to be made to attain a higher level of safety in nuclear power plant operation.

ACKNOWLEDGEMENTS

The author would like to express his thanks to E. Türkcan of ECN, the Netherlands, for his contribution to this study. He is also grateful to R. Oguma of Studsvik AB for many valuable discussions. He extends his gratitude to Professor K. Sugiyama of Tohoku University for his continuous support throughout this study.

TABLE 9-1 Typical PWR design parameters

```
-------------------------------------------------------------------
```

Thermal power	2440 (MW)
Electric power	826 (MWe)
Primary Pressure	157 (Kg/cm^2)
Coolant flow rate	45000 (Ton/hr)
Coolant temperature (inlet)	287 ($^\circ$C)
Coolant temperature (outlet)	322 ($^\circ$C)
Core diameter (effective)	3.04 (m)
Core height (effective)	3.66 (m)
Number of fuel assemblies	157
Fuel rods per assembly	15x15
Steam pressure (turbine inlet)	52.1 (kg/cm^2)

```
-------------------------------------------------------------------
```

TABLE 9-2   Typical BWR design parameters

```
-------------------------------------------------------------------
```

Thermal power	2380 (MW)
Electric power	780 (MWe)
Primary Pressure	70.7 (Kg/cm^2)
Coolant flow rate	33800 (Ton/hr)
Coolant temperature (inlet)	278 ($^\circ$C)
Coolant temperature (outlet)	286 ($^\circ$C)
Core diameter (effective)	4.03 (m)
Core height (effective)	3.66 (m)
Number of fuel assemblies	548
Fuel rods per assembly	7x7
Steam pressure (turbine inlet)	66.8 (kg/cm^2)

```
-------------------------------------------------------------------
```

TABLE 9-3   VR values obtained by the MAR model.

```
------------------------------------------------------------

    Signal                            VR-value
------------------------------------------------------------

Pressure (P)                          7.6x10⁻²
Steam flow (Wₛ)                       2.3x10⁻²
Bypass valve opening (BPR)            1.3x10⁻³
------------------------------------------------------------
```

Signal	VR-value
Pressure (P)	7.6×10^{-2}
Steam flow (W_s)	2.3×10^{-2}
Bypass valve opening (BPR)	1.3×10^{-3}

Chapter Ten

ANALYTIC REDUNDANCY MANAGEMENT
FOR SYSTEMS WITH
APPRECIABLE STRUCTURAL DYNAMICS

Raymond C. Montgomery and Jeffrey P. Williams

This chapter deals with analytic redundancy management for the class of systems that have appreciable structural dynamics and require active control. Specific systems in this class are large and light-weight spacecraft with control systems incorporating many distributed sensors and actuators. Two subjects relevant to the control system design problem are addressed. The first is the pre-operation spacecraft design problem of incorporating reliability via component placement. The second is the problem of designing operational on-line fault detection, identification, and control system reconfiguration (FDI&R) algorithms. In this chapter, the NASA Langley Research Center flexible grid apparatus is a laboratory system for applying the principles of the theory developed. For clarity, the applications are integrated into the theoretical expositions; hence, the apparatus is described in Section 10.2. Section 10.3 deals with the laboratory prototype design whilst Section 10.4 deals with on-line FDI&R.

10.1 INTRODUCTION

Future space missions may involve very large and highly flexible spacecraft that require active structural dynamics control. As an example, Wright (1981), considers a microwave radiometer over 100 metres in diameter. Therein, many actuators and sensors were needed to meet radiometry requirements. Also, to make the mission economically feasible, a mission life of 20 years or more with occasional revisits for repair and replacement was needed. Another example is the manned space station which is projected to operate indefinitely. These long life requirements, large numbers of sensors and actuators, and heavy dependence on the proper operation of the control system dictate the design ground rule that the control system must operate acceptably in the presence of component faults. The option of hardware duplication may not be possible when a large number of physically distributed components are attached to a light-weight, highly flexible spacecraft. Hence, sensors which measure dynamically different quantities and actuators which have different effects on the system need to be managed using analytic models. This *analytic redundancy* management is the subject of this chapter.

To incorporate reliability effectively into the design of space-craft control systems, both the preliminary laboratory prototype design and the on-orbit operation of the system must be considered. Preliminary design studies must be made on the effects of component placement on the probability of mission success. Also, on-line automatic fault detection, identification, and control system reconfiguration (FDI&R) algorithms based on analytical models must be designed since hardware duplication may not be viable. With this in mind, the theoretical work presented in this chapter is divided into two main parts, the first addresses preliminary design, viz., component location on the spacecraft. The other addresses on-line FDI&R. The concepts and theories presented are applied to the same example, the NASA Langley Research Center flexible grid apparatus. For clarity, the applications are integrated into the theoretical expositions, hence, the apparatus is described first, in Section 10.2. This is followed by the treatment of the preliminary design problem in Section 10.3 and by the on-line FDI&R in Section 10.4.

10.2 THE NASA LANGLEY RESEARCH CENTER FLEXIBLE GRID APPARATUS

The grid apparatus is a test facility designed to have appreciable low frequency structural dynamics, inertial sensors and actuators, and microprocessor-based distributed computing components. These

design features were considered because of, 1) *mass-to-stiffness trade-offs for the construction of large space structures,* 2) *the spacecraft 'free-free' environment for operation of sensors and actuators,* **and,** 3) *new potentials for space qualified computers.*

Figure 10-1: A photograph of the grid structure

Figure 10-1 is a photograph of the flexible grid experimental apparatus. The test article is a 7 ft x 10 ft planar structure made by overlaying aluminium bars of rectangular cross section and suspended by a cable at two locations on the top horizontal bar (Figure 10-2). The bars are centred every foot so that there are 8 vertical and 11 horizontal bars.

The sensor and actuator complement of the apparatus consists of eight accelerometers, six rate gyros, and six inertia wheel actuators. All sensors and actuators are inertial devices mounted to the test structure and are characteristic of those available for spacecraft. The only attachment of the apparatus to Earth ground is via the cables shown in Figure 10-2.

RATE GYRO LOCATIONS ON GRID

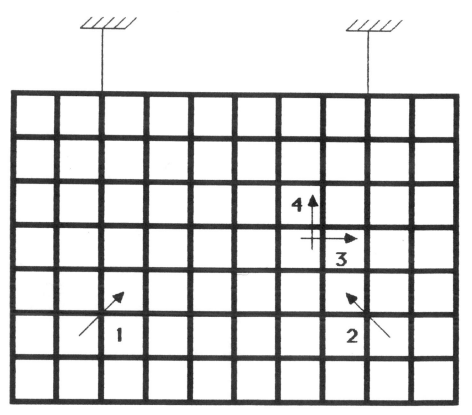

Figure 10-2: Locations of the rate gyro sensors on the grid

Sensor data in this chapter are taken primarily from the rate gyros. They measure rotational rate about their output axis with a range of 20 degrees/second and are packaged in 4 units which may be mounted at any of the nodes of the grid. Two units are 3-axis gyros and the remaining two are single-axis units. This packaging allows for collocated rate feedback for 4 of the actuators, but, because two units are 3-axis units which cannot be separated, collocated rate feedback is not possible for each of the 6 actuators.

A candidate location set for the gyro units was selected and simulation studies were made to determine if the locations were suitable. To facilitate the simulator, a finite element model of the apparatus was used. This model included the suspension cables, the test structure, and the masses of the attached sensor and actuators, and the gravity loading. Finite element analysis nodes were placed at each overlapping joint of the grid, the Earth ground attachment points of the cable, and every 1/2 ft along the 2 ft suspension cables. The finite elements of the grid were modelled as bending elements whereas the cable elements were modelled as two-force members. A total of 165 elements were, thus, included in the model. Four degrees of freedom appropriate for motion normal to the plane of the grid were considered but no damping was included. A modal analysis of this model was conducted using thirty modes. The frequencies of the first ten modes are listed in Table 10-1. The first three modes are the pendulum modes, the fourth is the first flexible or bending mode, and the fifth is the first torsional second flexible mode.

Mode Number	Frequency Hz
1	.364
2	.625
3	1.398
4	2.29
5	3.07
6	4.791
7	5.933
8	6.297
9	7.337
10	10.352

Table 10-1: Modal frequencies obtained from the finite element analysis of the grid

The simulation studies used the first 15 modes of the analysis. A sampling rate of 32 Hz was used to create a difference equation model of the structure. Modes were simulated with initial conditions that resulted in a maximum out-of-plane structural vibration of about 1 inch peak-to-peak which is the level expected in the apparatus.

Visual examination of the sensor outputs was made to determine if the six lowest frequency flexible modes had an acceptable output in the gyro units. The locations shown in Figure 10-2 were found suitable to observe these modes.

Control torques of up to 20 oz.-in.(0.014 kg-m) each can be applied by reaction of six grid-mounted, dc motors against the inertia of wheels attached to their rotors. The motors are powered by high bandwidth current amplifiers. Figure 10-3 shows their location as well as the orientation of their output axis. Each actuator is controlled by a *byte-oriented digital controller* that communicates through an asynchronous serial data communications channel to a host computer at 19,200 baud. The controller for each inertia wheel actuator is an INTEL 8751 based microcontroller that must be programmed in assembly language. Figure 10-4 illustrates schematically the signal flow for the system with one inertia wheel actuator.

TORQUE ACTUATORS

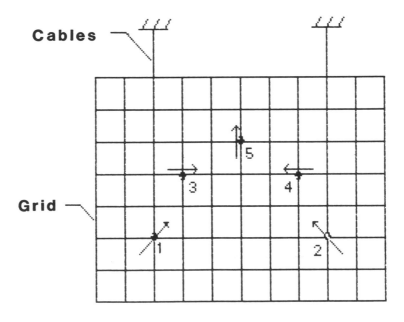

Figure 10-3: Locations of the inertia wheel actuators on the grid

"SMART" ACTUATOR

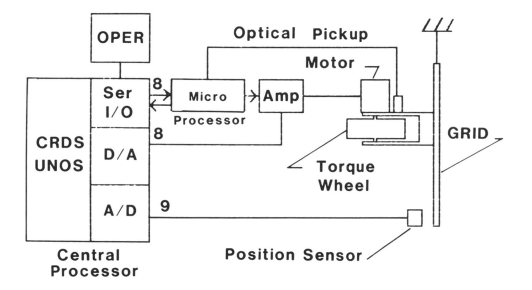

Figure 10-4: Schematic diagram of the communications interfaces for the apparatus

The host computer is a Charles River Data System (CRDS) computer with a UNIX-like operating system (UNOS) that can be programmed in either the FORTRAN 77 or C programming languages. Real-time programming of the host allows sampling the A/D converters and sending commands to the actuators up to approximately 2000 samples per second. The maximum sampling frequency depends on the amount of computation undertaken during the sampling interval. As an indication of this, a tenth order Kalman filter and regulator can be implemented using all of the rate gyro signals as sensors and the six inertia wheel actuators at 10 samples per second.

10.3 RELIABILITY CONSIDERATIONS IN PRE-OPERATION SPACECRAFT DESIGN

The goal of this part is to provide the spacecraft designer with a

method of evaluating different sets of control system component
locations using a total mission performance index that accounts for
component reliability. This was the subject of Montgomery et al.
(1981) wherein the static shape control of a beam was the primary
goal of the control system. In Montgomery et al. (1981) a performance
measure incorporating mission-oriented performance as well as
reliability was formulated and optimal locations of force actuators
were found. It was shown that reliability has substantial impact on
the selection of the optimal locations of the actuators. The approach
of Montgomery et al. (1981) was applied to static shape control of
the grid in Akpan (1982). Results of Akpan (1982) favoured clustering
the force actuators for long life missions (since any one may
substitute for the others) and distributing them physically for
improved performance for short life missions (or missions where
reservicing can be conducted frequently).

The results of this part follows the approach of Montgomery and
Vander Velde (1985) which extended Akpan (1982) to dynamic vibration
suppression and applied the theory developed to the NASA Langley
flexible grid. Other approaches to construct reliability performance
measures are available. Carignan and Vander Velde (1982), for
example, construct a measure of controllability and suggest
optimising the average value of that measure taken over the design
mission life. Similarly, other approaches for component placement are
available (e.g. Horner (1983)). However, to the knowledge of the
authors, only Montgomery and Vander Velde (1985) incorporate both
component placement and reliability.

Herein, as in Montgomery and Vander Velde (1985), vibration
suppression is treated using optimal regulator theory. First, the
general discrete regulator theory is overviewed. Then, reliability of
the components is considered and the fault characteristics of the
components are modelled. Finally, results of the analysis are presented which compare the performance of two actuator location sets for a
grid. Optimality is shown to depend on mission life or reservicing
interval.

10.3.1 Summary of essential regulator theory

Herein, consistent with current engineering practice, a finite
element model is assumed given for the purpose of control system

design. Such a model can be put in the form:

$$\dot{x} = Ax + Bu \qquad (10\text{-}3.1)$$

where x is an n-dimensional vector consisting of pairs of modal amplitude/velocity elements and u is an m-dimensional input vector consisting of the forces or moments.

Most modern control systems use a digital computer to generate torque and moment commands to actuators; hence, a difference equation form of model is most appropriate:

$$x_{k+1} = \Phi x_k + \Gamma u_k \,, \qquad (10\text{-}3.2)$$

One measure of performance for vibration suppression is:

$$J = \int_o^T (x'Qx + u'Ru)dt \qquad (10\text{-}3.3)$$

with matrices $Q = Q' > 0$ and $R = R' > 0$. This performance equation can be written in a form convenient to use of the difference equation model:

$$J = \sum_{k=0}^N (x_k'\hat{Q}x_k + x_k'Wu_k + u_k'\hat{R}u_k) \qquad (10\text{-}3.4)$$

The term involving W can be eliminated using the control variable transformation $u_k = -Fx_k + v_k$ and appropriately selecting the matrix F. This results in the model equation:

$$x_{k+1} = \Phi x_k + \Gamma v_k \qquad (10\text{-}3.5)$$

together with the performance measure equation:

$$J = \sum_{k=0}^N (x_k'Qx_k + v_k'\hat{R}v_k) \qquad (10\text{-}3.6)$$

The discrete optimal linear regulator problem is the problem of determining the control sequence v_k that minimises J. In our application, the only element of concern is the resulting optimal performance measure, J^*, which is of the form:

$$J^* = x_o'X_o x_o \qquad (10\text{-}3.7)$$

where x_o is the initial value of the state at t=0 and X_o is the solution of the discrete Ricatti equation at stage 0. That is, with

$X_n = 0$,

$$X_k = \hat{\phi}_k' X_{k+1} \hat{\phi}_k + \hat{F}_k' \tilde{R} \hat{F}_k + \tilde{Q}_k \qquad (10\text{-}3.8)$$

wherein:

$$F_k = (\tilde{R} + \Gamma' X_k \Gamma)^{-1} \Gamma' X_{k+1} \tilde{\phi} \quad \text{and} \quad \hat{\phi}_k = \tilde{\phi} - \Gamma \hat{F}_k \qquad (10\text{-}3.9)$$

Equation (10-3.7) provides a measure of the best realisable performance of a control system assumming that the m actuators making up u are functional. Computational algorithms for constructing the coefficient matrices needed in equation (10-3.4) and (10-3.6) and for solving for X_0 in (10-3.7) can be found in the ORACLS computer-aided design package, Armstrong (1980).

10.3.2 Effects of reliability on component placement

If one does not consider possible faults, the locations of control system components that produce the best system responses is the set of locations that minimises the function J^*. Reliability becomes an issue when the components of the control system are used for durations approaching their mean time to failure, t_m. In order to obtain a methodology for selecting the best locations for control system components, it is necessary to specify the class of faults to be considered and the action to be taken in response to the fault. Here, it is assumed that the fault is detected as soon as it occurs and that the faulty components are shut off and produce a zero system input. Also, it is assumed that the remaining working actuators cannot be relocated and will be used optimally; hence, equation (10-3.7) still describes the system performance index computed with a modified B matrix.

In summary, for any fault case one may compute the optimal realisable performance of the control system by constructing the appropriate B matrix of equation (10-3.1) and performing the calculations leading to the X_0 matrix of equation (10-3.7).

To examine the effect of reliability on component placement, consider the two sets of actuator locations on the grid shown in Figure 10-5. For this example, each set consists of five actuators that produce force normal to the plane of the grid and the Q and R matrices are identity. The best realisable performance of the control system, taken over a 5-sec time interval with a sampling time interval of 1/32 sec and calculated using ORACLS (Armstrong, (1980)) is shown in Table 10-2. The performance function was based on initial conditions of unity in the velocity of each modal coordinate. It shows that, under some fault modes, configuration 1 is better than configuration 2, whereas for others the opposite is true.

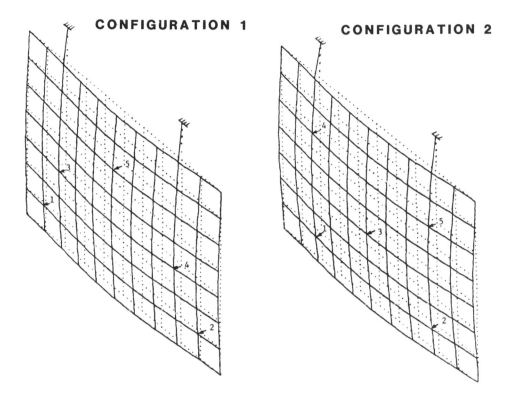

Figure 10-5: Two sets of actuator locations considered

The control system designer has the problem of selecting a configuration using the information in Table 10-2. One approach to this problem is to construct another function from the table that indicates how the system will most probably perform during a mission. This requires modelling the fault characteristics of the control system components as a function of the mission time, t. For this example, the fault characteristics are assumed to be identical for each component and described by an exponential probability density function for the time of fault as a function of mission time:

$$p(t) = \exp(- t/t_m)/t_m \qquad\qquad (10\text{-}3.10)$$

FAULT MODE	ACTUATOR* STATE	OPTIMAL COSTS CONFIGURATION 1	OPTIMAL COSTS CONFIGURATION 2
1	0 0 0 0 0	16.305	16.274
2	1 0 0 0 0	16.557	16.519
3	0 1 0 0 0	16.555	16.518
4	0 0 1 0 0	16.401	16.324
5	0 0 0 1 0	16.402	16.423
6	0 0 0 0 1	16.435	16.424
7	1 1 0 0 0	16.533	16.582
8	1 0 1 0 0	16.704	16.805
9	1 0 0 1 0	16.703	16.683
10	1 0 0 0 1	16.657	16.682
11	0 1 1 0 0	16.705	16.806
12	0 1 0 1 0	16.705	16.684
13	0 1 0 0 1	16.659	16.683
14	0 0 1 1 0	16.902	16.919
15	0 0 1 0 1	16.837	16.919
16	0 0 0 1 1	16.837	16.790

* 0   Indicates a functioning component

  1   Indicates a faulty component

Table 10-2: Summary of the best performance achievable for the fault modes considered.

One method to construct a performance index that reflects both reliability and system performance is to weigh the J^*'s listed in Table 10-2 according to the probability that the fault mode corresponding to each element will be in effect at the end of the mission. Figure 10-6 shows the variation of this weighted performance indicator as a function of t/t_m. Note that at $t/t_m = 0$ the weighted cost function is equal to that in Table 10-2 for the mode with no faults. Also, note that the weighted cost goes to zero as t/t_m increases. This occurs because the set of fault modes considered by the designer is not complete. It is possible that three, four, or even five simultaneous *or common mode faults* may occur. These may be considered as catastrophic faults since the rigid body dynamics cannot be controlled. They should not be considered in the design because the selection of actuator configuration should not be biased with catastrophic fault modes. Figure 10-7 is a graph of the incremental cost of configuration 2 over 1. The notable point is that configuration 2 is preferable to 1 for short missions whereas 1 is preferable to 2 for long missions.

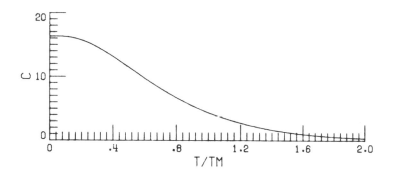

Figure 10-6: Variation of the weighted cost with t/t_m configuration 1

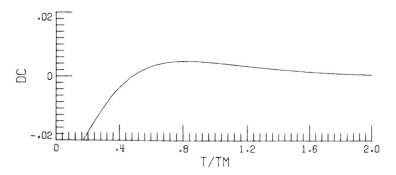

Figure 10-7: Cost of configuration 2 less cost of configuration 1

10.3.3 Summary

Section 10.3 has described a methodology that allows the control system designer to select from among a set of possible actuator locations the one which is best for vibration suppression considering the reliability of the components. The method has been applied to a grid structure as an example. In that case it was shown that optimal locations depend on the design mission life. The method generally involves, for each candidate location set, determining the best achievable performance of the control system for all fault modes that the designer wishes to consider. These values of performance are then used to construct the criterion function by taking the probability-weighted sum of the performance measures for each fault mode. The actuator locations are then chosen to minimise this cost criterion.

10.4 ON-LINE FDI&R IN STRUCTURAL DYNAMIC SYSTEMS USING ANALYTIC REDUNDANCY

Reconfiguration can generally be accomplished by switching to a controller designed for a specific fault mode if it can be identified. The detection and identification of the particular fault is the fundamental problem. Popular methods for detecting faults involve signals from three or more identical devices which sense the same physical quantity. This allows detection and identification of faults at the expense of added components - an option that may not be available when the components are physically distributed over a highly flexible structure. In some cases, the sensor set consists of signals from components whose generic type or location is different. Therefore, it is a redundant set only if the analytic relationship between the signals can be adequately modelled. This enables an *analytic redundancy management* (ARM) among sensors that have generically different outputs.

A significant step in analytic redundancy management (ARM) was made in the NASA Digital Fly by Wire F-8 aircraft, Deckert (1978). There, a system using ARM was developed and demonstrated in flight. The system used two hardware-redundant sensors to determine if a fault had occurred. The ARM feature came from a third analytic signal derived from kinematic and dynamic relations between other system sensors. This signal was used to identify which of the two hardware-redundant sensors has failed.

The F8 ARM was developed using rigid body kinematics and dynamics of the aircraft. The literature is rich in reports relating to theory applicable to ARM for this class of problems. Willsky (1976) provides a comprehensive literature survey. For large flexible spacecraft, ARM must both detect and identify faults of physically distributed sensors which produce signals related through more complex structural dynamics.

One of the early uses of optimal decision theory to accomplish ARM has been reported by Montgomery (1981). The system described was computationally intensive, requiring banks of Kalman filters. It still cannot be implemented with the current state of the art in computer technology. To gain an appreciation of this limitation, the grid apparatus, described in the previous section, uses a micro-computer-based system that has a computational capability similar to space qualified control computers. As presently configured the system can process a single steady-state LQG algorithm that uses 5 rate gyro sensors to estimate the five lowest frequency modes of motion of the grid and commands 5 torque wheel actuators at a sample rate of only 0.1 sec. This is indeed disappointing to the modern control theorist, but, it is, in fact, a realistic situation.

Unfortunately, high computational requirements are characteristic of optimal decision theoretic ARM. Thus, practical ARM fault detection and identification systems are suboptimal. Optimal decision theory is important, however, since it provides the basis of evaluation of the performance of various suboptimal ARM systems. We will now consider a suboptimal ARM system designed for and tested on the flexible grid.

10.4.1 A Suboptimal on-line fault detection system example

The suboptimal system reported in Montgomery and Williams (1985) and Williams and Montgomery (1985) has been selected for presentation and application to the grid in this part. Therein, the residual sequence of a single operating Kalman filter is used to detect and identify faults. After a fault has been identified, a Kalman filter previously designed for the remaining sensor set is used. The residual sequence of this filter is then processed to identify any further faults. The technique is suboptimal since only the residual from one Kalman filter is used and the fault hypotheses are tested sequentially. Also, herein, only bias type sensor faults are considered. The technique is amenable, however, to handling any sensor fault that has a recognisable effect on the *innovations* (or estimated residuals) of the Kalman filter.

Figure 10-8 is a schematic diagram of the system. The sensor measurements are sampled by analogue-to-digital (A/D) converters and processed by one of N Kalman filters selected by the fault state estimate H_i. The filters are designed for each anticipated fault condition. The decision as to whether or not a fault has occurred is made by processing the innovations sequence of the selected filter using Wald's sequential probability ratio test, SPRT, Wald (1947).

Wald's SPRT algorithm was designed to determine which of two statistical processes generated the input to the algorithm. Thus, it is a binary hypothesis test wherein the hypotheses (for our case) correspond to *fault* and *no fault*. At a specific sample time, insufficient data may have been accumulated to make a good decision. Thus, the possible outputs of the SPRT algorithm are: *fault detected, no fault detected*, and *no decision*.

Returning to Figure 10-8, since a fault of a single sensor affects more than one sensor innovations sequence, an interpreter is required to examine the innovations of the operating filter for the appropriate fault signature. Thus, after a fault detection is made the SPRT output interpreter produces the estimate of the fault mode. The remainder of this part is devoted to a detailed development of the theory underlying this suboptimal FDI&R strategy and presentation of examples to illustrate details of implementation. It is organised as follows: first, the elements of Kalman filter theory needed to implement the FDI&R strategy are presented; this is followed by a presentation and discussion of Wald's sequential probability ratio test; finally, experiments on the grid, intended to illustrate the practical implementation of the strategy, are presented and discussed.

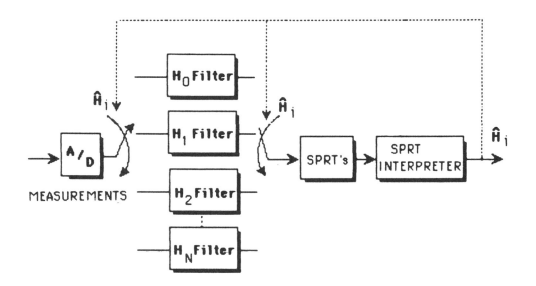

Figure 10-8: Overview of the fault detection logic

10.4.2 A Summary of essential Kalman filter theory

Now consider the estimation of the state x required to implement the regulator control law. With additive noise, the state model for designing the estimator is of the form:

$$x_{k+1} = \Phi x_k + \Gamma u_k + n_k \qquad (10\text{-}4\text{-}1)$$

with a measurement model:

$$y_k = H x_k + v_k \qquad (10\text{-}4.2)$$

where n_k and v_k are zero-mean Gaussian white noise sequences with nontrivial variances N and V, respectively. The filter used herein is the optimal Kalman filter that minimises the function

$$J = \lim_{k \to \infty} E(e_k' \, e_k)$$

where the estimation error is defined as $e_k = y_k - H\hat{x}_k$. The filter state $\hat{x}_k$ satisfies:

$$\hat{x}_{k+1} = \Phi\hat{x}_k + G(y_k - H\hat{x}_k) \tag{10-4.3}$$

wherein the filter gain G is given by:

$$G = \Phi PH'(R + HPH') \tag{10-4.4}$$

and P satisfies:

$$P = (\bar{\Phi}P\bar{\Phi})^{-1} + GRG' + N \tag{10-4.5}$$

and:

$$\bar{\Phi} = \Phi - GH \tag{10-4.6}$$

The matrix P represents the steady state variance of the innovation sequence of the filter. The mean-square of the innovation, trace (P), is a measure of the expected performance of the filter. Computational algorithms for determining the matrix performance indicators, χ and P, are included in the ORACLS computer aided design package, Armstrong (1980). They have been used herein to predict the performance of an advanced redundancy management control system for the grid apparatus discussed above.

10.4.3 Wald's sequential probability ratio test

Sequential testing of a sample was developed by Wald during World War II as a means to economise on the number of observations required in a test procedure. The method's primary feature is that the number of observations required to make the decision is not determined *a priori*. Whether a decision can be reached on the nth sample is based on the outcomes of the previous observations. The number of observations required to reach a decision is on the average less than that required for a similiar test with a fixed number of observations.

The SPRT algorithm decides in favour of one hypothesis, H_0 over another, H_1 by sequentially calculating the ratio of the probability of the input data sequence assumming one hypothesis to that assuming the other, Thus:

$$L_k = \frac{P(r_1, r_2, \ldots, r_k | H_1)}{P(r_1, r_2, \ldots, r_k | H_0)} \tag{10-4.7}$$

where $r = (r_1, r_2, \ldots, r_k)$ is the sample of residuals associated with one sensor. To determine the most probable hypothesis for the sequence, an inequality is formed as follows:

$$B < L_k < A \qquad\qquad (10\text{-}4.8)$$

As sampled data are used, the magnitude of the SPRT decision variable, L_k, is checked. If it is less than a predetermined threshold B, a decision is made in favour of the *no fault-hypothesis* H_0 and the test is terminated. If it is greater than a threshold A, the decision is made in favour of *fault-hypothesis* H_1 and, again, the test is terminated. No decision can be made as long as L_k is between A and B. In that case more data is required to make a decision and the test continues. The thresholds A and B are selected by the designer and reflect his concern over the risk involved of missing a developing fault and the nuisance created by sounding a false alarm.

In the implementation described here, the no-fault hypothesis, H_0 is characterised by the residual signal being Gaussian having a zero mean with a variance σ^2, whereas the fault hypothesis, H_1 is characterised by the residual signal being Gaussian having mean m with the same variance, σ^2. Such a fault condition is likely to be based on the electrical characteristics of available control system components. Assuming the sensor noise is Gaussian, the residual sequence of the Kalman filter will also be Gaussian. Under these assumptions the sequential probability ratio test becomes:

$$B < \frac{\exp\left[-1/2\sigma^2 \sum_{i=1}^{k}(r_i - m)^2\right]}{\exp\left[-1/2\sigma^2 \sum_{i=1}^{k} r_i^2\right]} < A \qquad\qquad (10\text{-}4.9)$$

The value of the test mean m is the designer's choice and may reflect robustness considerations.

For real-time analysis, one would prefer to eliminate the need for exponentiation. Wald notes that, by taking the logarithm of the last equation, the above simplifies to:

$$\ln B < \frac{-m}{2\sigma^2} \sum_{i=1}^{k}\left(m-2r_i\right) < \ln A \qquad\qquad (10\text{-}4.10)$$

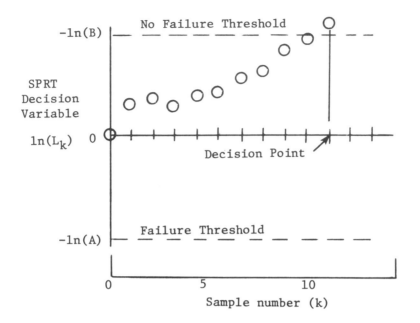

Figure 10-9: Behaviour of the SPRT decision variable

The threshold constants, A and B, are selected based on how certain one wants to be of making the correct decision. The formulas relating the probability of missed detection, P_m, the probability of false alarm P_f, and the decision thresholds, A and B, are $A \simeq (1-P_m)/P_f$ and $B \simeq P_m/(1-P_f)$, Deckert (1978). Figure 10-9 illustrates the mechanisation of the SPRT algorithm (note sign change so that no-fault decision is indicated by passage above the -ln B threshold).

10.4.4 Experimental results

The objective of the experimental implementation was to demonstrate the ability to identify component bias fault and in real-time using realistic sensing, computing and structural hardware. Specific questions to be answered are: How much do the faults affect the other residuals? What decision logic should be used and, when should the SPRT be started? Only four rate gyros are used and no closed-loop results are available. The data presented will involve primarily first mode vibrations. In this pendulum mode, the rate gyro sensitivities are proportional to $2\sqrt{2}$, $2\sqrt{2}$, 1, and 0, respectively (see Figure 10-2). Since the third sensor has a design variance one-hundred times less than the others, it is the major contributor to the estimate of the first modes motion. In fact, the Kalman gains on the first two sensors are such that their signals tend to cancel.

The effect of sensor bias fault of 0.01 rad/second on the Kalman filter residuals is shown in Figure 10-10. It is noted that the fault of the third component results in a high variance and low mean error in the residual for the first two sensors, but, a low variance and high mean error in the residual of the third. This result is caused by the Kalman filter attempting to filter the zero frequency (or sensor bias) signal since it is not included in the filter model. This pattern of residual means is desirable since the SPRT is designed to detect mean faults and the faulty sensor residual will be driven to the negative (fault) threshold the fastest.

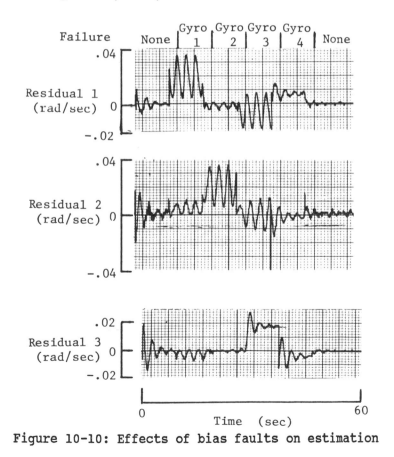

Figure 10-10: Effects of bias faults on estimation

Figure 10-11 shows the behaviour of the SPRT decision variable for each residual sequence as the individual sensors, in sequence, are failed, with 0.01 rad/second constant bias and zero variance, and later restored to normal operation. The decision variable is calculated continuously from time zero. Note that although the slope of the decision variable signal changes for all residuals for each of the first three sensors, only the one for the failed sensor tends toward the fault decision. Thus, it can be concluded that the required identification logic is the trivial case that the defective component is the one with a negative slope decision variable. The characteristics of the decision variable indicate some need for operational considerations of the algorithm. In Deckert (1978), the

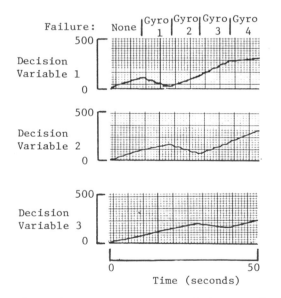

Figure 10-11: A time history of SPRT decision variables obtained from the grid experimental apparatus

SPRT was initiated upon recognition that two redundant sensors gave different readings. In this scheme, redundant sensors are not used, hence, the dilemma is: 'When should the SPRT be started?' One could start the SPRT at time zero and allow the variable to grow without bound until a fault occurred. After the fault onset, the decision variable would be driven back towards the fault detection threshold.

Starting from time zero is unacceptable because the decision variable is biased to the no-fault case over a very long time interval and this bias must be removed before a fault decision can be made. One might propose that the bias be ignored and the slope of the decision variable be used as an indication of the presence of a fault. This detection logic may not be used since the decision variable does not, in general, approach the thresholds monotonically, and the confidence in the decision expressed by the probabilities of missed detection, P_m, and false alarm, P_f, cannot be realised unless the fault state is constant over the test interval. The dilemma is solved by starting the test at time zero and then restarting it every time a decision is reached. Note that this solution will corrupt the confidence measures, but some degree of confidence in the test is maintained. The average number of samples required to reach a decision can be approximated by the formulae $\bar{k} = (\ln A) / (m^2/2 \cdot 2)$ if H_0 is true, and $\bar{k} = (\ln B) / (-m^2/2 \cdot 2)$ if H_1 is true. A typical decision variable time history is sketched in Figure 10-12. Note that the fault decision time will be extended unless the fault occurs immediately after a no fault decision has been made. The length of this extension may be approximated by $2 \cdot \ln(A)/m^2/2$. This time may be used to determine whether a closed-loop controller will cause damage before reconfiguration of the filter.

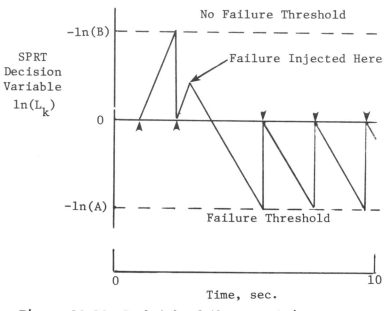

Figure 10-12: A sketch of the SPRT trigger process

   Implementation of the FDI&R scheme is accomplished in a straight-forward manner given the characteristics of the SPRT test for the mode of excitation studied. The filter design is based on the hypothesis of the most recent decision. As a sensor fault is uniquely reflected in the decision variable associated with that sensor, detection and identification may be accomplished simultaneously based on the outcome of a single decision variable. Reconfiguration of the filter to accommodate a fault of the third sensor is demonstrated in Figure 10-13. The probability for missed detection was set at 0.0001 and the probability for false detection was set to 0.01. These values were selected on the judgement that to miss a fault is more detrimental to the system than to generate a false detection. A bias fault is injected into the rate gyro signal at an arbitrary point in time and the residual immediately reflects the new mean-value. The fault decision is made within 2.1 seconds. The reconfiguration of the filter occurs in the same sample period and the estimate again converges in accordance with Kalman filter theory to produce zero mean residuals.

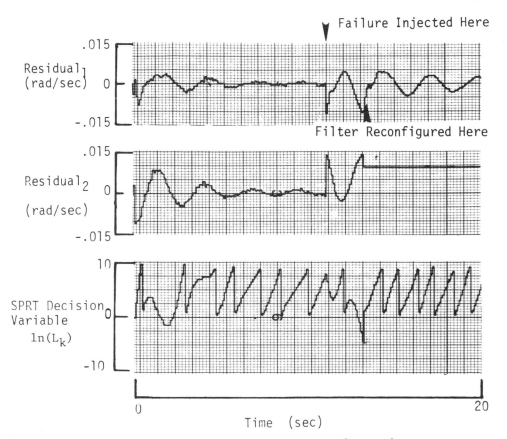

Figure 10-13: Reconfiguration of the Kalman filter in response to a fault

10.4.5 Concluding remarks

The system presented and illustrated in this part enables on-line detection and rapid reconfiguration of the Kalman filter for a structural dynamics system. The FDI&R system involves sequential testing of the innovations of a single, active, Kalman filter to detect any anomolous operation. A fault is then isolated by detecting its signature on the innovations. Individual faults obviously affect all innovations, but, the nature of the filter and SPRT decision process allows trivial detection of a fault.

The success of the method is conditioned on whether the theoretical zero-mean character of the innovations sequence can be relied upon as an indicator of a component fault. Unfortunately, component faults are not the only source of corruption to the

innovations sequence. Indeed, model error is another source of corruption. Hence, unless one can certify the model and the character of
the innovations sequence before attempting on-line fault detection,
the scheme will fail. Note that this does not require the model to be
exact, only that the statistics of the nominal Kalman filter be
known, and that the statistics of the innovations resulting from the
fault be sufficiently different from the nominal.

Another concern conditioning the ultimate success of the method
is the possibility of undetected faults which could, in turn, easily
become full component failures. The effect of undetected faults on a
closed-loop system can be assessed by determining the maximum
decision time if the fault occurs when the decision variable is about
to reach the no decision threshold. The control action caused by the
incorrect sensor reading must then be analysed. If it is determined
that the control action over this interval is not acceptable, the
decision time may be shortened somewhat by restarting the SPRT at a
point somewhat past the analytic time to make a no-fault decision.

Chapter Eleven

A FAULT DIAGNOSTIC APPROACH
TO SIGNAL PROCESSING
IN NON-DESTRUCTIVE TEST INSTRUMENTATION

P.A. Gorton and J.O. Gray

Eddy current non-destructive test instrumentation is widely used to detect cracks or surface flaws in fabricated metal samples. In this chapter an equivalent circuit model of the eddy current test phenomenon is posed and used as the basis for a digital signal processing procedure based on the Kalman filter theory. The objective is to evolve a practical method for the automatic classification of sample defects during the inspection process which can be realised in an economical way using concurrent processing techniques. The work is concerned primarily with the study and identification of defects in tubular metal samples using circular inspection coils. Simulation results are presented to verify the equivalent circuit model and to confirm the validity of the approach in identifying fault conditions.

11.1 INTRODUCTION

11.1.1 General

Modern engineering practice demands that materials be free, as far as possible, from major structural flaws. As a consequence, a range of techniques has been developed to locate flaws without testing materials to destruction. Such tests are used to provide a measure of quality assurance during the manufacture of engineering materials and for in-service testing of engineering components. The methods are collectively known as Non-Destructive Testing (NDT), Non-Destructive Inspection (NDI) or Non-Destructive Evaluation (NDE) and exploit a wide range of phenomena including electromagnetism, ultrasonics and X-rays etc., (see, for example, McMaster (1963), De Graaf and De Rijk (1979)). Recently, NDT problems in the Nuclear Power industry have provided a major stimulus for the development of increasingly sensitive, reliable and sophisticated techniques in all branches of NDT, (Natesh (1978)).

In this Chapter, we discuss the electromagnetic NDT of electrically conducting, cylindrical bars and tubes by means of induced eddy currents. The test may be applied to bars or tubes several centimetres in diameter or to wires less than one millimetre in diameter by choosing suitable testing conditions.

We regard the *detection* of defects by this technique to be a solved problem and our interest will focus on the *interpretation* of information from eddy current NDT instrumentation.

11.1.2 Historical background

The first known eddy current test was performed in 1879 by D. E. Hughes (1879) who used a so-called *Induction Balance* to distinguish various properties of test objects. The Induction Balance consisted of two identical coils into which the various objects under test could be placed. The coils were excited by a microphone placed close to a ticking clock and the measurement made by listening to the ticks with a telephone receiver and adjusting the coils until the sound disappeared. The adjustments for each test object were recorded. Hughes managed to distinguish forged sovereigns from genuine sovereigns by this method.

More recent eddy current NDT equipment is based on the work of Friedrich Forster and co-workers (Forster (1952)) who systematised the theory and practice of eddy current NDT by providing a theory based on the solution of the Helmholtz Equation and carrying out a series of experiments to relate the response of eddy current NDT instruments to the shape, size and location of defects in the sample under test. The technique is referred to as *Phase Analysis* and has led to a wide range of commercial equipment, (see, for example, Forster (1970), (1974), Stumm (1979)). In such equipment, the output is usually presented in the form of a polar display. The shape of the locus of a point on the display is related to the nature of the defect in the sample under test.

11.1.3 Eddy current NDT in practice

The conventional arrangement for conducting the eddy current test is shown in Figure 11-1. The material under test is passed through encircling inspection coils or the inspection coils are passed through the inside of tubular samples to inspect the inside walls and avoid the disassembly of equipment. Eddy currents are induced in the material by the application of a sinusoidal excitation signal at a frequency chosen to give a suitable distribution of current within the sample. Defects in the material (e.g. cracks, pits, dents and porous regions etc.) may be detected by monitoring signals derived from the inspection coils and it is possible for a skilled and experienced operator to classify defects according to type and depth beneath the surface of the material by interpreting the standard polar display.

A great deal of experimental work has been carried out to aid the interpretation of the polar display. The most useful results have been obtained by testing artificial samples consisting of mercury filled glass tubes with non-conducting test pieces introduced to simulate sample defects. The results are widely available in the literature (see, for example, McMaster (1963), Forster (1952), Hochschild (1961), Aldeen and Blitz (1979), Blitz and Rowse (1964)) and may be used to aid the interpretation of the polar display. The task of relating the polar display to the results of the experimental work is difficult and prone to human error and to the effect of random disturbances in the measurement circuitry caused by surface roughness of the sample, vibration and other effects.

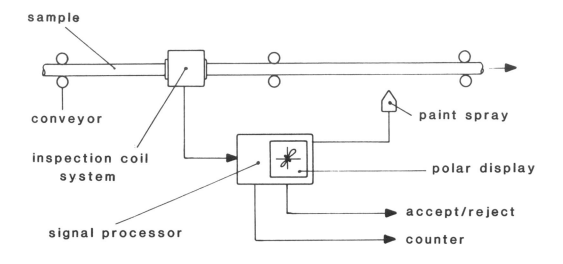

Figure 11-1: Conventional eddy current NDT equipment

Although the majority of commercially available instruments make use of a single frequency sinusoidal excitation signal, the usefulness of other signals has long been recognised and increasingly, use is being made of pulsed excitation, (Morris (1975) and Wittig (1977)), and multi-frequency excitation, (Davis (1978) and Larsson (1978)), in which the sum of several sinusoids at different frequencies is applied. Although these excitation signals are more *information-rich* than the single frequency sinusoidal excitation signal, the interpretation of the resulting output signal has remained problematic. As a result, sophisticated pattern recognition algorithms (processed off-line), (see, for example, McLean (1978) and Dan (1979)), and other data reduction techniques have been brought to bear on the problem of interpreting data from the various types of eddy current NDT instruments.

11.1.4 Automatic interpretation of eddy current NDT measurements

It is a highly desirable aim to classify the defects detected by eddy current NDT equipment automatically. In order to provide some motivation for this statement, we give the following example of a typical problem encountered in eddy current NDT. Copper wire, typically 0.5 cm in diameter is used as the raw material in the production of fine gauge copper wire less than 1mm in diameter (the fine gauge wire is subsequently used to manufacture inductors, solenoids and relays etc.). The raw material is drawn (stretched) to reduce the diameter and wound onto reels which are supplied in several standard weights. The heaviest reels require the production of the longest, unbroken lengths of wire. If the wire breaks during the drawing process the reel is lost and the material has to be reprocessed in some way. In order to increase the efficiency of the production process by minimising breakages, eddy current NDT is sometimes used to grade the raw material so that the material with the fewest defects may be used in the production of the heaviest reels. However, it is necessary to test the raw material at such a high rate that there is insufficient time for an operator to interpret the test and defects have to be registered according to the amplitude (and possibly phase) of the signals detected by the eddy current NDT equipment. Unfortunately, for physical reasons, eddy current NDT gives rise to signals with small amplitudes for sub-surface defects. (This problem may be alleviated to some extent by the choice of the excitation signal and is one of the reasons for the use of multi-frequency excitation. However, as we have pointed out, the interpretation of the resulting signals by conventional techniques remains problematic.) It may happen, therefore, that dents and other surface effects which may cause no difficulties in the drawing process may be registered as defects whilst an internal porous region or even an internal piece of slag which would certainly lead to a breakage may not be registered at all. Consequently, the grading of the raw material may be fairly coarse in practice.

The above example is typical of the eddy current NDT problems that are encountered in high throughput manufacturing applications.

We may summarise the main requirements for the automatic classification of sample defects as follows:

(a) to classify automatically defects by type and hence eliminate the need for a skilled operator,

(b) to eliminate as far as possible the misinterpretation of signals due to noise etc., and

(c) to increase the effectiveness of eddy current NDT by classifying defects with any form of excitation signal.

It is important that the classification technique can be applied to samples of various materials of differing dimensions at high throughput rates. Furthermore the implementation costs associated with the solution should be compatible with the cost of the related eddy current NDT equipment.

In this Chapter, we discuss the concept of *hypothesis testing* as a means of automatically interpreting signals from eddy current NDT instruments. The basis of the idea is to compare the response obtained from the instrument in the presence of a defect with a set of hypothetical responses and then to determine which hypothetical response corresponds most closely (in some sense) to the actual response. The problem of developing an automatic classifier based on the hypothesis testing concept may be resolved in two issues:

(a) the formulation of suitable models with which to predict the responses associated with sample defects, and

(b) the development of a means of measuring the correspondence between the predicted responses and the actual response of the instrument.

Consequently, the material in the remainder of this Chapter mainly adresses these two issues.

11.1.5 Modelling eddy current NDT phenomena

The development of suitable models of eddy current NDT phenomena is a difficult undertaking for the following reasons:

(a) The signals which arise from a defect depend on the location of the defect, the excitation signal, the dimensions and bulk electrical properties of the material under test.

(b) The theoretical and experimental results in the literature are generally only valid for sinusoidal excitation signals.

(c) The theoretical results that are available are usually applicable only to a highly idealised situation in which the test coil and sample are assumed to be of infinite length. In practice, multiple inspection coils, arranged in a bridge configuration are deliberately made short to give good sensitivity to transverse defects.

In order to automate the interpretation of signals from eddy current NDT equipment it is necessary to formulate models which encompass the *a-priori* information available to the operator. On this basis we may identify the following requirements for models of eddy current NDT upon which to base a classification scheme:

(a) The models must represent the basic underlying electromagnetic phenomena for all the conditions under consideration (at least approximately).

(b) The models must represent the effect of sample defects.

To some extent, these requirements are satisfied by methods reported in the literature. For example, more elaborate analytic solutions to the field equations encountered in the design of eddy current NDT transducers than those developed by Forster and his co-workers are given by Dodd (1977), and the theory has been extended to deal with idealised sample cracks, see, for example, Kahn et al (1977) and Spal and Kahn (1977)). Furthermore, the finite element method has been shown to be useful in modelling sample defects, with sample defects being represented by breaks in the finite element mesh, (see Polanisamy and Lord (1979) and Kincaid and Chari (1979)). However, the possibility of incorporating such approaches into an automatic classifier appears to be a daunting one if only for computational reasons.

It should be pointed out at this stage, that the problem of modelling sample defects is difficult since, by definition, defects are random disturbances in the structure of the sample under test. We therefore reject the idea of modelling specific sample defects and concentrate our efforts on developing a model with which we may represent the behaviour of broad classes of defects, for example, surface defects and sub-surface defects. The model we describe is a simple, lumped parameter approximation of the behaviour of the sample and inspection coils which gives results which are in close agreement with the exact results of the conventional theory. We will treat the sample and inspection coil as a *dynamical system* with the excitation signal as input. Sample defects may be modelled by varying the parameters of the model and so to some extent we are able to treat the problem of classifying sample defects as one of *parameter identification*.

11.1.6 Multiple model estimation techniques

In order to determine the correspondence between the hypothetical defects and the actual defect in the sample under test we make use of an adaptive estimation scheme first proposed by Magill (1965), which makes use of a bank of elemental estimators. The algorithm has come to be known as *the Multiple Model Estimation Algorithm* (MMEA), (see Athans and Chang (1975) and Anderson and Moore (1979)). In the present application, each elemental estimator is matched to a specific hypothesised defect and the residuals behaviour of the estimators monitored via a simple nonlinear calculation to find the estimator which gives the closest correspondence to the actual defective sample and inspection coil system. The approach has a pleasing similiarity to the conventional method of interpreting signals from eddy current transducers.

11.1.7 Organisation of the chapter

The remainder of this Chapter is organized in the following way. In Section 11.2, a lumped parameter model of the eddy current NDT of cylindrical, conducting test samples is described. The model is firstly evaluated under simplifying assumptions to give results which may be compared with the theoretical and experimental results published in the literature. The model is then extended to deal with a practical transducer arrangement known as a Felici Bridge. An experiment to validate the Felici Bridge model is then described. In Section 11.3, the automatic classification of sample defects by means of the Multiple Model Estimation Algorithm is discussed. Simulations are presented which demonstrate the automatic classification of sample defects. Final conclusions are drawn in Section 11.4.

11.2. LUMPED PARAMETER MODEL

11.2.1 General

In this section, we describe the lumped parameter model which we will use to form the hypothesis models required for the classification scheme described in Section 11.3. Exact solutions for the impedance of the inspection coil and distribution of induced current within the sample may be found by solving the appropriate partial differential equations. However, as we pointed out, such an approach is not appropriate and we adopt a simplified approach in which the sample under test is represented as an assembly of concentric, tubular elements in which the induced current is assumed to flow in *current sheets*. The electrical properties of the elements are determined by standard methods.

11.2.2 Single inspection coil and sample

Consider the arrangement shown in Figure 11-2 in which a tubular sample with inner radius a_{in} and outer radius a_{ou}, conductivity c and length d is concentric with a single inspection coil of radius a_{pr} and length d_{pr}. Assume that the inductance of the inspection coil is l_{pr}, the resistance of the primary circuit is r_{pr} and that the relative permeability of the material is unity.

Now suppose that the sample is comprised of N concentric tubular elements and that the sample and inspection coil may be represented by the equivalent circuit shown in Figure 11-2. Assume that the induced current in the n-th sample element is x_n and that the n-th sample element has an equivalent resistance r_n and self-inductance l_n. Let the mutual inductance of the inspection coil and the n-th sample element be m_{prn} and the mutual inductance of the n-th and m-th sample elements be m_{mn}. We may then assemble the following vectors and matrices to represent the properties of the sample and inspection coil:

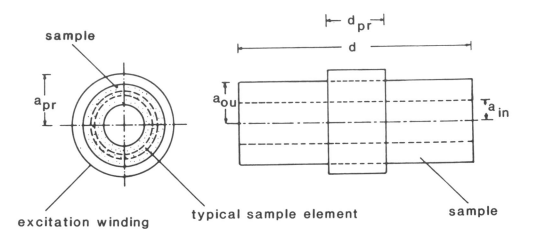

Figure 11-2: Sample and inspection coil

$$Q = \begin{bmatrix} m_{pr1}, \ldots, m_{prN} \end{bmatrix}' \qquad (11\text{-}2.1)$$

$$x = \begin{bmatrix} x_1, \ldots, x_N \end{bmatrix}' \qquad (11\text{-}2.2)$$

where Q is an N-vector of mutual inductances of the inspection coil and sample elements, x is an N-vector of currents induced in the sample elements,

$$R = \begin{bmatrix} r_1 & & O \\ & \ddots & \\ O & & r_N \end{bmatrix} \qquad (11\text{-}2.3)$$

where R is an N x N diagonal matrix of equivalent resistances, and

$$M = \begin{bmatrix} l_1 & \cdots & m_{1N} \\ \vdots & \ddots & \vdots \\ m_{N1} & \cdots & l_N \end{bmatrix} \qquad (11\text{-}2.4)$$

where M is an N x N symmetric matrix of self and mutual inductances of the sample elements.

   Kirchhoff's laws may be applied to the circuit of Figure 11-3 to give:

$$\left.\begin{array}{l} (r_p + sl_p)i_p(s) + sQ'x(s) = u(s) \\[2mm] (R + sM)x(s) + sQi_p(s) = 0 \end{array}\right\} \qquad (11\text{-}2.5)$$

The impedance of the inspection coil and sample may be obtained as a function of angular frequency by eliminating $x(s)$ from equation (11-2.5) and substituting $s = j\omega$ to give:

$$Z(j\omega) = r_p + j\omega l_p + \omega^2 Q'(R + j\omega M)^{-1}Q \qquad (11-2.6)$$

Assuming that $(R + j\omega M)$ may be inverted, i.e.

$$(R + j\omega M)^{-1} = P + jS \qquad (11-2.7)$$

then the following expressions for the real and imaginary parts of the inspection coil are obtained:

$$\left. \begin{array}{l} \text{Re } [Z(j\omega)] = r_p + \omega^2 Q'PQ \\[2mm] \text{Im } [Z(j\omega)] = \omega l_p + \omega^2 Q'SQ/l_p \end{array} \right\} \qquad (11-2.8)$$

On further assuming that the resistance in the primary circuit is negligible (i.e. $r_p = 0$) and on carrying out a conventional normalisation (i.e. dividing throughout by ωl_p) we obtain the following expressions for the normalised impedance of the inspection coil:

$$\left. \begin{array}{l} z_{re} = \omega Q'PQ/l_p \\[2mm] z_{im} = 1 + \omega Q'SQ/l_p \end{array} \right\} \qquad (11-2.9)$$

The equivalent resistances of the elements may be calculated from the dimensions of the sample elements and the conductivity of the material. The self-inductances and mutual inductances may be calculated by means of the standard inductance formulae given by Grover (1946). Note that on substituting the inductance formulae into equation (11-2.9) the expressions for the normalised impedance become independent of the number of turns on the inspection coil. Furthermore, when the inspection coil and sample are long in comparison to their diameters, the inductance formulae reduce to simple expressions.

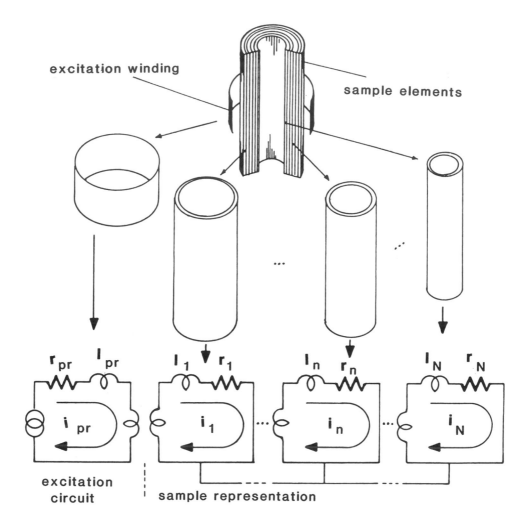

Figure 11-3: Model of eddy current test

The locus of the normalised impedance as frequency varies may be plotted on an Argand Diagram as shown in Figure 11-4 for a solid cylindrical sample. In the figure, the locus of the normalised impedance has been evaluated for several values of N using the sample and inspection coil characteristics given in Table 11-1. The locus of the normalised impedance calculated by solving the Helmholtz equation is shown for comparison.

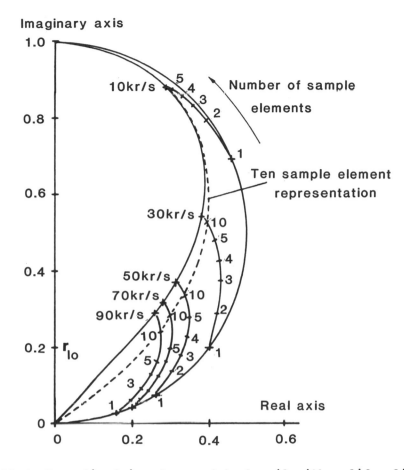

Figure 11-4: Normalised impedance of test coil with solid cylindrical sample showing the effect of increasing the number of sample elements in the model

11.2.3 Distribution of induced current

The current flowing in the sample elements may be determined directly from equation (11-2.5):

$$x(s) = - s(R + sM)^{-1}Qi_p(s) \qquad (11\text{-}2.10)$$

which may be evaluated by substituting $s = j\omega$ and using equation (11-2.7) to give:

$$x(j\omega) = - j\omega(P + jS)Q \qquad (11\text{-}2.11)$$

for unit excitation current. Equation (11-2.11) is evaluated using the sample and inspection coil parameters given in Table 11-1 with N=20 and plotted in Figure 11-5 for several values of frequency. Figure 11-5 illustrates *skin effect* whereby the induced current tends to be concentrated at the surface of the sample at high excitation frequencies.

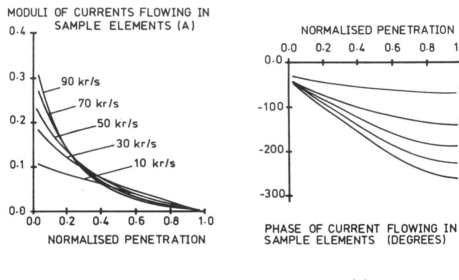

(a)   (b)

Figure 11-5: Current   flowing   in   solid   cylindrical   sample
            (20 elements, points joined by straight lines)
            (a) Moduli of currents
            (b) Phase of currents (relative to excitation current)

11.2.4 Modelling sample defects

The effect of sample defects may be modelled by adjusting the elements of the matrix of equivalent resistances to produce a new matrix $\tilde{R}$.

The departures from the operating point on the normalised impedance plane which correspond to adjustments of the elements of R are given by:

$$z_{re} = \omega Q'(P - \tilde{P})Q/l_{pr}$$
$$z_{im} = \omega Q'(S - \tilde{S})Q/l_{pr}$$

(11-2.12)

where $\tilde{P}$ and $\tilde{S}$ are given by:

$$(\tilde{R} + j\omega M)^{-1} = \tilde{P} + j\tilde{S} \qquad\qquad (11\text{-}2.13)$$

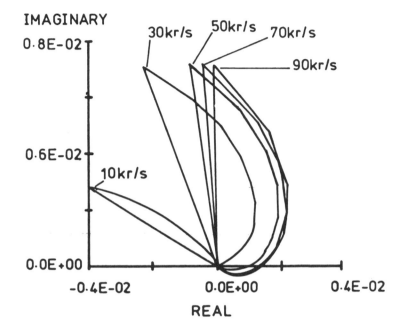

Figure 11-6: Departures from operating point on the normalised impedance plane as resistance of sample elements are increased sequentially

Plots of the departures from the operating point on the normalised impedance plane which result from increasing the equivalent resistance of each of the sample elements sequentially at several different excitation frequencies are shown in Figure 11-6 using the sample and inspection coil parameters given in Table 11-1 and N=20. The results show a close similarity to the experimental results widely reported in the literature. This demonstrates the usefulness of the model for predicting the signals which arise from sample defects at given excitation frequencies. Similar results may be obtained for tubular samples.

Sample Material	Stainless Steel
Sample Resistivity	72$\mu\Omega$ cm
Sample Diameter	2.5 cm
Sample Length	25 cm
Inspection Coil Diameter	2.5 cm
Inspection Coil Length	25 cm

Table 11-1: Characteristics of sample and inspection coil

11.2.5 Felici Bridge

In order to detect the small signals that arise from practical sample defects, an arrangement known as a Felici Bridge is often used in eddy current NDT instrumentation. The Felici Bridge is illustrated in Figure 11-7. Note that the secondary windings are arranged to cancel out the direct inductive coupling between the primary and secondary windings.

In order to model the Felici Bridge, we assume that the test and reference samples are represented by assemblies of N concentric elements as before. We may then assemble the N-vectors x_t, x_r, Q_t, and Q_r and the N x N matrices R_t, R_r, M_t and M_r for the test and reference samples with the elements defined as in equations (11-2.1)-(11-2.4). Further N-vectors Φ_r and Φ_t may be defined to represent the inductive coupling between the samples and secondary windings:

$$\left.\begin{array}{l} \Phi_r = \begin{bmatrix} m_{sr1}, \ldots, m_{srN} \end{bmatrix}' \\[2mm] \Phi_t = \begin{bmatrix} m_{st1}, \ldots, m_{stN} \end{bmatrix}' \end{array}\right\} \qquad (11-2.14)$$

where m_{srn} is the mutual inductance of the n-th element of the reference sample and the reference secondary winding and m_{stn} is the mutual inductance of the n-th element of the test sample and test secondary winding.

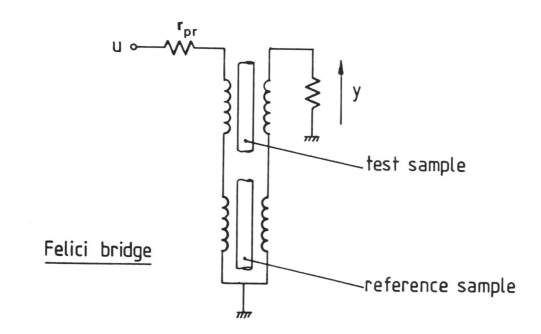

Figure 11-7: Felici bridge

We may now assemble the following $(2N + 2)\times(2N + 2)$ matrices:

$$
R_f = \begin{bmatrix}
\begin{matrix} r_p & 0 \\ 0 & r_s \end{matrix} & O & O \\
\hline
O & R_r & O \\
\hline
O & O & R_t
\end{bmatrix}
\tag{11-2.15}
$$

where R_f is a diagonal matrix of resistances and:

$$
M_f = \begin{bmatrix}
\begin{matrix} 1_p & 0 \\ 0 & 1_s \end{matrix} & \begin{matrix} Q'_r \\ \Phi'_r \end{matrix} & \begin{matrix} Q_t \\ -\Phi_t \end{matrix} \\
\hline
\begin{matrix} Q_r & \Phi_r \end{matrix} & M_r & O \\
\hline
\begin{matrix} Q_t & -\Phi_t \end{matrix} & O & M_t
\end{bmatrix}
\tag{11-2.16}
$$

and M_f is a symmetric matrix of self and mutual indutances r_p, r_s, l_p and l_s are the total resistances and inductances in the primary and secondary circuits. Kirchhoff's laws may be applied to give:

$$(R_f + sM_f)x_f(s) = bu(s) \qquad (11-2.17)$$

where $x_f(s)$ is a $(2N + 2)$ - vector of currents given by:

$$x_f = \left[i_p, i_s \mid x'_r \mid x'_t \right]' \qquad (11-2.18)$$

and where b is a $(2N + 2)$ - vector given by:

$$b = [\ 1,0,\ldots,0\ ]' \qquad (11-2.19)$$

Equation (11-2.17) may be expressed in state-space form:

$$\left.\begin{array}{l} sx_f(s) = -M^{-1}R_fx_f(s) + M^{-1}bu(s) \\[2mm] y(s) \quad = c'x_f(s) \end{array}\right\} \qquad (11-2.20)$$

where C is an $(2N + 2)$ - vector given by:

$$c = \left[0, r_{1o}, 0, \ldots, 0\right]' \qquad (11-2.21)$$

In order to validate the model, a practical Felici Bridge transducer was used to measure the response to a pulsed excitation signal with two specially constructed tubular test samples with machined grooves on their inner and outer surfaces, respectively. The appropriate sample and inspection coil parameters were then used to form the vectors and matrices in equation (11-2.20) and a dynamical system simulation program used to integrate the equations numerically. The result is shown in Figure 11-8. Close agreement between the simulation and experimental results was obtained.

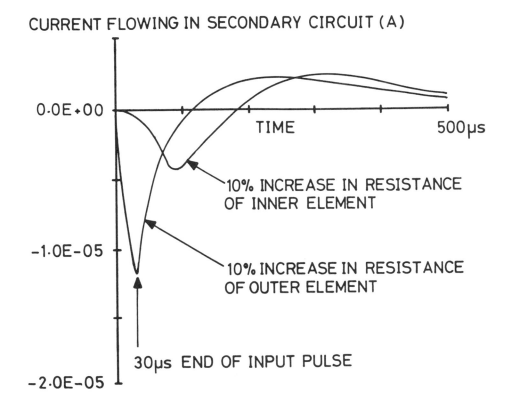

CURRENT FLOWING IN SECONDARY CIRCUIT (A)

Figure 11-8: Simulation of response of felici bridge to pulsed excitation

11.3. AUTOMATIC CLASSIFICATION OF SAMPLE DEFECTS

11.3.1 General

One way of making use of the model described in the previous section to classify sample defects would be to design a matched filter for a set of expected signals associated with various defect types. The filters could be designed to maximise the Signal to Noise Ratio and would be specific to the selected excitation signal, sample and inspection coil parameters. The defect could then be classified according to which filter most closely *matches*, in some sense, the signal arising from the sample defect. However, as we have pointed out, a desirable feature of any potential classification scheme is that it should be valid for any excitation signal. Furthermore, we have associated the occurrence of sample defects with changes in the dynamical behaviour of the sample and inspection coils. We therefore turn to a scheme which makes use of a bank of elemental state estimators (Kalman Filters) known as the Multiple Model Estimation Algorithm, (see Magill (1965), Athans and Chang (1975) and Anderson and Moore (1979)). Each Kalman Filter is matched to a hypothesised sample defect and the residuals behaviour is monitored to determine which Kalman Filter gives the closest correspondence to the defective sample and inspection coils. The algorithm produces an estimate of the system state vector (i.e. the current induced in the sample) which is generated in the form of a weighted sum of the estimates produced by the individual Kalman Filters. The algorithm also produces a set of output signals which correspond to the *a-posteriori* probabilities of the individual Kalman Filters matching the actual sample defect.

11.3.2 The multiple model estimation algorithm

The continuous time state-space model developed in Section 11.2 may be discretised by means of a standard procedure and expressed in the following form:

$$
\left. \begin{array}{l} x(k+1) = F(L)x(k) + G(L)u(k) + w(k) \\[2mm] z(k) \quad = c'x(k) + v(k) \end{array} \right\} \qquad (11\text{-}3.1)
$$

where $x(k)$ is the state vector at the k-th instant and $F(L)$, $G(L)$ and c is a suitable discrete time representation of the sample and inspection coils arrangement. L is used to model sample defects and may take values $L = L_j \in \{L_0, \ldots, L_K\}$. ($L = L_0 = I$ is used to model a defect-free sample.) Note that the $(2N + 2) \times (2N + 2)$ system of equations in equation (11-2.20) may be used to produce hypothesis models of order $(N + 2)$ prior of discretisation. For convenience we write:

$$
\left. \begin{array}{l} F_j = P(L_j) \\[2mm] G_j = G(L_j) \end{array} \right\} \qquad (11\text{-}3.2)
$$

$w(k)$ and $v(k)$ are assumed to be white noise sequences with zero mean, i.e.

$$E[w(k)] = E[v(k)] = 0$$

$$E[w(m)w'(n)] = E[v(m)v'(n)] = 0 \; ; \; m \neq n$$

(11-3.3)

The covariances of $w(k)$ and $v(k)$ are given by:

$$E[w(k)w'(k)] = U$$

$$E[v(k)v'(k)] = R$$

(11-3.4)

We construct a Kalman filter for each value of $L_j \in \{L_1, \ldots, L_k\}$ so that:

$$\hat{x}_j(k|k) = \hat{x}_j(k|k-1) + K_j(k)(z(k) - c'\hat{x}_j(k|k-1))$$

$$K_j(k) = \Sigma_j(k|k-1)c(c'\Sigma_j(k|k-1)c+R)^{-1}$$

$$\Sigma_j(k|k) = \Sigma_j(k|k-1) - K_j(k)c'\Sigma_j(k|k-1)$$

(11-3.5)

where $\hat{x}_j(k|k)$ is the current state estimate and $\hat{x}_j(k|k-1)$ is the best state estimate given information up to time k-1. $K_j(k)$ is known as the Kalman gain. $\Sigma_j(k|k)$ represents the estimation accuracy immediately after a measurement and $\Sigma_j(k|k-1)$ is valid just before a measurement. The time update equations are given by

$$\hat{x}_j(k+1|k) = F_j\hat{x}(k|k) + G_ju(k)$$

$$\Sigma_j(k+1|k) = F_j\Sigma_j(k|k)F_j' + U$$

(11-3.6)

Now let h be a hypothesis variable and let $(h_0, \ldots, h_K)$ be a set of events. The interpretation that is attached to the event $h = h_j$ is that (F_j, G_j, c) represents the sample and inspection coil arrangement. Denote the sequence of measurement $z(0), \ldots, z(k)$ by $Z(k)$. The probability that $h = h_j$ conditioned on the measurement $Z(k)$ may be written:

$$P(h=h_j|Z(k)) = P(h_j|Z(k))$$

(11-3.7)

in which $P(h_j|Z(k))$ is a convenient notation. It can be shown that (see Magill (1965)), for the scalar case:

$$P(h_j|Z(k)) = fP(h_j|Z(k-1))(\Omega(k|h_j))^{-1/2}$$

$$\times \exp\left\{-\frac{1}{2} \tilde{z}(k|h_j)^2/\Omega(k|h_j)\right\}$$

(11-3.8)

where $\tilde{z}(k|h_j)$ and $\Omega(k|h_j)$ are the innovations sequence and the

variance of the innovations sequence associated with the j-th Kalman
filter at the k-th sampling instant, f is a normalising constant
independent of h_j chosen to ensure that $\sum_{j=0}^{K} P(h|Z(k)) = 1$. The computation is organised as shown in Figure 11-9.

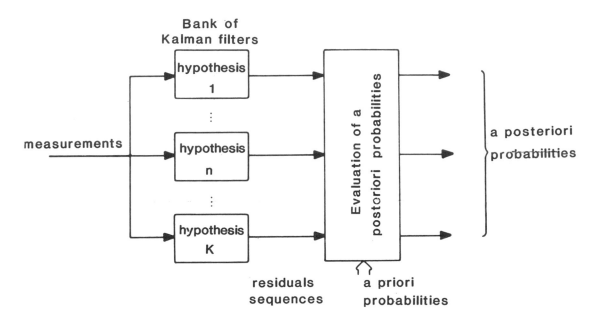

Figure 11-9: Organisation of multiple model estimation algorithm

11.3.3 Simulation results

Simulations have been performed in which the discrete time MMEA is
used to classify signals derived from a continuous time simulation of
a Felici Bridge eddy current NDT transducer. An example of the
evolution of the *a-posteriori* probabilities is shown in Figure 11-10.
The classification of sample defects by means of the MMEA has been
simulated extensively. From the results we make the following
observations about the general behaviour of the algorithm:

(a) The algorithm converges on the correct hypothesis more rapidly
    with multi-frequency and pulsed excitation than with single
    frequency excitation.

(b) The algorithm converges more rapidly with outer surface defects
    than with internal sample defects.

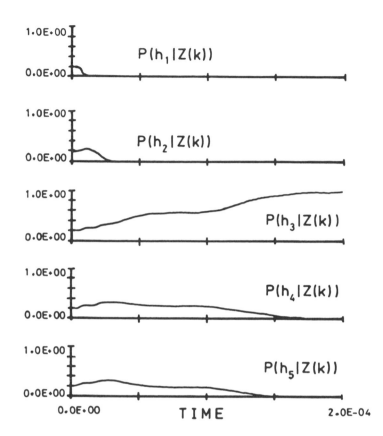

Figure 11-10: Example of evolution of *a-posteriori* probabilities

11.4. DISCUSSION AND CONCLUSIONS

A lumped parameter approximate model of a practical (Felici Bridge)
eddy current NDT transducer has been developed. It has been shown
that the model may be used to represent defective samples by
comparing the behaviour of the model with widely published experimental results. It is proposed that the model may be used to classify
sample defects automatically by incorporating it into a well known
parameter identification algorithm known as the Multiple Model
Estimation Algorithm. Simulation results show that the algorithm is
capable of discriminating between signals from a Felici Bridge trans-

ducer with various commonly used excitation signals.

The equivalent circuit model of eddy current test phenomena and its state-space representation given here have several advantages over the conventional representation based on the solution of Helmholtz's equation. These include the ability to approximately model fault conditions within the sample volume and the ability to investigate analytically the effects of a range of excitation waveforms of arbitrary complexity. In addition, by incorporating both the excitation and detection coil impedances within the model, simple optimisation strategies can be adopted to maximise the performance of a transducer configuration relative to the dimensions and electromagnetic properties of any test cylinder. A state-space representation allows the use of modern algorithms such as the Multiple Model Estimation Algorithm which offers the possibility of an entirely novel method of signal processing for eddy current NDT.

The procedure is being used primarily as a design tool with experimental realisation being restricted to off-line analysis due to bandwidth considerations. However, the Multiple Model Estimation Algorithm is an ideal candidate for parallel decomposition and studies are now being undertaken to exploit the possibilities presented by the recently available VSLI processors (which are optimised for concurrent operation) and to examine the economics of engineering a new generation of eddy current test instrumentation based on completely new methods of digital signal processing.

Chapter Twelve

A MULTIPLE MODEL ADAPTIVE FILTERING APPROACH TO FAULT DIAGNOSIS IN STOCHASTIC SYSTEMS

Keigo Watanabe

A new approach to instrument fault detection and identification (FDI) problems is developed for stochastic dynamical systems. A decentralised mutiple model adaptive filtering (MMAF) technique is first derived and then applied to detecting the sensor faults and estimating the state under the fault mode. The proposed FDI system has two-fold decision making which consists of generating the local and global probabilities for the hypothesised faults. Furthermore, it has an advantage of simply computing such probabilities, compared with the traditional MMAF approach for a case of multiple sensors. The method is applied to a simulation for the lateral dynamics for a VSTOL aircraft, in which two sensors may malfunction or fail independently or simultaneously.

12.1 INTRODUCTION

In order to design a reliable, fault-tolerant, digital control system or estimation system, it is important that system faults (e.g., sensor or actuator faults) be promptly detected and identified so that appropriate remedies can be applied. Several surveys exist which are concerned with numerous approaches to the problem of FDI in stochastic or deterministic dynamical systems. Willsky (1976) is a classic paper, discussing many of the techniques available at the time. Pau (1981), Isermann (1984) and Nakamizo (1985) survey many methods for FDI in various theories or applications. The interested reader may like to refer to Himmelblau (1976) for the fault detection and diagnosis in chemical and petrochemical processes. More recently, Basseville and Benveniste (1985) have edited a book comprising a number of papers on the detection of abrupt changes in dynamical systems.

In a variety of approaches to the design of FDI system, we can classify, for example, as follows:

(a) *Fault-sensitive filters* (Beard (1971); Jones (1973))

(b) *Innovation-based detection (and diagnosis) systems*: (i) a chi-squared test (Willsky et al. (1974, 1975); Watanabe et al. (1979, 1981); Usoro et al. (1985)), (ii) a sequential probability ratio test (SPRT) (Gai and Curry (1976); Chien and Adams (1976); Soeda et al. (1978); Yoshimura et al. (1979); Speyer and White (1984)), (iii) a generalised likelihood ratio (GLR) test (Willsky and Jones (1976); Ono et al. (1984); Kumamaru (1984))

(c) *Multiple model adaptive filters* (MMAFs) (Willsky et al. (1974, 1975); Gustafson et al. (1978a, 1978b); Willsky et al. (1980); Okita et al. (1985))

(d) *Two-stage bias correction filters* (Friedland (1979, 1981, 1982); Friedland and Grabousky (1982); Caglayan and Lancraft (1983); Watanabe (1985))

(e) *Analytical redundancy systems* (Deckert et al. (1977); Frank and Keller (1980); Clark and Setzer (1980); Watanabe and Himmelblau (1982); Chow and Willsky (1984); Patton et al. (1987)).

In this chapter we focus on the class (c). Such an FDI system involves the use of a bank of linear Kalman filters - one for each of the hypothesised fault modes and one for the normal mode. This method has the advantage of simply making a decision logic, as well as directly providing fault identification information. Moreover, it can readily compensate the control or estimation under the fault modes, apart from the substantial increases in the computation burden.

Our objective is to describe a new approach to instrument FDI problems for a system with multiple sensors by applying a decentralised multiple model adaptive filtering (MMAF) technique. The proposed FDI system has twofold decision making which consists of generating the local and global *a-posteriori* probabilities for the hypothesised faults. In addition, it has an advantage of simply computing such probabilities, compared with the usual multiple model adaptive filtering method (Lainiotis (1971, 1976)). In other words, the present method can deal with a scalar innovation process in the local processing of the parallel Kalman filters for a general case of multiple sensors.

This chapter is organised as follows. In Section 12.2, the problem statement and the essential assumptions are presented. In Section 12.3, the MMAF method is described for a local model with a single sensor, in which the sensor may take hypothesised fault modes at unknown time. This result is used in Section 12.4 to construct a decentralised MMAF and to derive the global probabilities for a linear stochastic system with multiple sensors. A numerical example is presented in Section 12.5 to illustrate the effectiveness of the proposed approach.

12.2 PROBLEM STATEMENT

Consider the following discrete-time stochastic system:

$$x(k+1) = Ax(k) + Gw(k) \qquad\qquad (12\text{-}2.1)$$

$$z_i(k) = C_i(\theta_i)x(k) + v_i(k), \quad i=1,\ldots,m \qquad\qquad (12\text{-}2.2)$$

where $x(k) \epsilon R^n$, $w(k) \epsilon R^p$, $z_i(k)$, $v_i(k) \epsilon R^l$, and A, G and C_i are assumed to have appropriate dimensions. The noises $[w(k), v_1(k), \ldots, v_m(k)]$ are assumed to be zero-mean white Gaussian processes with covariance:

$$E \left[\begin{bmatrix} w(k) \\ v_1(k) \\ \vdots \\ v_m(k) \end{bmatrix} [w^T(j), v_1(j), \ldots, v_m(j)] \right] = \text{diag }(Q, R_1, \ldots, R_m)\delta_{kj}$$

$$\qquad\qquad\qquad\qquad\qquad\qquad\qquad\qquad\qquad (12\text{-}2.3)$$

where $Q > 0$, $R_i > 0$, $i=1,\ldots,m$, δ_{kj} is the Kronecker delta function, and to be independent of the initial state $x(0)$. For the simplicity of the problem, it is also assumed that the central processor has the *a-priori* information $x(0) \sim N(0, P(0))$ and the local processors have the same initial conditions. In equation (12-2.2), θ_i denotes the unknown status of the ith sensor, where it is assumed that θ_i takes $M^{(i)}$ status so that:

$$\Omega^{(i)} \overset{\Delta}{=} [\theta_i : \theta_i = 1, \ldots, M^{(i)}] \qquad\qquad (12\text{-}2.4)$$

which is an index set. Let us introduce an m-tuple index set:

$$\Theta \overset{\Delta}{=} \{j : j = \Theta_1 \times \Theta_2 \times \ldots \times \Theta_i \times \ldots \times \Theta_m; \; \Theta_i \epsilon \; \Omega^{(i)}\} \qquad (12\text{-}2.5)$$

Notice that under these conditions Θ is the index for the permutation of unknown status in all sensors. In the following, *two* hypotheses representing unknown sensor status are defined.

(a) $\bar{H}_j$, $j \epsilon \Theta$: It is a *global* hypothesis since it concerns the status of all sensors.

(b) $H^i_{j(i)}$, $j(i) \equiv \Theta_i \epsilon \; \Omega^{(i)}$:   It is a *local* hypothesis since it only concerns the status of the i-th sensor.

Clearly, the global hypothesis can be specified by local hypotheses, i.e.,

$$H_j = (\bar{H}^1_{j(1)}, \; \bar{H}^2_{j(2)}, \ldots, \; \bar{H}^m_{j(m)}) \qquad (12\text{-}2.6)$$

where it is assumed that we can use the local *a-priori* probabilities $P^{(i)}_r(\bar{H}^i_{j(i)})$ and the global *a-priori* probabilities $P_r(H_j)$.

The purpose of this paper is as follows: Given all measurements data $Z_k \overset{\Delta}{=} \{Z^1_k, \ldots, Z^m_k\}$, where $Z^i_k \overset{\Delta}{=} \{z_i(j), 0 \leqslant j \leqslant k\}$, we desire to develop a decentralised MMAF and to apply the result for detecting sensor faults. A schematic diagram of the FDI system considered here appears in Figure 12-1.

The present approach is based on the following assumptions.

A1) The pairs (A,G) and $(C_i(\Theta_i),A)$ are stabilisable and detectable, respectively.

A2) The pairs (A,G) and $(C(j),A)$ are stabilisable and detectable, respectively, where

$$C(j) \overset{\Delta}{=} \begin{bmatrix} C_1(j(1)) \\ \vdots \\ C_m(j(m)) \end{bmatrix}, \quad \text{for } j \; \epsilon \; \Theta \qquad (12\text{-}2.7)$$

and A is nonsingular.

A3) The scalar measurements $z_i(k)$, $i=1,\ldots,m$ are mutually independent. This means that $z^1_k, \ldots, z^m_k$ are mutually independent and also $z_i(k)$ and z^j_k are mutually independent, where $i \neq j$.

A4) The scalar measurements $z_i(k)$, $i=1,\ldots,m$ are also conditionally independent given H_j and Z_{k-1}.

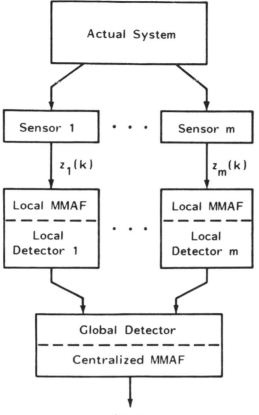

Figure 12-1: An FDI system based on a decentralised MMAF

12.3 LOCAL MULTIPLE MODEL ADAPTIVE FILTERING

12.3.1 Introduction

In this section we discuss the local MMAF for the system (2-1), (2-2). The derivation of such a filter is routine, and we refer the reader to Lainiotis (1971), Lainiotis (1976), Maybeck (1982) and Watanabe et al. (1981) and the references cited therein for a detailed development of the technique.

12.3.2 Parallel bank of Kalman filters

The conditional estimate of the state at local station i given Z_k^i and

$\bar{H}^i_{j(i)}$, i.e.,

$$\hat{x}^{(i)}(k/k,j(i)=\ell) \triangleq E[x(k)/Z^i_k,\bar{H}^i_{j(i)}], \quad i=1,\ldots,m \qquad (12\text{-}3.1)$$

can be obtained by the well-known Kalman filter:

Measurement update:

$$\hat{x}^{(i)}(k/k,\ell) = \hat{x}^{(i)}(k/k-1,\ell) + K^{(i)}(k,\ell)[z_i(k)$$

$$- C_i(\ell)\hat{x}^{(i)}(k/k-1,\ell)]$$

$$\hat{x}^{(i)}(0/-1,\ell) = 0, \quad \ell=1,\ldots,M^{(i)} \qquad (12\text{-}3.2)$$

$$K^{(i)}(k,\ell) \triangleq P^{(i)}(k/k-1,\ell)C^T_i(\ell)/[C_i(\ell)P^{(i)}(k/k-1,\ell)C^T_i(\ell) + R_i],$$

$$P^{(i)}(0/-1,\ell) = P(0) \qquad (12\text{-}3.3a)$$

or:

$$K^{(i)}(k,\ell) \triangleq P^{(i)}(k/k,\ell)C^T_i(\ell)/R_i \qquad (12\text{-}3.3b)$$

$$P^{(i)-1}(k/k,\ell) = P^{(i)-1}(k/k-1,\ell) + C^T_i(\ell)C_i(\ell)/R_i \qquad (12\text{-}3.4a)$$

or:

$$P^{(i)}(k/k,\ell) = [I - K^{(i)}(k,\ell)C_i(\ell)]P^{(i)}(k/k-1,\ell) \qquad (12\text{-}3.4b)$$

where:

$$\hat{x}^{(i)}(k/k-1,\ell) \triangleq E[x(k)/Z^i_{k-1},\bar{H}^i_{j(i)}] \qquad (12\text{-}3.5)$$

$$P^{(i)}(k/k-1,\ell) \triangleq E\{[x(k) - \hat{x}^{(i)}(k/k-1,\ell)][x(k) - \hat{x}^{(i)}(k/k-1,\ell)]^T$$

$$/Z^i_{k-1},\bar{H}^i_{j(i)}\} \qquad (12\text{-}3.6)$$

and:

$$P^{(i)}(k/k,\ell) \triangleq E\{[x(k) - \hat{x}^{(i)}(k/k,\ell)][x(k) - \hat{x}^{(i)}(k/k,\ell)]^T$$

$$/Z^i_k,H^i_{j(i)}\} \qquad (12\text{-}3.7)$$

Time update:

$$\hat{x}^{(i)}(k+1/k,\ell) = A\hat{x}^{(i)}(k/k,\ell) \qquad (12\text{-}3.8)$$

$$P^{(i)}(k+1/k,\ell) = AP^{(i)}(k/k,\ell)A^T + GQG^T \qquad (12\text{-}3.9)$$

Thus, the local processing of station i with a single sensor consists of a bank of $M^{(i)}$ parallel Kalman filters generating the scalar innovation processes:

$$\nu_i(k,\ell) \triangleq z_i(k) - C_i(\ell)\hat{x}^{(i)}(k/k-1,\ell) \qquad (12\text{-}3.10)$$

with variances:

$$\tilde{P}^{(i)}(k,\ell) \triangleq C_i(\ell)P^{(i)}(k/k-1,\ell)C_i^T(\ell) + R_i \qquad (12\text{-}3.11)$$

The local *a-posteriori* probability, which is adopted as a *local detector* for ith sensor,

$$P_r^{(i)}(\ell/k) \triangleq P_r^{(i)}(\bar{H}_{j(i)}^i = \ell/Z_k^i) \qquad (12\text{-}3.12)$$

is obtained recursively by using the outputs of the bank of Kalman filters, as:

$$P_r^{(i)}(\ell/k) = \frac{L^{(i)}(\ell/k)P_r^{(i)}(\ell/k-1)}{\displaystyle\sum_{j=1}^{M^{(i)}} L^{(i)}(j/k)P_r^{(i)}(j/k-1)} \qquad (12\text{-}3.13)$$

where:

$$L^{(i)}(\ell/k) \triangleq (2\pi)^{1/2}P_r^{(i)}(z_i(k)/Z_{k-1}^i, \bar{H}_{j(i)}^i) \quad \text{and:}$$

$$L^{(i)}(\ell/k) = (\tilde{P}^{(i)}(k,\ell))^{-1/2}\exp\left\{-\frac{1}{2}\nu_i^2(k,\ell)/\tilde{P}^{(i)}(k,\ell)\right\} \qquad (12\text{-}3.14)$$

It is well-known (Kucera (1972)) that there is a unique positive semidefinite stabilising solution to $P^{(i)}(k/k-1,\ell)$ (or $P^{(i)-1}(k/k-1,\ell)$) as $k \to \infty$ if and only if the assumption A1) in section 12.2 is satisfied. From this point of view, we assume that the local Kalman filters have converged prior to their use in the FDI system. Then, we can substantially reduce the computation burden of the $M^{(i)}$ parallel Kalman filters, when using the steady-state version of the above algorithm. In this case, the Kalman filter gains for the local station i become:

$$K^{(i)}(\ell) = P^{(i)}(\ell)C_i^T(\ell)/[C_i(\ell)P^{(i)}(\ell)C_i^T(\ell) + R_i] \qquad (12\text{-}3.15)$$

where $P^{(i)}(\ell)$ is a unique solution to the following algebraic Riccati equation:

$$P^{(i)}(\ell) = A[P^{(i)}(\ell) - P^{(i)}(\ell)C_i^T(\ell)P^{(i)}(\ell)$$

$$/[C_i(\ell)\ P^{(i)}(\ell)\ C_i^T(\ell) + R_i]]A^T + GQG^T \qquad (12\text{-}3.16)$$

The innovation variances also reduce to:

$$\tilde{P}^{(i)}(\ell) = C_i(\ell)P^{(i)}(\ell)C_i^T(\ell) + R_i \qquad (12\text{-}3.17)$$

12.3.3 Local decision-making

The local decision making can be simply constructed by choosing the local hypothesis whose probability exceeds a threshold (we have used in this paper $P_r^{(i)}(\ell/k) = 0.9$).

To avoid that one of the probabilities becomes very small (or very large), or to facilitate that the local detector responds quickly after the fault has occurred, we must set a bound on the probability, Gustafson et al. (1978a), Gustafson et al. (1978b) and Willsky et al. (1980). In this paper, a lower bound of 0.01 on $P_r^{(i)}(\ell/k)$ is set so that the upper bound results in:

$$\text{Upper bound of } P_r^{(i)}(\ell/k) = 1.0 - 0.01(M^{(i)} - 1) \qquad (12\text{-}3.18)$$

Additionally, the *a-priori* probabilities $P_r^{(i)}(\bar{H}_{j(i)=\ell}^i\ /\text{-}1) \equiv P_r^{(i)}(\bar{H}_{j(i)}^i)$, $i=1,\ldots,m$, can be taken as:

$$P_r^{(i)}(\ell/\text{-}1) = 1/M^{(i)}, \quad \ell=1,\ldots,M^{(i)} \qquad (12\text{-}3.19)$$

if there are no *a-priori* information.

12.4 DECENTRALISED MULTIPLE MODEL ADAPTIVE FILTERING

In this section, we show how the results indicated in the previous section are incorporated to construct the global filtered estimate. That is, we develop here a new decentralised multiple model adaptive filtering technique. First, the following theorem is stated.

Theorem 1 **Let:**

$$\hat{x}(k/k,j) \triangleq E[x(k)/Z_k^1,\ldots,Z_k^m, H_j] \qquad (12\text{-}4.1)$$

be the centralised estimate of x(k) given $[Z_k^1, \ldots, Z_k^m, H_j]$. Then:

$$\hat{x}(k/k,j) = \sum_{i=1}^{m} r^{(i)}(k,j) + v^{(i)}(k,j)\hat{x}^{(i)}(k/k,j(i)) \qquad (12\text{-}4.2)$$

where:

$$v^{(i)}(k,j) \triangleq P(k/k,j)P^{(i)-1}(k/k,j(i)) \qquad (12\text{-}4.3)$$

$$P^{-1}(k/k,j) = P^{-1}(k/k-1,j) + \sum_{i=1}^{m} C_i^T(j(i))C_i(j(i))/R_i,$$

$$P(0/-1,j) = P(0) \qquad (12\text{-}4.4)$$

$$P(k+1/k,j) = AP(k/k,j)A^T + GQG^T \qquad (12\text{-}4.5)$$

in which $P(k/k,j)$ and $P(k/k-1,j)$ are the measurement updated and time updated centralised estimation error covariance conditioned on H_j. Furthermore, the compensator $r^{(i)}(k,j)$ is subject to:

$$r^{(i)}(k,j) = A*(k,j)r^{(i)}(k-1,j) + K*^{(i)}(k,j)\hat{x}^{(i)}(k/k-1,j(i)),$$

$$r^{(i)}(-1,j) = 0 \qquad (12\text{-}4.6)$$

where:

$$A*(k,j) \triangleq P(k/k,j)P^{-1}(k/k-1,j)A \qquad (12\text{-}4.7)$$

$$K*^{(i)}(k,j) \triangleq A*(k,j)P(k-1/k-1,j)P^{(i)-1}(k-1/k-1,j(i))A^{-1}$$

$$- P(k/k,j)P^{(i)-1}(k/k-1,j(i)) \qquad (12\text{-}4.8)$$

Proof: The proof is adapted from the result of usual decentralised Kalman Filter (KF), Speyer (1979) and Watanabe (1986). Given all measurements data $[Z_k^1, \ldots, Z_k^m]$ and the hypothesis H_j, the global filtered estimate $\hat{x}(k/k,j)$ is given by:

$$\hat{x}(k/k,j) = \hat{x}(k/k-1,j) + P(k/k,j) \sum_{i=1}^{m} C_i^T(j(i))[z_i(k)$$

$$- C_i(j(i))\hat{x}(k/k-1,j)]/R_i,$$

$$\hat{x}(0/-1,j) = 0 \qquad (12\text{-}4.9)$$

$$P^{-1}(k/k,j) = P^{-1}(k/k-1,j) + \sum_{i=1}^{m} C_i^T(j(i))C_i(j(i))/R_i,$$

$$P(0/-1,j) = P(0) \qquad (12\text{-}4.10)$$

$$\hat{x}(k+1/k,j) = A\hat{x}(k/k,j) \qquad (12\text{-}4.11)$$

$$P(k+1/k,j) = AP(k/k,j)A^T + GQG^T \qquad (12\text{-}4.12)$$

The corresponding information type filter, i.e., $d(k,j) \triangleq P^{-1}(k/k,j)$ $\hat{x}(k/k,j)$ reduces to:

$$d(k,j) = [I - D(k)G^T]A^{-T}d(k-1,j) + \sum_{i=1}^{m} C_i^T(j(i))z_i(k)/R_i,$$

$$d(-1,j) = 0 \qquad\qquad (12-4.13)$$

where:

$$D(k) \triangleq NG(G^T N G + Q^{-1})^{-1}, \quad N \triangleq A^{-T}P^{-1}(k-1/k-1,j)A^{-1} \qquad (12-4.14)$$

The local information type filter corresponding to (12-3.2) is also reduced to:

$$d^{(i)}(k,j(i)) = [I - D^{(i)}(k)G^T]A^{-T}d^{(i)}(k-1,j(i))$$

$$+ C_i^T(j(i))z_i(k)/R_i,$$

$$d^{(i)}(-1,j(i)) = 0 \qquad\qquad (12-4.15)$$

where:

$$D^{(i)}(k) \triangleq N^{(i)}G(G^T N^{(i)} G + Q^{-1})^{-1}$$

$$N^{(i)} \triangleq A^{-T}P^{(i)-1}(k-1/k-1,j(i))A^{-1} \qquad (12-4.16)$$

Replacing the closed-loop matrix $[I - D^{(i)}(k)G^T]$ in (12-4.15) by the global closed-loop one $[I - D(k)G^T]$ and superposing it and (12-4.13), it is easily seen that:

$$q(k,j) = d(k,j) - \sum_{i=1}^{m} d^{(i)}(k,j(i)) \qquad\qquad (12-4.17)$$

which satisfies:

$$q(k,j) = [I - D(k)G^T]A^{-T}q(k-1,j) + \sum_{i=1}^{m} [D^{(i)}(k) - D(k)]$$

$$\times G^T A^{-T}d^{(i)}(k-1,j(i)),$$

$$q(-1,j) = 0 \qquad\qquad (12-4.18)$$

Defining:

$$r(k,j) \triangleq P(k/k,j)q(k,j) \qquad\qquad (12-4.19a)$$

$$= \hat{x}(k/k,j) - P(k/k,j) \sum_{i=1}^{m} P^{(i)-1}(k/k,j(i))\hat{x}^{(i)}(k/k,j(i))$$

$$(12-4.19b)$$

and substituting (12-4.18) into (12-4.19a) yields:

$$r(k,j) = P(k/k,j)[I - D(k)G^T]A^{-T}P^{-1}(k-1/k-1,j)r(k-1,j)$$
$$+ P(k/k,j) \sum_{i=1}^{m} [D^{(i)}(k) - D(k)]G^TA^{-T}P^{(i)-1}(k-1/k-1,j(i))$$
$$\times \hat{x}^{(i)}(k-1/k-1,j(i)) \qquad (12\text{-}4.20)$$

The equation (12-4.2) is apparent from (12-4.19b), where the compensator $r(k,j)$ is written as $r(k,j) = \sum_{i=1}^{m} r^{(i)}(k,j)$, in which $r^{(i)}(k,j)$ consisting of equations (12-4.6)-(12-4.8) can be derived by some manipulations of (12-4.20) (see Watanabe (1986) for the details).

When the assumptions A2) are satisfied, there exists a unique stabilising solution to $P(k/k-1,j)$ (or $P^{-1}(k/k-1,j)$) as $k \to \infty$. Therefore, taking account of assumptions A1) and A2) we get the following corollary.

Corollary 1: If assumptions A1) and A2) of Section 12.2 are satisfied, matrices $V^{(i)}(k,j)$, $A*(k,j)$ and $K*^{(i)}(k,j)$ for $i=1,\ldots,m$ and $Vj \in \Theta$ result in

$$V^{(i)}(j) = P_f(j)P_f^{(i)-1}(j(i)) \qquad (12\text{-}4.21)$$

$$A*(j) = P_f(j)P^{-1}(j)A \qquad (12\text{-}4.22)$$

$$K*^{(i)}(j) = A*(j)V^{(i)}(j)A^{-1} - P_f(j)P^{(i)-1}(j(i)) \qquad (12\text{-}4.23)$$

where $P_f(j)$ and $P_f^{(i)}(j(i))$ are unique solutions to:

$$P_f(j) = [P^{-1}(j) + \sum_{i=1}^{m} C_i^T(j(i))C_i(j(i))/R_i]^{-1} \qquad (12\text{-}4.24)$$

$$P_f^{(i)}(j(i)) = [P^{(i)-1}(j(i)) + C_i^T(j(i))C_i(j(i))/R_i]^{-1} \qquad (12\text{-}4.25)$$

Here, $P(j)$ is a unique stabilising solution to the following algebraic Riccati equation:

$$P(j) = A \left\{ P^{-1}(j) + \sum_{i=1}^{m} C_i^T(j(i))C_i(j(i))/R_i \right\}^{-1}A^T + GQG^T \qquad (12\text{-}4.26)$$

We further obtain algorithms for computing the global (centralised) filtering estimate and the global *a-posteriori* probability.

Theorem 2: Given all measurement data $[Z_k^1, \ldots, Z_k^m]$, the centralised estimate $\hat{x}(k/k) \triangleq E[x(k)/Z_k^1, \ldots, Z_k^m]$ is given by:

$$\hat{x}(k/k) = \sum_{j \in \Theta} \hat{x}(k/k,j) P_r(j/k) \qquad (12\text{-}4.27)$$

and the associated error covariance matrix $P(k/k) \triangleq E\{[x(k) - \hat{x}(k/k)] [x(k) - \hat{x}(k/k)]^T / Z_k^1, \ldots, Z_k^m\}$ is also provided by:

$$P(k/k) = \sum_{j \in \Theta} \{P(k/k,j) + [\hat{x}(k/k,j) - \hat{x}(k/k)][\hat{x}(k/k,j) - \hat{x}(k/k)]^T\}$$

$$\times P_r(j/k) \qquad (12\text{-}4.28)$$

where the global *a-posteriori* probability $P_r(j/k) \triangleq P_r(H_j/Z_k)$, which is adopted as the *global detector*, can be reconstructed using the recursive algorithm:

$$P_r(j/k) = \frac{\prod_{i=1}^{m} L^{(i)}(j(i)/k) P_r(j/k-1)}{\sum_{\ell \in \Theta} \prod_{i=1}^{m} L^{(i)}(\ell(i)/k) P_r(\ell/k-1)} \qquad (12\text{-}4.29)$$

Proof: The results of equations (12-4.27) and (12-4.28) follow directly from those of the well-known MMAF (Lainiotis (1971, 1976); Maybeck (1982); Watanabe et al. (1981), etc.). Therefore, we prove only the result of (12-4.29). Using the Bayes' rule we obtain:

$$P_r(H_j/Z_k) = P_r(H_j, Z_k)/P_r(Z_k)$$

$$= \frac{P_r(z_1(k), \ldots, z_m(k)/H_j, Z_{k-1}) P_r(H_j/Z_{k-1})}{P_r(z_1(k), \ldots, z_m(k)/Z_{k-1})}$$

$$= \frac{P_r(z_1(k), \ldots, z_m(k)/H_j, Z_{k-1}) P_r(H_j/Z_{k-1})}{\sum_{\ell \in \Theta} P_r(z_1(k), \ldots, z_m(k)/H_\ell, Z_{k-1}) P_r(H_\ell/Z_{k-1})} \qquad (12\text{-}4.30)$$

Now, taking account of the assumption A4) gives:

$$P_r(z_1(k), \ldots, z_m(k)/H_j, Z_{k-1}) = \prod_{i=1}^{m} P_r^{(i)}(z_i(k)/H_j, Z_{k-1}) \qquad (12\text{-}4.31)$$

Furthermore, using the assumption A3), it follows that:

$$P_r(z_1(k),\ldots,z_m(k)/H_j,Z_{k-1}) = \prod_{i=1}^{m} P_r^{(i)}(z_i(k)/H_{j(i)}^i,Z_{k-1}^i)$$

$$= (2\pi)^{-m/2} \prod_{i=1}^{m} L^{(i)}(j(i)/k) \qquad (12\text{-}4.32)$$

Substituting (12-4.32) into (12-4.30) completes the proof. It should be noted that the present algorithm for computing the global (or centralised) probabilites is much simpler than that suggested in recently by Castanon and Teneketzis (1985) or Teneketzis et al. (1985). This is due to utilising the probability density function $L^{(i)}(j(i))$ for the local innovation under the assumptions A3) and A4), instead of using the local *a-posteriori* probabilities $P_r^{(i)}(j(i)/k)$ to obtain the global *a-posteriori* probability.

Global decision making

The global decision making used in this paper consists of two stages: *preparatory decision* and *final decision*. The preparatory decision rule is simply to choose the global hypothesis whose probability exceeds a threshold (e.g., $P_r(j/k) = 0.9$). This can be written as follows:
1) The hypothesis H_j is true if $P_r(j/k) \geqslant 0.9$.
The final decision rule can be written as:
2) The hypothesis H_j is true only if $P_r(j/k) \geqslant 0.9$ and all $P_r^{(i)}(j(i)/k) \geqslant 0.9$, $i = 1,\ldots,m$.
This final decision rule is sufficient for detecting and identifying a fault or failure among the m sensors. In practice, however, there may exist a case when the final decision rule is not accomplished. Furthermore, there may exist a time-delay between the preparatory decision and the final decision times.

As previously discussed in the problem of the local MMAF, assuming a lower bound of $P_r(j/k)$ is 0.01, we have:

$$\text{Upper bound of } P_r(j/k) = 1.0 - 0.01(M-1) \qquad (12\text{-}4.33)$$

where M denotes the total number of global hypotheses. In addition, if we have no *a-priori* information, we can choose

$$P_r(j/-1) \equiv P_r(H_j) = 1/M, \quad j = 1,\ldots,M \qquad (12\text{-}4.34)$$

12.5 SIMULATION STUDIES

In this section we apply the proposed FDI scheme to the estimation of a VSTOL aircraft (Wagdi (1982)) whose discrete-time lateral dynamics may be represented by (12-2.1), where:

$$A = \begin{bmatrix} 0.9949 & 0.0 & -0.1 & 0.0238 \\ -0.0737 & 0.08665 & 0.0369 & 0.001 \\ 0.001 & 0.0107 & 0.9668 & 0 \\ 0 & 0.1 & 0.0 & 0 \end{bmatrix},$$

$$G = \begin{bmatrix} 1 & 0 & 0 \\ 0 & 1 & 0 \\ 0 & 0 & 1 \\ 0 & 0 & 0 \end{bmatrix}, \quad x^T = [\beta, \ p, \ r, \ \Phi] \qquad (12\text{-}5.1)$$

β[rad] is the sideslip angle, p[rad/sec] is the roll velocity, r[rad/sec] is the yaw velocity and Φ[rad] is the roll angle. The observation system consists of two sensors: a single-axis roll rate gyro and a vertical gyro. In such a case, each observation matrix becomes:

$C_1 = [0 \ \ 1 \ \ 0 \ \ 0],$

$C_2 = [0 \ \ 0 \ \ 0 \ \ 1] \qquad\qquad\qquad\qquad\qquad\qquad (12\text{-}5.2)$

The system noise covariance matrix is taken as $Q = \text{diag}(0.01, 0.008, 0.01)$ and the measurement noise variances are $R_1 = R_2 = 0.0001$. The initial state of the system is also assumed to be $x^T(0) = [0, 0, 0, 0.1]$.

For the present simulation, it is assumed that each sensor takes two statuses: a normal status and a fault status. In this case, we set:

$\bar{H}_1^1 : C_1(\theta_1{=}1) = [0 \ \ 1 \ \ 0 \ \ 0], \quad \bar{H}_2^1 : C_1(\theta_1{=}2) = [0 \ \ 1.5 \ \ 0 \ \ 0]$

$\bar{H}_1^2 : C_2(\theta_2{=}1) = [0 \ \ 0 \ \ 0 \ \ 1], \quad \bar{H}_2^2 : C_2(\theta_2{=}2) = [0 \ \ 0 \ \ 0 \ \ 2] \quad (12\text{-}5.3)$

so that the global hypotheses become:

$$H_1 = (\bar{H}_1^{-1}, \bar{H}_1^{-2})$$

$$H_2 = (\bar{H}_2^{-1}, \bar{H}_1^{-2})$$

$$H_3 = (\bar{H}_1^{-1}, \bar{H}_2^{-2})$$

$$H_4 = (\bar{H}_2^{-1}, \bar{H}_2^{-2}) \qquad\qquad (12\text{-}5.4)$$

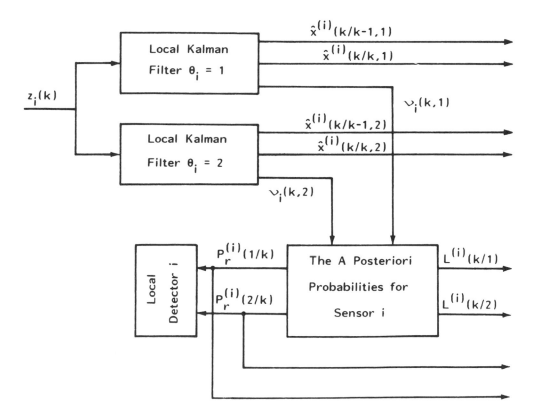

Figure 12-2(a): Block diagram of a local MMAF for the sensor i in the given example, where i = 1, 2

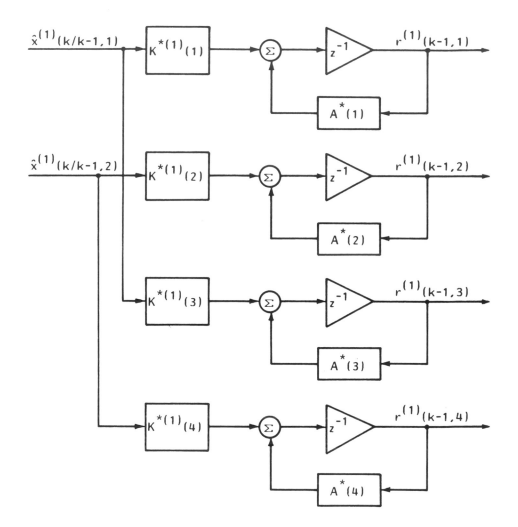

Figure 12-2(b): Block diagram of a compensator 1 in the given example

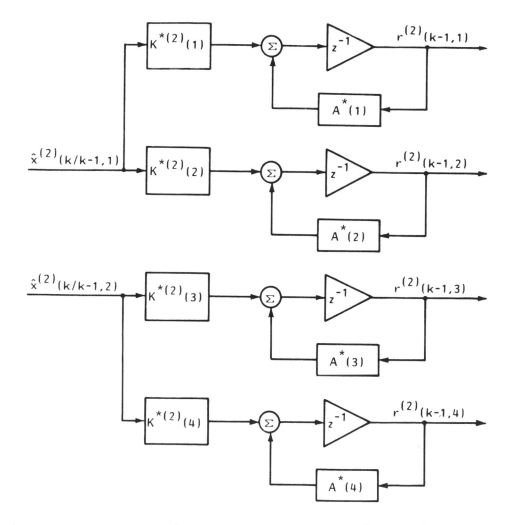

Figure 12-2(c): Block diagram of a compensator 2 in the given example

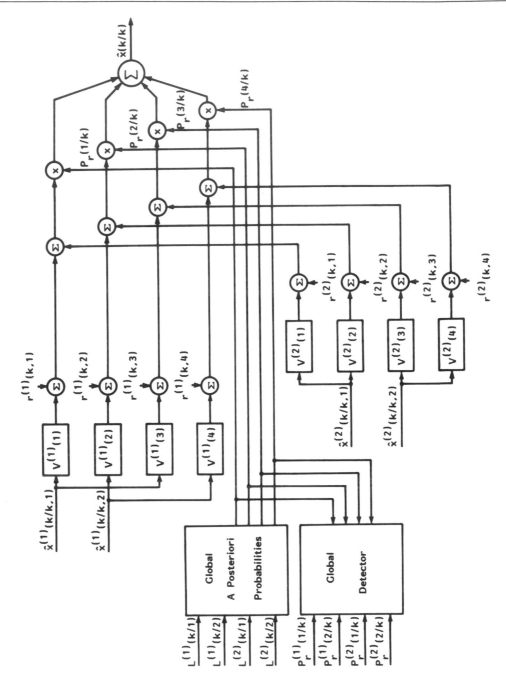

Figure 12-2(d): Block diagram of a decentralised MMAF in the given example

Figure 12-2 shows the block diagram of the decentralised MMAF for this example. Note that the pair (A, G) is stabilisable, and all pairs ($C_i(\ell)$, A) and (C(j), A) are detectable, where $i = 1,2$, $\ell = 1,2$ and $j = 1,\ldots,4$, because the dynamic system matrix A is asymptotically stable (see Watanabe (1986)). Therefore, we see that all stabilising solutions exist uniquely for all one-step predicted or filtered error covariances, and hence we can use the steady-state version of the theory presented in the preceding sections. We adopted here the doubling algorithm (Sidhu and Bierman (1977); Watanabe and Iwasaki (1983)) to obtain such steady-state solutions. Thus we have the following steady-state one-step predicted error covariances and gains for local filters:

$$P^{(1)}(1) = \begin{bmatrix} 0.15779 & -0.12245\times10^{-1} & -0.41983\times10^{-1} & -0.13895\times10^{-4} \\ & 0.90872\times10^{-2} & 0.58347\times10^{-2} & 0.20737\times10^{-5} \\ & & 0.99111\times10^{-1} & 0.62326\times10^{-5} \\ & \text{symmetric} & & 0.98912\times10^{-6} \end{bmatrix}$$

$$P^{(1)}(2) = \begin{bmatrix} 0.15727 & -0.12195\times10^{-1} & -0.41834\times10^{-1} & -0.61919\times10^{-5} \\ & 0.90821\times10^{-2} & 0.58187\times10^{-2} & 0.92549\times10^{-6} \\ & & 0.99044\times10^{-1} & 0.27809\times10^{-5} \\ & \text{symmetric} & & 0.44228\times10^{-6} \end{bmatrix}$$

$$P^{(2)}(1) = \begin{bmatrix} 0.25926 & -0.22380\times10^{-1} & -0.70272\times10^{-1} & -0.20397\times10^{-2} \\ & 0.10156\times10^{-1} & 0.89416\times10^{-2} & 0.25956\times10^{-3} \\ & & 0.11045 & 0.77488\times10^{-3} \\ & \text{symmetric} & & 0.98156\times10^{-4} \end{bmatrix}$$

$$P^{(2)}(2) = \begin{bmatrix} 0.19803 & -0.16923\times10^{-1} & 0.55885\times10^{-1} & -0.14855\times10^{-2} \\ & 0.96667\times10^{-2} & 0.75893\times10^{-2} & 0.20965\times10^{-3} \\ & & 0.10510 & 0.63340\times10^{-3} \\ & \text{symmetric} & & 0.92940\times10^{-4} \end{bmatrix}$$

$$K^{(1)}(1) = [-1.3328,\ 098912,\ 0.63509,\ 0.22572\times10^{-3}]^T$$

$$K^{(1)}(2) = [-0.89083,\ 0.66342,\ 0.42504,\ 0.67605\times10^{-4}]^T$$

$$K^{(2)}(1) = [-10.293,\ 1.3099,\ 3.9104,\ 0.49535]^T$$

$$K^{(2)}(2) = [-6.2975,\ 0.88879,\ 2.6852,\ 0.39401]^T$$

Similarly, we can obtain the steady-state matrices, $V^{(i)}(j)$, $A*(j)$ and $K*^{(i)}(j)$, $i = 1,2$ and $j = 1,\ldots,4$, which are necessary for constructing the centralised estimates, but we omit the presentation of those results because of the limit of space.

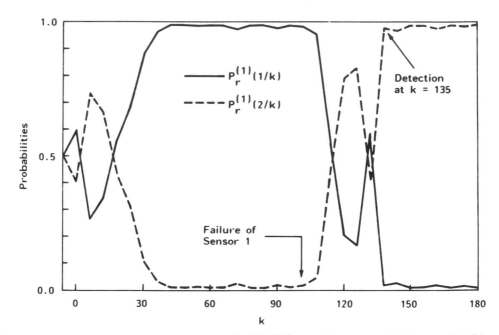

Figure 12-3: Local *a-posteriori* probabilities of sensor 1 for a fault of sensor 1

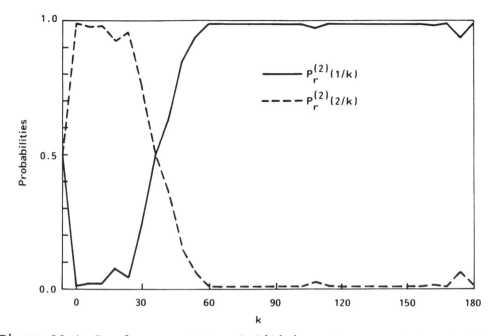

Figure 12-4: Local *a-posteriori* probabilities of sensor 2 for fault of sensor 1

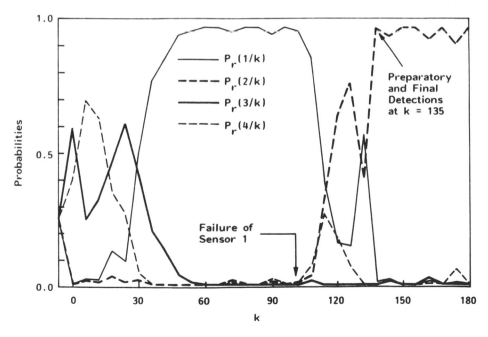

Figure 12-5: Global *a-posteriori* probabilities for the fault of sensor 1

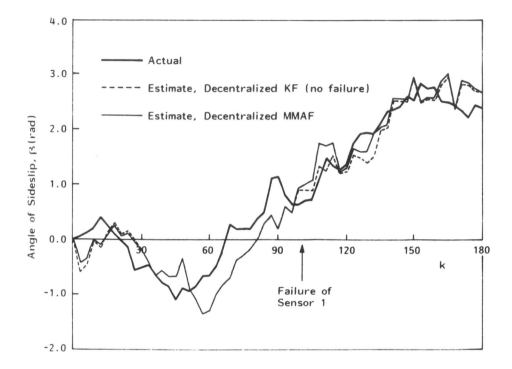

Figure 12-6: Estimate of angle of sideslip for the fault of sensor 1

Figures 12-3 - 12-5 show the results with a malfunction in sensor 1
at k = 100. The fault is simultaneously detected and identified by
both the local and global detectors at step k = 135. In addition, it
is seen from Figure 12-6 that the estimate due to the decentralised
MMAF promptly converges to that due to the decentralised Kalman
filter (KF) for a case of no faults, after the fault has occurred.

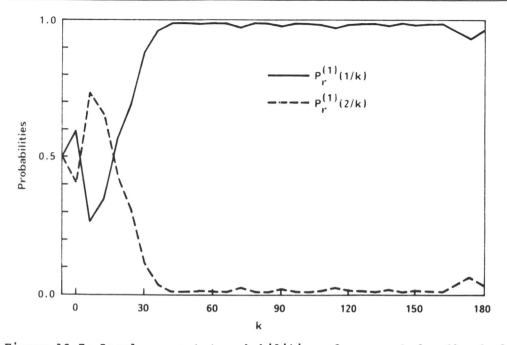

Figure 12-7: Local *a-posteriori* probabilities of sensor 1 for the fault
    of sensor 2

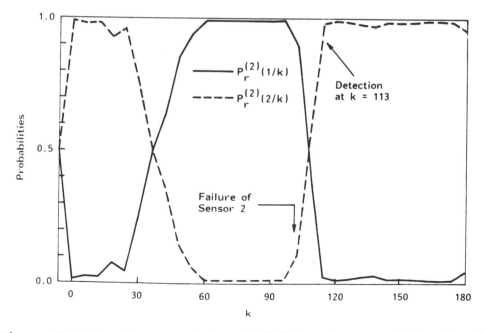

Figure 12-8: Local *a-posteriori* probabilities of sensor 2 for the fault
    of sensor 2

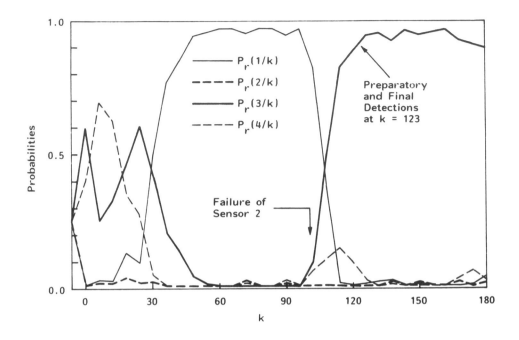

Figure 12-9: Global *a-posteriori* probabilities for the fault of sensor 2

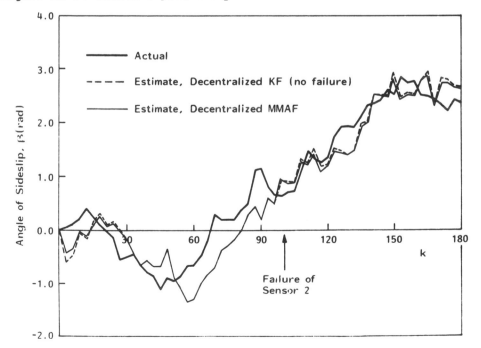

Figure 12-10: Estimate of angle of sideslip for the fault of sensor 2

The results obtained for a case when the sensor 2 failed at k = 100 are given in Figures 12-7 - 12-9, and show that the local detector 2 detects the fault at step k = 113, but the preparatory and final decisions in the global detector are made simultaneously at step k = 123. The result for the estimate of sideslip angle is shown in Figure 12-10.

The results in Figures 12-11 to 12-14 were obtained for a case when the two sensors had a simultaneous fault onset at k = 100. For this case, it should be noted that there exists a time-delay between the preparatory decision and the final one in the global detector: the local detector 2 detects the fault at step k = 113 and the preparatory decision in the global detector is made at the same time, whereas the local decision of sensor 1 and the final decision of the global detector are made simutaneously at step k = 135.

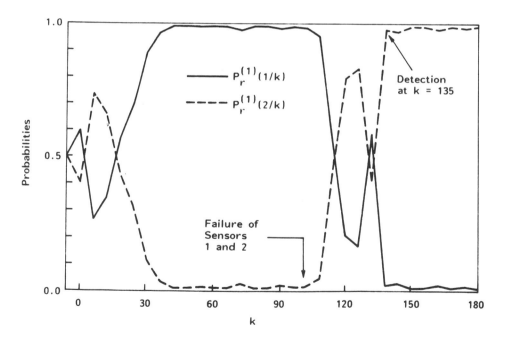

Figure 12-11: Local *a-posteriori* probabilities of sensor 1 for the fault of sensor 1 and 2

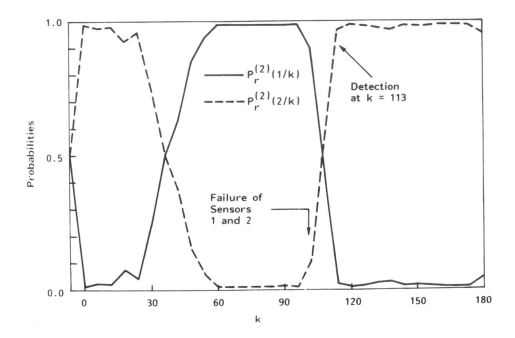

Figure 12-12: Local *a-posteriori* probabilities of sensor 2 for the fault of sensors 1 and 2

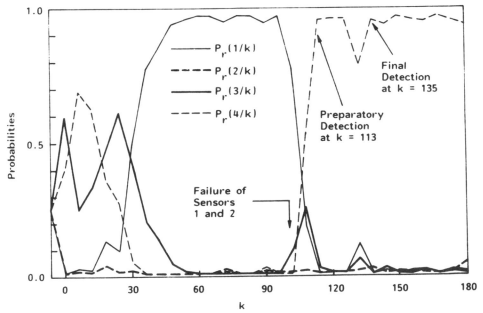

Figure 12-13: Global *a-posteriori* probabilities for the fault of sensors 1 and 2

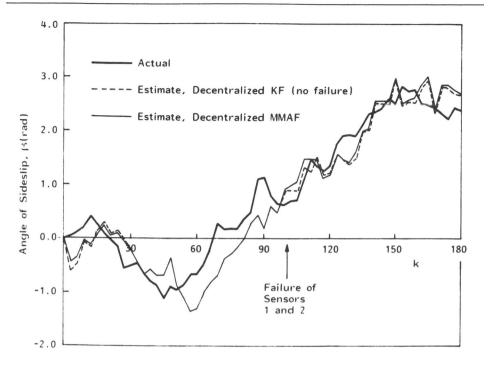

Figure 12-14: Estimate of angle of sideslip for the fault of sensors
1 and 2

12.6 CONCLUSIONS

A new approach has been introduced for fault detection and
identification (FDI) problems in stochastic dynamical systems. A
decentralised multiple model adaptive filtering technique has been
used to derive a decision rule for checking the operational modes of
multiple sensors and to estimate the state under the normal and fault
modes. The proposed methodology has been applied to a practical
example, and the results indicate that it is a useful algorithm for
FDI problems.

Although we confined our attention to the problem of sensor fault
detection and identification for a stochastic system with no
controllers, we can extend the result to a case of detecting actuator
and/or sensor faults or failures for a stochastic system with some
controllers. Moreover, note that the present approach can be applied
naturally to the FDI problems for essentially distributed systems and

that the *unknown parameter* θ_i can be modelled as a Markov chain with known or unknown transition statistics (Tugnait (1982); Matthews and Tugnait (1983)).

ACKNOWLEDGEMENTS

The author wishes to express his gratitude to Professor Takashi Soeda, President, and Professor Toshio Yoshimura of the University of Tokushima for their valuable encouragement.

Chapter Thirteen
SOME STATISTICAL METHODS
FOR FAULT DIAGNOSIS
FOR DYNAMICAL SYSTEMS

K. Kumamaru, S. Sagara and T. Söderström

In this paper two statistical methods for fault diagnosis are
proposed based on the state-space and input-output models, respectively. For systems with known dynamics a hierarchical two-level
diagnosis technique is developed. In the first level a parameter
monitoring scheme for on-line fault detection is introduced. After a
fault declaration in the first level diagnosis, the faulty signal
contents are examined in the second level using a generalised
likelihood ratio (GLR) test. The fault detection and identification
(FDI) can be implemented in a systematic way for black-box type
systems, general input-output model structures are used for system
identification. A fault in the system will be reflected as a
difference between two identified models by using data from distinct
time intervals. The Kullback discrimination information (KDI) is
introduced for model discrimination. Several new criteria for fault
detection are then derived from the KDI analysis. Both approaches to
statistical fault diagnosis are discussed in detail. Several
simulation studies on some low order systems are carried out to
demonstrate the effectiveness of the methods.

13.1 INTRODUCTION

An appropriate diagnosis is essential for system maintenance in many
control applications. From a technical viewpoint, it seems that the
realisability of a practical diagnosis system has been fairly
enhanced by the remarkable development of computer processing
technology. Until now many kinds of fault diagnosis methods have been
proposed (see the survey papers Willsky (1976), Isermann (1984) and

the book by Basseville and Benveniste (1985)).

The design scheme of fault diagnosis system has a different aspect depending on what kinds of models are used for system and fault mode descriptions. One way of system representation is the state-space modelling for the object with known dynamics including real physical parameters. In such a system most of the fault modes might be modelled by abrupt changes in the system configuration parameters and/or inputs. Some filtering approaches can be applied to information processing for joint detection, identification and estimation of the fault. The observer design method (Hertel and Clark (1982), Frank and Keller (1980)) and the generalised likelihood ratio (GLR) test (Willsky and Jones (1976)) are considered to be typical ones out of many more based on such model. Recently a redundancy analysis has been developed for robust fault diagnosis (Chow and Willsky (1984), Lou et al. (1986), Desai and Ray (1984)). Anyway when fault diagnosis is based on state-space modelling, *it is a crucial problem to develop a systematic fault detection and identification which can be implemented as an on-line procedure.* On the other hand, for black-box type systems some input-output models (like ARMAX models) are used for system description. Such models are commonly used in process control and other applications where lack of insight prevents use of a state-space model parametrised with physical parameters. According to this type of modelling, the fault will be reflected in the model parameter changes and can be estimated by using system identification. However, it is no longer possible to relate explicitly the changing modes to real fault or failure phenomena, since the malfunction due to a change in one physical component will usually induce several model parameters' changes. Thus the diagostic function for such a system becomes only a fault detecting one for the alarm declaration. In this case, *an important subject is how to construct an effective detection index which is sensitive to the fault.* From these points of view we propose here some statistical methods for fault diagnosis of dynamical systems based on state-space and input-output modellings, respectively.

Firstly, for systems which can be described explicitly by a state-space model, we develop a hierarchical diagnosis technique for systematic fault detection and identification (FDI). The diagnosis system is composed of a two-level decision structure. To solve the difficulties in recognising the faulty component among many candidates, a parameter monitoring scheme is proposed for on-line fault detection in the first level diagnosis. After the fault detection the faulty signal contents are examined in the second level diagnosis using a GLR test based on *an oriented fault modelling* (Kumamaru et al. (1983)). This fault modelling approach has made it possible to perform the GLR test in the framework of linear filtering theory. The second level diagnosis may be carried out by an off-line procedure, because a rather large amount of computational labour is required for a precise diagnosis. These two-level diagnosis structures are combined in a hierarchical way by using, in the second level, the information obtained from first level diagnosis results.

Next, a method of fault detection is proposed for systems with *unknown dynamics*. General input-output model structures are considered for system identification. A fault in the system will then be reflected in the change of identified models. Thus the fault detection problem leads to model discrimination which is performed as a batch procedure. For this purpose we introduce an effective detection index using Kullback discrimination information (KDI). The KDI is known as a distortion measure for comparing two probability density functions, see Kullback (1959). This index has extensive applications to information processing for Gaussian systems. Applications have been made for seismic, speech, and EEG data (Shumway and Unger (1974), Gray et al. (1980), Ishii et al. (1980)). We modify the KDI to be applied to likelihood functions for two models identified by using input-output data from distinct time intervals. In order to evaluate the KDI in a feasible way for finite but fairly large data sets, a recursive calculation algorithm is derived based on a Bayesian approach. As a result several new criteria for model discrimination are obtained. These criteria give meaningful information about the model structures and the noise statistics. They can hence be used effectively for fault detection as well as for model cross-validation (Söderström and Kumamaru (1985a), (1985b)) in the thresholding approach. Furthermore, for a reasonable selection of a threshold value, the statistical properties of the criteria are analysed based on the asymptotic properties of model parameter estimates.

The chapter consists of three main sections organised as follows. In Section 13.2, we will discuss the statistical background which are introduced in our approach to fault diagnosis. Section 13.3 is concerned with the hierarchical diagnosis of linear systems described by state-space models. The system and the fault modelling are given in Section 13.2 for the statement of the problem. Section 13.3 is devoted to design of a hierarchical diagnosis system with new ideas, which include a parameter monitoring scheme and an oriented fault modelling. In Section 13.4 simulation results for a second order system are given. In Section 13.4, we propose a fault detection method using the KDI. Some basic notation and assumptions are introduced in section 13.4.1. Section 13.4.2 contains an evaluation of the KDI and several new criteria for fault detection are derived. The statistical properties of the criteria are analysed in section 13.4.3. In section 13.4.4 simulation studies on a damped oscillator are shown and discussed. Finally both approaches are summarised in the conclusions.

13.2 STATISTICAL BACKGROUND

In this part we will discuss the statistical background which is introduced in our approach to fault diagnosis. We will also frequently use the concept of hypothesis testing. For this subject and its use in statistical decision problem, see Fukunaga (1979).

13.2.1 Kullback discrimination information

The Kullback discrimination information (KDI) is basically a measure to compare two probability density functions $p_1(x)$ and $p_2(x)$. It is given by (see Kullback (1959)):

$$I[1,2] = \int p_1(x) l_n \frac{p_1(x)}{p_2(x)} \, dx \geqslant 0 \qquad\qquad (13\text{-}2.1)$$

where equality holds if and only if $p_1(x) = p_2(x)$. We can hence use the value of $I[1,2]$ as a measure of how much $p_1(x)$ deviates from $p_2(x)$. We can also introduce a symmetric form by considering $J[1,2] = I[1,2] + I[2,1]$, which is called *divergence*. When $p_i(x)$ is an n-dimensional Gaussian distribution function with mean vector M_i and covariance matrix $\Sigma_i (i=1,2)$:

$$p_i(x) = N(M_i, \Sigma_i) \tag{13-2.2}$$

we have the form of I [1,2] as:

$$I[1,2] = \frac{1}{2}\mathrm{tr}(\Sigma_2^{-1}\Sigma_1 - I) + \frac{1}{2}(M_1 - M_2)^T \Sigma_2^{-1}(M_1 - M_2)$$

$$- \frac{1}{2}\ln(\det\Sigma_1/\det\Sigma_2) \tag{13-2.3}$$

13.2.2 Jensen difference measure

Consider n-dimensional multinominal distributions x, y:

$$x = (x_1, x_2, \ldots, x_n), \quad x_i \geqslant 0, \quad \sum_{i=1}^{n} x_i = 1$$

$$\tag{13-2.4}$$

$$y = (y_1, y_2, \ldots, y_n), \quad y_i \geqslant 0, \quad \sum_{i=1}^{n} y_i = 1$$

Introduce the Shannon entropy function:

$$H_n(x) = - \sum_{i=1}^{n} x_i \ln(x_i) \tag{13-2.5}$$

Then the Jensen difference is defined by:

$$J_n(x,y) = H_n(\frac{x+y}{2}) - \frac{1}{2}[H_n(x) + H_n(y)] \tag{13-2.6}$$

$J_n(x,y)$ is non-negative and vanishes if and only if x=y. Thus it provides a natural measure of divergence between the distributions x and y. A natural extension of equation (13-2.6) to a mixture of k distributions y_1, y_2, $\ldots$, y_k with a vector of *a priori* weights π = $(\pi_1, \pi_2, \ldots, \pi_k)$ is given by:

$$J_n^\pi(y_1, y_2, \ldots, y_k) = H_n(\sum_{i=1}^{k} \pi_i y_i) - \sum_{i=1}^{k} \pi_i H_n(y_i) \tag{13-2.7}$$

where y_1, y_2, $\ldots$, y_k are the n-dimensional distributions as defined in (13-2.4). J_n^π is called *generalised Jensen difference*. For detailed discussion on the properties of the Jensen difference measure, see Burbea and Rao (1982).

13.2.3 Parallel Kalman filters

Consider a discrete-time linear system:

$$x_{k+1} = F(\theta)x_k + w_k \tag{13-2.8}$$

$$y_k = H(\theta)x_k + v_k \qquad k=1,2,\ldots \tag{13-2.9}$$

where x_k is the nx1 state vector, y_k is the mx1 observation vector (m≤n), w_k and v_k are the mutually independent white Gaussian noises with:

$$w_k \in N(0,Q_k), \quad v_k \in N(0,R_k) \tag{13-2.10}$$

$F(\theta)$ and $H(\theta)$ are the nxn transition and the mxn observation matrices, which depend on θ, the *scalar-valued* unknown parameter. The admissible region of θ is assumed to be known as:

$$\theta \in [\theta_-, \theta_+] \tag{13-2.11}$$

Consider joint parameter and state estimation for the system (13-2.8), (13-2.9). It leads to a nonlinear estimation problem. Here we will introduce a parallel filtering approach to perform the estimation in the framework of linear filtering theory.

First let us quantitise the admissible region (13-2.11) into a set of discrete elements:

$$\Theta = \{\theta_1, \theta_2, \ldots, \theta_\ell\} \text{ with } \theta_1 = \theta_- \text{ and } \theta_\ell = \theta_+ \tag{13-2.12}$$

If the set is sufficiently dense, we can expect that the true value of θ, say $\theta*$, is approximately included in this set. We hence assume that:

$$\theta^* \in \Theta\{\theta_1, \theta_2, \ldots, \theta_\ell\} \tag{13-2.13}$$

Next, consider the *a-posteriori* probability density function of x_k based on Y^k : $p(x_k/Y^k)$, where $Y^k = \{y_1, y_2, \ldots, y_k\}$ is the measurement sequence up to k. Using Bayes' chain rule we have:

$$p(x_k/Y^k) = \sum_{j=1}^{\ell} p(x_k/\theta_j,Y^k)Pr(\theta_j/Y^k) \qquad (13-2.14)$$

Assuming $\theta = \theta_j$ in the system description (13-2.8), (13-2.9), we can express the conditional probability densities by Gaussian distributions as follows:

$$p(x_k/\theta_j,Y^k) = N(\hat{x}_k(j),\hat{S}_k(j)) \qquad (13-2.15)$$

$$p(x_{k+1}/\theta_j,Y^k) = N(\bar{x}_{k+1}(j),\bar{S}_{k+1}(j)) \qquad (13-2.16)$$

where $\hat{x}_k(j)$ and $\bar{x}_{j+1}(j)$ are the filter estimate of x_k and the prediction of x_{k+1} based on Y^k for $\theta = \theta_j$, respectively. $\hat{S}_k(j)$ and $\bar{S}_{k+1}(j)$ are the corresponding error covariance matrices. These are obtained by the parallel Kalman filters as follows:

$$\hat{x}_{k+1}(j) = \bar{x}_{k+1}(j) + K_{k+1}(j)[y_{k+1}-H(\theta_j)\bar{x}_{k+1}(j)] \qquad (13-2.17)$$

$$\bar{x}_{k+1}(j) = F(\theta_j)\hat{x}_k(j) \qquad (13-2.18)$$

$$K_{k+1}(j) = \bar{S}_{k+1}(j)H^T(\theta_j)[H(\theta_j)\bar{S}_{k+1}(j)H^T(\theta_j)+R_{k+1}]^{-1} \qquad (13-2.19)$$

$$\hat{S}_{k+1}(j) = \bar{S}_{k+1}(j) - K_{k+1}(j)H(\theta_j)\bar{S}_{k+1}(j) \qquad (13-2.20)$$

$$S_{k+1}(j) = F(\theta_j)\hat{S}_k(j)F^T(\theta_j) + Q_k \qquad (13-2.21)$$

$$k=1,2,\ldots \quad j=1,2,\ldots,\ell$$

On the other hand $Pr(\theta_j/Y^k)$ denotes a learned parameter distribution based on the data Y^k and it can be calculated in the recursive form:

$$Pr(\theta_j/Y^{k+1})= \frac{Pr(\theta_j/Y^k) \int p(y_{k+1}/\theta_j,x_{k+1})p(x_{k+1}/\theta_j,Y^k)dx_{k+1}}{\sum_{j=1}^{\ell} (\text{Numerator})} \qquad (13-2.22)$$

From equations (13-2.9), (13-2.10) we have:

$$p(y_{k+1}/\theta_j,x_{k+1}) = N(H(\theta_j)x_{k+1},R_{k+1}) \qquad (13-2.23)$$

Substitution of (13-2.6) and (13-2.23) into (13-2.22) and analytical integration yield:

$$\Pr(\theta_j/Y^{k+1}) = \Pr(\theta_j/Y^k)q_j(y_{k+1})/\sum_{j=1}^{\ell} \text{ (Numerator)} \qquad (13\text{-}2.24)$$

where $q_j(y_{k+1})$ is the probability density function of Gaussian distribution given by:

$$q_j(y_{k+1}) = N(H(\theta_j)\bar{x}_{k+1}(j), \ H(\theta_j)\bar{S}_{k+1}(j)H^T(\theta_j)+R_{k+1}) \quad (13.2.25)$$

Initial values of $\Pr(\theta_j/Y^0)$ are usually set to be uniform distribution; $\Pr(\theta_j/Y^0) = 1/\ell$ ($j = 1, 2, \ldots, \ell$) when there is no *a-priori* information about θ.

Finally the state estimation is obtained as the weighted sum of each parallel filtering result as follows,

$$\hat{x}_k = E[x_k/Y^k] = \sum_{j=1}^{\ell} \hat{x}_k(j)\Pr(\theta_j/Y^k) \qquad (13\text{-}2.26)$$

$$\hat{S}_k = \text{Cov}[x_k/Y^k] = \sum_{j=1}^{\ell} \Pr(\theta_j/Y^k)[\hat{S}_k(j)+(x_k-\hat{x}_k(j))$$

$$(x_k-\hat{x}_k(j))^T] \qquad (13\text{-}2.27)$$

In a similar way we have for the prediction of x_{k+1}:

$$\bar{x}_{k+1} = E[x_{k+1}/Y^k] = \sum_{j=1}^{\ell} \bar{x}_{k+1}(j)\Pr(\theta_j/Y^k) \qquad (13\text{-}2.28)$$

$$\bar{S}_{k+1} = \text{Cov}[x_{k+1}/Y^k] = \sum_{j=1}^{\ell} \Pr(\theta_j/Y^k)[\bar{S}_{k+1}(j)+(x_{k+1}-\bar{x}_{k+1}(j))$$

$$(x_{k+1}-\bar{x}_{k+1}(j))^T] \qquad (13\text{-}2.29)$$

If the *identifiability condition*:

$$p(x_k/\theta_i,Y^k) \neq P(x_k/\theta_j,Y^k) \text{ for a.e. } x_k \text{ and for } \theta_i \neq \theta_j \quad (13\text{-}2.30)$$

is fulfilled, the convergence of the parallel filtering is guaranteed in the sense of Liporace (1971):

$$\Pr(\theta_j/Y^k) = 1 \text{ for } \theta_j = \theta^* \in \Theta \qquad (13\text{-}2.31)$$
$$\lim k \to \infty$$

In this way we can implement simultaneous parameter and state estimation by using a bank of Kalman filters. Note here that this approach is rather more effective than the nonlinear estimation scheme and is a practical way when the unknown parameter is *a scalar*. Otherwise it becomes impractical due to *the curse of dimensionality*. Furthermore we know from experience on numerical simulations that this approach gives a reasonable result even when $\theta*$ is not exactly included in θ. In such a case the probability distribution of $\hat{\theta}$, the element of θ which is nearest to $\theta*$, may instead converge to 1.

13.3 HIERARCHICAL FAULT DIAGNOSIS OF DYNAMIC SYSTEMS

13.3.1 Statement of the problem

13.3.1.1 System description
As the object to be diagnosed, consider a linear system with a transfer function $G(s,p)$. The system is written in the continuous-time state-space form:

$$\dot{x}(t) = A(p)x(t) + B(p)u(t) + w(t) \qquad (13\text{-}3.1)$$

$$y(t) = C(p)x(t) + D(p)u(t) + v(t) \qquad (13\text{-}3.2)$$

where $x(t)$ is the $n \times 1$ state vector, $u(t)$ the $r \times 1$ input vector, $y(t)$ the $m \times 1$ observation vector, $w(t)$ and $v(t)$ the system and the observation noise vectors. p is the $N \times 1$ system configuration parameter vector. It specifies system faults in the physical sense. In other words, when a fault occurs, one component of the vector p will change. We will write the vector p as:

$$p = [p^1 \ p^2 \ \dots \ p^N]^T \qquad (13\text{-}3.3)$$

In (13-3.1) and (13-3.2), $A(p)$, $B(p)$, $C(p)$ and $D(p)$ are matrices with appropriate dimensions. They depend on the parameter vector p. We will assume that in the normal situation the parameter vector p takes a nominal value, say $\bar{p}$. The continuous system can be sampled which gives for an arbitrary sampling period T,

$$x_{k+1} = F(\bar{p})x_k + G(\bar{p})u_k + w_k \qquad (13\text{-}3.4)$$

$$y_k = C(\bar{p})x_k + D(\bar{p})u_k + v_k \qquad k=1,2,\dots \qquad (13\text{-}3.5)$$

where the noise vectors w_k, v_k are assumed to be mutually independent white Gaussian distributed variables with:

$$w_k \in N(0, Q_k), \quad v_k \in N(0, R_k) \tag{13-3.6}$$

The matrices $F(\bar{p})$ and $G(\bar{p})$ are calculated by:

$$F(\bar{p}) = e^{A(\bar{p})T} \tag{13-3.7}$$

$$G(\bar{p}) = \int_0^T e^{A(\bar{p})(T-t)} dt \, B(\bar{p}) \tag{13-3.8}$$

13.3.1.2 Fault modelling

Often most of the fault modes of dynamic systems can be modelled by abrupt changes in the system parameters. Assume that more than two sets of components of the vector p never jump simultaneously. We can then express the fault by:

$$\bar{p}^i \rightarrow \bar{p}^i + \Delta p^i . I_{k,\theta} \quad i \in \{1, 2, \ldots, N\} \tag{13-3.9}$$

where the index i denotes a faulty component and the symbol $I_{k,\theta}$ is the unit step function. Hence θ stands for a fault onset time. Δp^i is the jump quantity in parameter p^i. The variables $(\theta, i, \Delta p^i)$ are all unknowns to be determined by a fault diagnosis procedure. In this way the fault diagnosis problem leads to the joint problem of: *detection* $(\hat{\theta})$, *recognition* $(\hat{i})$ and *estimation* $(\Delta \hat{p}^i)$ based on the statistical processing of the data obtained from the system. There are, however, some difficulties in performing such a fault detection and identification. One of the prominent difficulties is how to determine the faulty component i from many possible candidates. To solve this problem we will propose a *hierarchical diagnosis* using a parameter monitoring scheme.

13.3.2 Design of hierarchical diagnosis system

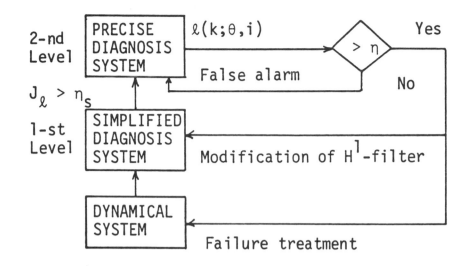

Figure 13-1: Structure of the hierarchical diagnosis system

The structure of the hierarchical diagnosis system is shown in Figure 13-1. This system is composed of two decision levels. It has a simplified diagnosis in the first level and a precise diagnosis in the second level. The former is for on-line fault detection and the latter is for fault identification which may be implemented as an off-line procedure. In the following we will discuss details of the design method for each diagnosis system.

13.3.2.1 Simplified diagnosis; first level
(A Parameter Monitoring Scheme)

We will first discuss the basic idea of a parameter monitoring scheme for on-line fault detection (Kumamaru et al. (1986)). Suppose that a fault occurred in the system due to a change in a component of the vector p at an instant of past stage. We intend to detect the fault occurrence without a knowledge of the component that changed. For this purpose, choose one component of p, say p^j, as a monitored parameter α given by:

$$\alpha = p^j \in \{p^1, p^2, \ldots, p^N\} \tag{13-3.10}$$

Note here that the chosen parameter α does not necessarily coincide with the really faulty component. When the system is normal, the value of α is known as $\alpha = \bar{p}^j$, by our assumption. Instead we treat it as an unknown parameter. Perform a joint parameter (α) and state estimation for the system (13-3.4), (13-3.5), where the vector p is set to be $p=\bar{p}$ except for the component $\alpha = p^j$. Then a fault effect will appear as a change in the estimate of α from its nominal value, even if the fault is not due to a change in α itself. When there is no fault, the estimate will converge to the nominal value. In this way, such a result of information processing can be used as an index for fault detection.

Now let us describe the implementation of the parameter monitoring scheme. Consider *two* kinds of data sets Y_R^k and Y_M^k up to the current stage k. Here Y_R^k denotes the real data set obtained from the system. Y_M^k is a simulated data set which can be calculated from the model with $p = \bar{p}$, where the initial value of x_1 is replaced with its estimate $\hat{x}_1$ and the noise sequences sampled from the distributions (13-3.6) are used. Then these two data sets can be seen as statistically equivalent when the system is normal. On the other hand they will differ in a statistical sense when a fault has occurred. Let us introduce a set of discrete elements for an admissible region of α, i.e.:

$$\alpha \in \quad \{\alpha_1, \alpha_2, \ldots, \alpha_\ell\} \tag{13-3.11}$$

Perform a simultaneous parameter (α) and state estimation by using these two data sets. The parallel Kalman filter approach introduced in Section 13.2.3 can then be applied to this estimation. We thus have two estimated parameter distributions $Pr(\alpha_j/Y_R^k)$ and $Pr(\alpha_j/Y_M^k)$, $j=1,2,\ldots,\ell$. The difference between the distributions can be calculated by using the Jensen Difference Measure (JDM) (see (13-2.6)),

$$J_\ell(M,R;k) = H_\ell(\frac{1}{2} (P_r^k(M)+P_r^k(R))) - \frac{1}{2} [H_\ell(P_r^k(M))+H_\ell(P_r^k(R))] \tag{13-3.12}$$

where $P_r^k(R)$ and $P_r^k(M)$ are the ℓ-dimensional multinominal distributions obtained for Y_R^k and Y_M^k, respectively. As a result, the fault detection leads to the following decision problem,

$$J_\ell(M,R;k) \overset{\geq}{\underset{<}{}} \eta_s \rightarrow \begin{cases} \text{Fault} \\ \text{No fault} \end{cases} \qquad (13\text{-}3.13)$$

Remarks

(a) *An appropriate choice of the monitored parameter* α *is crucial in this approach.* The standard way will be to choose it by considering the parameter sensitivity in the system and observation equations. However, to choose one with high sensitivity as a candidate of monitored parameter is not necessarily optimal in the sense that the resulting detection index J_ℓ becomes sensitive to the fault. Such a sensitivity depends on which component of $p^1,\ldots,p^N$ has really changed in the fault, as well as the choice of α. If some *a-priori* information about the system and the fault characteristics is available, it may suggest a proper way of choosing the parameter α.

(b) *The choice of the threshold value* η_s *is also important for satisfactory fault detection.* The value of η_s which gives a specified significance level of the decision can usually be determined from statistical properties of the index (see Fukunaga (1979) for a detailed discussion). However it can be shown that the JDM introduced here is bounded within a narrow region as follows:

$$0 \leqslant J_\ell(M,R;k) \leqslant \ln(2) \qquad (13\text{-}3.14)$$

The upper bound in (13-3.14) is obtained by considering an ideal case of:

$$\text{Pr}(\alpha = \alpha_i/Y_R^k) = 1, \ \text{Pr}(\alpha = \alpha_j/Y_M^k) = 1, \ i \neq j \qquad (13\text{-}3.15)$$

in (13-3.12). Therefore an appropriate value of η_s can easily be determined from this region.

(c) *The proposed method for fault detection can be implemented on-line by using a bank of Kalman filters.* Note, however, that since the parameter α is a temporary one chosen for monitoring the system operating mode, it does not necessarily coincide with a really faulty component of p. Thus after the fault detection, the true fault mode which has occurred should be examined by a fault detection and identification procedure. This is done in the following second level diagnosis for precise decision making.

13.3.2.2 Precise diagnosis; second level
(oriented fault modelling)

The fault effect due to a change in the system parameter can be expressed as an additive *jump* in the linear system equations (13-3.4), (13-3.5). For instance when the parameter p^i changed from $\bar{p}^i$ to $\bar{p}^i + \Delta p^i$ at time θ, the resulting system equation can approximately be written by the form (Kumamaru et al. (1983)):

$$x_{k+1} = F(\bar{p})x_k + G(\bar{p})u_k + D_k^i(x_k, u_k, \bar{p})\Delta p^i + w_k \qquad (13-3.16)$$
$$k \geqslant \theta$$

where $D_k^i(.)$ denotes a directional vector which specifies in the system equation an orientation of the fault effect due to the parameter change Δp^i. To obtain an explicit form of $D_k^i(.)$, let us apply a perturbation analysis to the original continuous system (13-3.1):

$$\dot{x}(t) = A(\bar{p})x(t) + B(\bar{p})u(t) + w(t) + \mu(t) \qquad (13-3.17)$$

$$\mu(t) = [A_p^i x(t) + B_p^i u(t)]\Delta p^i, \quad t \geqslant kT \qquad (13-3.18)$$

where the following linearisations are used:

$$A(\bar{p} + \Delta p^i d^i) = A(\bar{p}) + A_p^i \Delta p^i, \quad B(\bar{p} + \Delta p^i d^i) = B(\bar{p}) + B_p^i \Delta p^i \qquad (13-3.19)$$
$$A_p^i, B_p^i : \text{the gradients w.r.t. } p^i \text{ at } p = \bar{p}$$

$$d^i = [\underbrace{0...0\ 1\ 0...0}_{i}]^T : \text{Nx1 indicator vector}$$

Regarding the term $\mu(t)$ as an alternative input and using the fact $u(t) = u_k$ for $t \in [kT, (k+1)T]$, we have the discrete form (13-3.16) of (13-3.17), (13-3.18), where $D_k^i(.)$ is given by:

$$D_k^i(x_k, u_k, \bar{p}) = \int_{kT}^{(k+1)T} e^{A(\bar{p})[(k+1)T - \tau]}[A_p^i x(\tau) + B_p^i u(\tau)]d\tau \qquad (13-3.20)$$

In the integration of (13-3.20), use the linear approximation of $x(\tau)$ given as:

$$x(\tau) = [I+A(\bar{p})(\tau-kT)]x_k + B(\bar{p})(\tau-kT)u_k \qquad (13\text{-}3.21)$$

We can then derive the following expression of $D_k^i(.)$, when $A(\bar{p})$ is nonsingular:

$$D_k^i(x_k,u_k,\bar{p}) = \Phi^i(\bar{p})x_k + \Gamma^i(\bar{p})u_k \qquad (13\text{-}3.22)$$

where:

$$\Phi^i(\bar{p}) = [F(\bar{p})-I]A^{-1}(\bar{p})[A_p^i+A^{-1}(\bar{p})A_p^iA(\bar{p})]-A^{-1}(\bar{p})A_p^iA(\bar{p})T \qquad (13\text{-}3.23)$$

$$\Gamma^i(\bar{p}) = [F(\bar{p})-I]A^{-1}(\bar{p})[B_p^i+A^{-1}(\bar{p})A_p^iB(\bar{p})]-A^{-1}(\bar{p})A_p^iB(\bar{p})T \qquad (13\text{-}3.24)$$

If $A^{-1}(\bar{p})$ does not exist, an alternative expression of $e^{A(p)}$ using, e.g. the Lagrange-Sylvester interpolation polynominal, may be applied for implementation of the integrals;

$$\int_0^T e^{A(\bar{p})t}dt \text{ and } \int_0^T e^{A(\bar{p})t}tdt$$

Note that if the sampling time T is small, (13-3.23) and (13-3.24) give:

$$\Phi^i(\bar{p}) = A_p^iT, \quad \Gamma^i(\bar{p}) = B_p^iT \qquad (13\text{-}3.25)$$

which coincide with the results obtained using a simple approximation of $F(\bar{p})=I+A(\bar{p})T$. To modify the approximation in (13-3.22) we consider an additional disturbance w_k' with normal distribution:

$$w_k' \in N(0,Q_k') \qquad (13\text{-}3.26)$$

where Q_k' $(>Q_k)$ is treated as an adjustable parameter determined by considering the approximation degree.

In a similar way the effect of the fault can be expressed in the observation equation as:

$$y_{k+1} = C(\bar{p})x_{k+1} + D(\bar{p})u_{k+1} + E^i_{k+1}(x_{k+1}, u_{k+1}, \bar{p})\Delta p^i + v'_k \qquad (13\text{-}3.27)$$

$$E^i_{k+1}(x_{k+1}, u_{k+1}, \bar{p}) = C^i_p x_{k+1} + D^i_p u_{k+1}, \qquad k \geqslant \theta+1 \qquad (13\text{-}3.28)$$

C^i_p, D^i_p: the derivatives of $C(p)$ and $D(p)$ w.r.t. p^i at $p=\bar{p}$

From the meanings of the directional vectors D^i_k and E^i_{k+1}, we will call the models (13-3.16), (13-3.27) *oriented fault models*.

Data Processing for Fault Detection and Identification (FDI)

Let us introduce the following *two* hypotheses according to the system operating modes:

$$H^0 \qquad ; \text{ Normal situation } \quad (k \leqslant \theta) \qquad\qquad (13\text{-}3.29)$$

$$H^1(\theta, i) \; ; \text{ Fault occurrence at } \theta \text{ in the parameter } p^i$$
$$\qquad\qquad (k\text{-}M \leqslant \theta \leqslant k\text{-}1) \qquad\qquad\qquad\qquad (13\text{-}3.30)$$

where k denotes the present stage and M is a data window. The number of M is selected by considering the trade-off between the allowable computational labour and the required decision accuracy. For further discussions on the choice of M, see Willsky (1976). The state estimation is performed under both of these two hypotheses. We can describe the corresponding filters in the following way.

(a)*(H⁰-Filter)*

For the state estimation under the hypothesis H^0, the standard Kalman filter can be applied to the system (13-3.4), (13-3.5) with a known nominal value of $\bar{p}$. However, we have already obtained similar estimation results in the simplified diagnosis. Strictly speaking, these results are not same as the H^0-filter when a fault has occurred, but for convenience we will use them as H^0-filter results and denote them by:

$$\{(\hat{x}^o_j, \bar{x}^o_{j+1}), (\hat{S}^o_j, \bar{S}^o_{j+1}); \; j=1,2,\ldots,k\} \qquad (13\text{-}3.31)$$

where the superscript o denotes estimation results under the hypothesis H^0.

(b) *(H¹(ө,i)-Filter)*

From the oriented fault model the system equation under the hypothesis H^1 (ө,i) can be written in the form:

$$x_{ө+1} = F(\bar{p})x_ө + G(\bar{p})u_ө + \hat{D}_ө^i \Delta p^i + w_ө' \qquad (13\text{-}3.32)$$

$$y_{ө+1} = C(\bar{p})x_{ө+1} + D(\bar{p})u_{ө+1} + \hat{E}_{ө+1}^i \Delta p^i + v_{ө+1}' \qquad (13\text{-}3.33)$$

where the directional vectors $D_ө^i$ and $E_{ө+1}^i$ are replaced with their estimates:

$$\hat{D}_ө^i = D_ө^i(\hat{x}_ө^o, u_ө, \bar{p}), \qquad \hat{E}_{ө+1}^i = E_{ө+1}^i(\bar{x}_{ө+1}^o, u_{ө+1}, \bar{p}) \qquad (13\text{-}3.34)$$

The expressions (13-3.34) would be valid since $\hat{x}_ө^o$ and $\bar{x}_{ө+1}^o$ which are obtained under the hypothesis H^o can be considered to be fairy good estimates at time k=ө and k=ө+1, respectively. From (13-3.32) - (13-3.34), the innovation can be written as:

$$\tilde{y}_{ө+1}^o = y_{ө+1} - \bar{y}_{ө+1}^o = L^i(ө,ө+1)\Delta p^i + \xi_{ө+1} \qquad (13\text{-}3.35)$$

where:

$$L^i(ө,ө+1) = C(\bar{p})\hat{D}_ө^i + \hat{E}_{ө+1}^i \qquad (13\text{-}3.36)$$

$$\xi_{ө+1} = C(\bar{p})F(\bar{p})\tilde{x}^o + w_ө' + v_{ө+1}' \qquad (13\text{-}3.37)$$

and the following state estimations under the hypothesis H^o are used:

$$\bar{x}_{ө+1}^o = F(\bar{p})\hat{x}_ө^o + G(\bar{p})u_ө \qquad (13\text{-}3.38)$$

$$\bar{y}_{ө+1}^o = C(\bar{p})\bar{x}_{ө+1}^o + D(\bar{p})u_{ө+1} \qquad (13\text{-}3.39)$$

$$\tilde{x}_ө^o = x_ө - \hat{x}_ө^o \in N(0, \hat{S}_ө^o) \qquad (13\text{-}3.40)$$

Regarding (13-3.35) as a linear equation of Δp^i with additive noise $\xi_{ө+1}$:

$$E[\xi_{ө+1}] = 0$$
$$\text{Cov}[\xi_{ө+1}] = M_{ө+1} = C(\bar{p})[F(\bar{p})\hat{S}_ө^o F^T(\bar{p})+Q_ө']C^T(\bar{p})+R_{ө+1}' \qquad (13\text{-}3.41)$$

we have the weighted least-square's estimate of Δp^i in the form:

$$\Delta \hat{p}^i(\theta) = [L^i(\theta,\theta+1)^T M_{\theta+1}^{-1} L^i(\theta,\theta+1)]^{-1} L^i(\theta,\theta+1)^T M_{\theta+1}^{-1} \tilde{y}_{\theta+1}^o$$

$$(13\text{-}3.42)$$

Now replace the parameter value $\bar{p}$ with $\hat{p}$:

$$\hat{p} = \bar{p} + \Delta \hat{p}^i(\theta) d^i \qquad\qquad (13\text{-}3.43)$$

Then we have new system equations with the parameter value $\hat{p}$. Apply a Kalman filter to this new system from the stage of $(\theta+1)$ to k by using:

$$\bar{x}_{\theta+1} = F(\hat{p})\hat{x}_\theta^o + G(\hat{p})u_\theta, \quad \bar{S}_{\theta+1} = F(\hat{p})\hat{S}_\theta^o F^T(\hat{p}) + Q_\theta \qquad (13\text{-}3.44)$$

as the initial estimate of $x_{\theta+1}$. We thus have the $H^1(\theta,i)$ filter denoted by:

$$\{(\hat{x}_j(\theta,i),\bar{x}_{j+1}(\theta,i)), (\hat{S}_j(\theta,i),\bar{S}_{j+1}(\theta,i));$$
$$j=\theta+1,\theta+2,\ldots,k\}$$

$$(13\text{-}3.45)$$

GLR Test for Fault Identification

Introduce the *Generalised Likelihood Ratio* (GLR) function defined by:

$$\ell(k;\theta,i) = 2 \ln \frac{p(y_{k-M+1},\ldots,y_k/H^1(\theta,i))}{p(y_{k-M+1},\ldots,y_k/H^0)} \qquad (13\text{-}3.46)$$

It can be reduced to the following form using *Bayes' chain rule*:

$$\ell(k;\theta,i) = \sum_{j=\theta+1}^{k} \{\|\tilde{y}_j^o\|_{(\Sigma_j^o)^{-1}}^2 - \|\tilde{y}_j(\theta,i)\|_{(\Sigma_j(\theta,i))^{-1}}^2$$
$$+ \ln [\det\Sigma_j^o/\det\Sigma_j(\theta,i)]\} \qquad (13\text{-}3.47)$$

where $\tilde{y}_j^o$ and $\tilde{y}_j(\theta,i)$ are the innovations, Σ_j^o and $\Sigma_j(\theta,i)$ are the corresponding covariance matrices under the hypotheses H^o and H^1 (θ,i), respectively. Calculate the maximum value of $\ell(k;\theta,i)$ for $k-M \leq \theta \leq k-1$ and for $i=1,2,\ldots,N$. The fault diagnosis is then performed by the following thresholding approach with a proper value of η_p.

$$\underset{\theta,i}{\text{Max}} \; \ell(k;\theta,i) = \ell(k;\hat{\theta},\hat{i}) \underset{<}{\overset{>}{\geqq}} \eta_p \rightarrow \begin{Bmatrix} H^1(\hat{\theta},\hat{i}) \\ H^0 \end{Bmatrix} \qquad (13-3.48)$$

This GLR test requires a fairly large amount of computational labour in the calculation of $\ell(k;\hat{\theta},\hat{i})$. Therefore such a precise diagnosis may be implemented by an off-line procedure after a fault declaration is concluded in the first level diagnosis. As to the choice of η_p, a χ^2-distribution of $\ell(k;\hat{\theta},\hat{i})$ may be used for approximately specifying a significance level under the assumption that $\tilde{y}_j^0$ and $\tilde{y}_j(\theta,i)$ have Gaussian distributions.

13.3.3. Numerical Examples

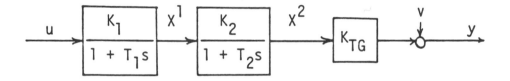

Figure 13-2: Transfer function of the second order system

To confirm the effectiveness of the method let us consider a second order linear system with the transfer function description given in Figure 13-2. Simulation studies were carried out for several choices of the monitored parameter α. Among them the simulation results with following specifications are shown in Figure 13-3 and Table 13-1.

k	i　θ	116	117	118	119	_120_
121	1	−0.24	−0.16	−0.09	−0.03	−0.01
	2	−6.11	−1.98	−9.29	5.94	_15.20_
	i　θ	117	118	119	_120_	121
122	1	−0.31	−0.20	−0.10	−0.07	−0.01
	2	−3.12	−14.00	8.53	_16.59_	4.38
	i　θ	118	119	_120_	121	122
123	1	−0.46	−0.30	−0.26	−0.08	−0.01
	2	−19.15	11.29	_19.56_	9.43	5.08
	i　θ	119	_120_	121	122	123
124	1	−0.53	−0.48	−0.20	−0.04	0.00
	2	12.74	_16.54_	10.97	6.33	1.90
	i　θ	_120_	121	122	123	124
125	1	−1.19	−0.64	−0.22	−0.04	−0.01
	2	_23.75_	17.97	13.61	7.57	7.69

$$\left[\begin{array}{l} i = 1 \;\rightarrow\; T_1 \\ i = 2 \;\rightarrow\; K_{TG} \end{array} \right]$$

Table 13-1: Calculated GLR. The data window M=5 and the underlined columns correspond to $\ell(k;\hat{\theta},\hat{i})$.

Note that: u=100 (constant input, T=0.01 sec., α=T_1, η_s=0.5,
Θ {$\alpha_1,\alpha_2,\ldots,\alpha_\ell$} = (2.0,2.8,3.6,...,18.0), ℓ=21, Fault parameter
K_{TG}:$K_{TG}\rightarrow K_{TG}$+0.025 at Θ=120, Nominal values of system parameters:
T_1=10.0, K_1=1.0, T_2=1.0, K_2=1.5151, K_{TG}=0.06

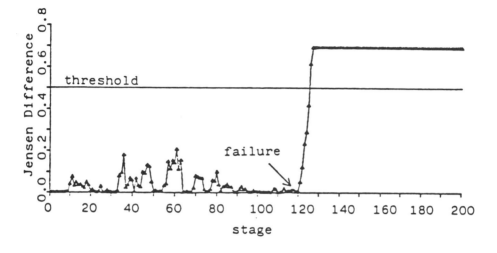

Figure 13-3: Calculated Jensen difference measure. The threshold is
set to be η_s = 0.5

It can be seen that the fault which occurred in K_{TG} is definitely
detected through the parameter monitoring scheme with the choice of α
= T_1 (see Figure 13-3). On the other hand the failure onset time (Θ
=120) and the faulty component (i=2, K_{TG}) are exactly detected and
identified in the second level diagnosis using the GLR test. This is
clear from the Table 13-1, since for each k the maximum value of
ℓ(k;Θ,i) is attained at $\hat{\Theta}$ = 120 and for $\hat{i}$ = 2. The obtained estimate
of Δp^1 was 0.02888. Note that it is essential for satisfactory
decision in the second level diagnosis to obtain good accuracy of
parameter jump estimation, because the accuracy directly effects the
performance of resulting H^1-filter results. Therefore in some cases
succeeding innovations $\tilde{y}^0_{\Theta+2}$, $\tilde{y}^0_{\Theta+3}$,..., $\tilde{y}^0_k$ should also be used in the
least-squares estimation in order to obtain better estimates of the
jump quantity. However, unlike the ideal case of a linear system with
pure additive jump as considered in Willsky and Jones (1976), the
succeeding innovations after the stage Θ+1 become nonlinear functions
of Δp^i, because the fault mode considered here is specified by a

change in the system configuration parameter. In addition, these innovations include more expression errors than the $\tilde{y}^0_{\theta+1}$. Therefore only one or two succeeding innovations will be available to obtain a rather good estimate of Δp^1.

13.4 ON THE USE OF KULLBACK DISCRIMINATION INFORMATION FOR FAULT DETECTION BASED ON INPUT-OUTPUT MODELS

13.4.1 Preliminaries

Consider a linear multivariable system given by:

$$S: y(t) = G_o(q^{-1})u(t) + H_o(q^{-1})e(t) \quad t=1,2,\ldots \qquad (13\text{-}4.1)$$

where $y(t)$ is the output of dimension ny at time t, $u(t)$ is the input of dimension nu and $e(t)$ is the white Gaussian noise of dimension ny

$$e(t) \in N(0,\Lambda_o) \qquad (13\text{-}4.2)$$

Further, $G_o(q^{-1})$ and $H_o(q^{-1})$ are matrix-valued rational functions of q^{-1}, the backward shift operator. We assume that:

(a) $G_o(q^{-1})$ and $H_o(q^{-1})$ are causal
(b) $G_o(0) = 0$, $H_o(0) = I$
(c) $H_o(q^{-1})$ and $H_o^{-1}(q^{-1})$ are asymptotically stable.

These conditions are natural and fairly weak, see the spectral factorisation theorem, Anderson and Moore (1979). When identifying the system we assume that parametric models of the form:

$$M : y(t) = G(q^{-1};\theta)u(t) + H(q^{-1};\theta)e(t) \qquad (13\text{-}4.3)$$

$$Ee(t)e^T(s) = \Lambda(\theta)\delta_{t,s}$$

are used. ARX and ARMAX models are only *two* possible alternatives out of many more. In what follows we will for generality deal with models of the form (13-4.3) without specifying any particular parametrisation. We assume though that for any value of θ the filters $G(q^{-1}:\theta)$ and $H(q^{-1}:\theta)$ fulfil the assumptions stated above for G_o and H_o.

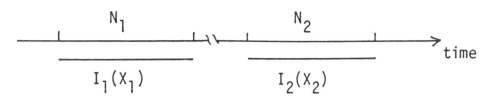

Figure 13-4: Distinct time intervals I_1 and I_2.

Assume then that data from the system are available from two distinct time intervals I_1 and I_2, see Figure 13-4. The number of data points in the interval I_i will be denoted by N_i. When no confusion is likely, we will number the data as $t=1,2,\ldots,N_i$ for each interval I_i. The experimental condition (e.g. the input characteristics like the spectral density $\Phi_u(\omega)$) will be denoted by X_i for the interval I_i. Now perform identification using data from these two intervals. The parameter estimate obtained from the interval I_i will be denoted by $\hat{\theta}_i$. To simplify notations introduce:

$$G_i = G(q^{-1};\hat{\theta}_i), \quad H_i = H(q^{-1};\hat{\theta}_i), \quad \Lambda_i = \Lambda(\hat{\theta}_i) \quad i=1,2 \qquad (13-4.4)$$

We also introduce the prediction error for the model corresponding to $\hat{\theta}_i$:

$$\epsilon(t,\hat{\theta}_i) = y(t)-\hat{y}(t/t-1,\hat{\theta}_i) = H_i^{-1}y(t)-H_i^{-1}G_iu(t) \qquad (13-4.5)$$

We now consider the problem to decide whether or not:

$$\hat{\theta}_1 = \hat{\theta}_2 \qquad (13-4.6)$$

reflecting the uncertainties in the data. We can apply it to a fault detection problem based on an input-output model in the following sense, Kumamaru et al. (1984), Kumamaru et al. (1985), Söderström and Kumamaru (1985a). The fault effect due to a physical parameter change in the system will be used to improve the model including parameter changes. Then if two models identified are significantly different we conclude that a fault has occurred between the intervals. True, the fault detection is often applied as an on-line method, see e.g. Hägglund (1983). However, in some cases it is sufficient to use it as a batch (repeated) method, see Kumamaru et al. ((1984) and (1985)). Then the fault detection can be stated just as the problem (13-4.6). Furthermore the problem can also be applied to a cross-validation for identified models (see Söderström and Kumamaru (1985a), (1985b), and Kumamaru and Söderström (1986)).

Clearly for such an approach to our stated problem it is necessary to have an effective measure of the difference between two models. For this purpose we introduce the Kullback Discrimination Information (KDI), see Section 13.2.1 for detailed discussion of the basic properties. Now we apply it to likelihood functions for the identified models. We thus have:

$$I_{N_1}[1,2] = \int p(Y_{N_1}/\hat{\theta}_1, U_{N_1}-1) \ln \frac{p(Y_{N_1}/\theta_1, U_{N_1}-1)}{p(Y_{N_1}/\hat{\theta}_2, U_{N_1}-1)} \, dY_{N_1} \qquad (13-4.7)$$

where capital symbols Y_{N1} and U_{N1} denote the data collections up to N_1 taken from the first interval I_1 defined by:

$$Y_{N_1} = [y(1)^T y(2)^T \ldots y(N_1)^T]^T, \quad U_{N_1} = [u(1)^T u(2)^T \ldots u(N_1)^T]^T$$
$$(13-4.8)$$

The index in (13-4.7) hence indicates how well the model using $\hat{\theta}_2$ describes the data in the interval I_1. In other words the criteria $I_{N1}[1,2]$ is an index for discriminating the models $M(\hat{\theta}_1)$ and $M(\hat{\theta}_2)$ via the difference in the corresponding likelihood functions. When applying the criteria (13-4.7) to model discrimination we must define a threshold η. We then will make the following decision:

$$I_{N_1}[1,2] \gtrless \eta \rightarrow \begin{cases} M(\hat{\theta}_1) \neq M(\hat{\theta}_2) \\ M(\hat{\theta}_1) = M(\hat{\theta}_2) \end{cases} \qquad (13-4.9)$$

The value of threshold η should be set according to the statistical properties of $I_{N1}[1,2]$ under various conditions. We will give some properties of this kind in Section 13.4.3. Note here that direct evaluation of (13-4.7) using the multidimensional (ny x N_1) Gaussian distribution of the likelihood functions requires a great computational labour due to large matrix operations for a finite but fairly large data set (see (13-2.3)). In order to solve this problem we will give a feasible way of evaluating the KDI. As a result some new criteria for fault detection are obtained.

13.4.2. Evaluation of the KDI

The index $I_{N1}[1,2]$ can be recursively computed. The derivation steps are as follows (see Söderström and Kumamaru (1985a) for details):

(a) *Apply Bayes' rule:*

$$p(Y_{k+1}/\hat{\theta}_i,U_k) = p(y(k+1)/\hat{\theta}_i,Y_k,U_k)p(Y_k/\hat{\theta}_i,U_k) \qquad (13\text{-}4.10)$$

(b) *Due to the assumption on Gaussian distributed disturbance we have*

$$p(y(k+1)/\hat{\theta}_i,Y_k,U_k) = N(m_{ki},\Lambda_i) \qquad (13\text{-}4.11)$$

Here the conditional mean m_{ki} is equal to one-step ahead prediction of $y(k+1)$, which can be found from (13-4.5)

(c) *Use of straightforward manipulation with the model equations, which are assumed to describe the data.*

The result of such a derivation is:

$$I_{N_1}[1,2] = I_0[1,2] + \sum_{j=1}^{3} I_{N_1}^{(j)}[1,2] \qquad (13\text{-}4.12)$$

where $I_0[1,2]$ is an initial value which might be neglected and other components are given as:

$$I_{N_1}^{(1)}[1,2] = \frac{N_1}{2}[\mathrm{tr}(\Lambda_2^{-1}\Lambda_1-I) - \ln(\det\Lambda_1/\det\Lambda_2)] \qquad (13\text{-}4.13)$$

$$I_{N_1}^{(2)}[1,2] = \frac{1}{2}\sum_{k=0}^{N_1-1} \|H_2^{-1}(G_2-G_1)u(k+1)\|^2_{\Lambda_2^{-1}} \qquad (13\text{-}4.14)$$

$$I_{N_1}^{(3)}[1,2] = \frac{N_1}{2}\mathrm{tr}[\Lambda_2^{-1}\frac{1}{2\pi i}\oint[H_2^{-1}(z)H_1(z)-I]\Lambda_1$$
$$[H_1^T(z^{-1})H_2^{-T}(z^{-1})-I]\frac{dz}{z}] \qquad (13.4.15)$$

These different terms have the following meanings:
The first term (13-4.13) expresses the deviation of Λ_2 from Λ_1. The second term (13-4.14) shows the difference between G_1 and G_2. Note that the difference in the deterministic output $Gu(t)$ is filtered with H_2^{-1}. In this way we relate the difference to the prediction error instead of the output. The last term (13-4.15) describes the difference between the noise filters H_1 and H_2. All of these terms become zero when $\hat{\theta}_1 = \hat{\theta}_2$ and otherwise positive from the property of the KDI construction. In particular the case $G_1 = G_2$ is the only case

when the second term becomes zero if the input is persistently exciting the system. We further have:

$$I_{N_1}^{(2)}[1,2] + I_{N_1}^{(3)}[1,2] = \frac{1}{2}\sum_{k=0}^{N_1-1} E[\|\hat{y}(k+1/k,\theta_2)-\hat{y}(k+1/k,\theta_1)\|_{\Lambda_2^{-1}}^2$$

$$/U_k,y(t)=G_1u(t)+H_1e(t)] \qquad (13\text{-}4.16)$$

Now we will describe some criteria that can be used for model discrimination. As we mentioned in section 13.4.1 we can apply them to (batch) fault detection and to model validation. First let us consider criteria based on the KDI. The following *three* criteria correspond to $I_{N_1}^{(j)}[1,2]$ for j=1,2,3

$$\hat{W}_1 = N_1[tr(\Lambda_2^{-1}\Lambda_1-I)-\ln(\det\Lambda_1/\det\Lambda_2)] \qquad (13\text{-}4.17)$$

$$W_1^0 = \sum_{t=1}^{N_1} \|H_2^{-1}(G_2-G_1)u(t)\|_{\Lambda_2^{-1}}^2 \qquad (13\text{-}4.18)$$

$$W_1^e = \frac{N_1}{2\pi i} tr\{\Lambda_2^{-1} \oint [H_2^{-1}(z)H_1(z)-I]\Lambda_1[H_1^T(z^{-1})$$
$$H_2^{-T}(z^{-1})-I]\frac{dz}{z}\} \qquad (13\text{-}4.19)$$

A *fourth* criterion can be derived from the expression (13-4.16) in which the expectation is replaced by the sample mean. In this way we obtain:

$$W_2^e = \sum_{t=1}^{N_1} \|\epsilon(t,\hat{\theta}_2)-\epsilon(t,\hat{\theta}_1)\|_{\Lambda_2^{-1}}^2 \qquad (13\text{-}4.20)$$

Note that this criterion depends explicitly on the output, not only the input. Let us also introduce a *traditional* criterion related to *standard* use of cross-validation:

$$W_3^e = N_1\det(\frac{1}{N_1}\sum_{t=1}^{N_1} \epsilon(t,\hat{\theta}_2)\epsilon^T(t,\hat{\theta}_2))$$

$$- N_1\det(\frac{1}{N_1}\sum_{t=1}^{N_1} \epsilon(t,\hat{\theta}_1)\epsilon^T(t,\hat{\theta}_1)) \qquad (13\text{-}4.21)$$

All criteria W introduced here can be used as indices for fault detection. They are used in the thresholding approach,

$$W \gtrless \eta \rightarrow \begin{cases} \text{Fault} \\ \text{No fault} \end{cases} \qquad (13\text{-}4.22)$$

where the value of threshold η should appropriately be determined under a specification of decision accuracy, e.g. a significance level for a false alarm rate. To do this we need to analyse the statistical properties of the criteria in the normal situation. In this concern we will discuss the asymptotic properties of the criteria under the stationary operation of the system.

13.4.3. Asymptotic properties of the criteria

To perform a detailed analysis of the statistical properties of the criteria, we must introduce some assumptions on the parameter estimates $\hat{\theta}_1$ and $\hat{\theta}_2$. Assume that a prediction error method (PEM) (Ljung (1976), Söderström and Stoica (1988)), is used for parameter estimation. Then we take the estimated parameter vector as:

$$\hat{\theta}_i = \arg \underset{\theta_i}{\text{Min}} [\det \frac{1}{N_i} \sum_{t=1}^{N_i} \epsilon(t,\theta_i), \epsilon^T(t,\theta_i)] \qquad (13\text{-}4.23)$$

The asymptotic parameter vectors θ_i^* are given by

$$\theta_i^* = \arg \underset{\theta_i}{\text{Min}} [\det E\epsilon(t,\theta_i)\epsilon^T(t,\theta_i)] \qquad (13\text{-}4.24)$$

θ_i^* does not depend on i when the system is included in the model structure. We will occasionally denote it by θ^* in such a case and call it the *true parameter vector*. The prediction error $\epsilon(t,\theta^*)$ becomes identical to the white noise sequence e(t) of the true system, see (13-4.1). The estimates have then the following asymptotic Gaussian distributions (Ljung and Caines (1979), Söderström and Stoica (1988)):

$$\sqrt{N_i}(\hat{\theta}_i - \theta^*) \xrightarrow{\text{dist}} N(0, P_i) \qquad (13\text{-}4.25)$$

$$P_i = [E\Psi^T(t)\Lambda_0^{-1}\Psi(t)]^{-1} \qquad (13\text{-}4.26)$$

where $\Psi(t)$ denotes the gradient of the prediction error with respect to the parameter vector:

$$\Psi(t) = \frac{\partial \epsilon(t,\theta)}{\partial \theta}\Big|_{\theta=\theta^*} \qquad (13\text{-}4.27)$$

Note that the expectation in (13-4.26) is to be carried out for the experimental condition X_i. Thus the covariance matrix P_i will depend on X_i.

Now let us consider the ideal case that the system is stationary and the model structure is sufficiently rich to describe the system. We can then expect that the estimated parameter vectors $\hat{\theta}_1$ and $\hat{\theta}_2$ are close to each other. They will not be identical, however, since the time intervals I_1 and I_2 are finite and there will always be some random fluctuations. It can be shown that all criteria are approximately expressed by:

$$W/N_1 \approx (\hat{\theta}_1-\hat{\theta}_2)^T Q (\hat{\theta}_1-\hat{\theta}_2) \qquad (13\text{-}4.28)$$

where the weighting matrix Q depends on the criteria and the model structure. By using the Taylor expansion, the Q matrices are given as follows (see Söderström and Kumamaru (1985a) for details of the derivation).

$$Q(W_1^{\Lambda})^{i,j} = \frac{1}{2}\,\text{tr}[\Lambda_0^{-1}\frac{\partial \Lambda}{\partial \theta_i}\Lambda_0^{-1}\frac{\partial \Lambda}{\partial \theta_j}\,_{\theta=\theta^*}] \qquad (13\text{-}4.29)$$

$$Q(W_1^{o})^{i,j} = E[H_0^{-1}(q^{-1})\frac{\partial G(q^{-1};\theta)}{\partial \theta_i}\,u(t)]^T\Lambda_0^{-1}$$

$$[H_0^{-1}(q^{-1})\frac{\partial G(q^{-1};\theta)}{\partial \theta_j}\,u(t)]\,_{\theta=\theta^*} \qquad (13\text{-}4.30)$$

$$Q(W_1^{e})^{i,j} = \frac{1}{2\pi i}\oint\text{tr}\Big\{\Lambda_0^{-1}H_0^{-1}(z)\frac{\partial H(z;\theta)}{\partial \theta_i}\Lambda_0$$

$$[\frac{\partial H(z^{-1};\theta)}{\partial \theta_j}]^T H_0^{-T}(z^{-1})\Big\}\,_{\theta=\theta^*}\frac{dz}{z} \qquad (13\text{-}4.31)$$

$$Q(W_2^{e}) = \frac{1}{N_1}\sum_{t=1}^{N_1}\Psi^T(t)\Lambda_0^{-1}\Psi(t) \approx P_1^{-1} \qquad (13\text{-}4.32)$$

$$Q(W_3^{e}) = \det\Lambda_0 P_1^{-1} \qquad (13\text{-}4.33)$$

Since $\hat{\Theta}_1$ and $\hat{\Theta}_2$ are independent, from (13-4.25) we have:

$$\sqrt{N_1}(\hat{\Theta}_1 - \hat{\Theta}_2) = \sqrt{N_1}(\hat{\Theta}_1 - \Theta^*) - \sqrt{\frac{N_1}{N_2}} \cdot \sqrt{N_2}(\hat{\Theta}_2 - \Theta^*) \xrightarrow{\text{dist}} N(0, \bar{P})$$

$$(13\text{-}3.34)$$

with:

$$\bar{P} = P_1 + (N_1/N_2)P_2 \qquad\qquad\qquad (13\text{-}4.35)$$

If in the special case the experimental conditions are equal $(X_1 = X_2)$, we have $P_1 = P_2$ and hence:

$$\bar{P} = P_1(1 + N_1/N_2) \qquad\qquad\qquad (13\text{-}4.36)$$

By inspection of (13-4.28) and (13-4.34) we can suggest that the asymptotic distribution of the criteria will be a χ^2-distributed if the matrices Q and $\bar{P}$ satisfy $Q = \bar{P}^{-1}$. This is true and the results are summarised in the Lemma.

Lemma: Assume that the model structure includes the true system and that the system is stationary in both intervals I_1 and I_2 with the same experimental condition $(X_1 = X_2)$.

(a) Assume that the model parametrisation is such that (13-4.3) can be written as:

$$M : y(t) = G(q^{-1}; \Theta_u)u(t) + H(q^{-1}; \Theta_e)e(t) \qquad (13\text{-}4.37)$$

$$Ee(t)e^T(s) = \Lambda(\Theta_\lambda)\delta_{t,s} \qquad\qquad (13\text{-}4.38)$$

where Θ_u, Θ_e and Θ_λ with dimensions $d_u = \dim(\Theta_u)$, $d_e = \dim(\Theta_e)$, $d_\lambda = \dim(\Theta_\lambda)$ do not have any common parameters. Then as N_1 and N_2 tend to infinity we have:

$$EW_1^\Lambda = d_\lambda(d_\lambda + 1), \text{ if } d_\lambda = 1 \text{ we have } W_1^\Lambda \xrightarrow{\text{dist}} \chi^2(1) \qquad (13\text{-}4.39)$$

$$W_1^O/(1 + N_1/N_2) \xrightarrow{\text{dist}} \chi^2(d_u) \qquad\qquad (13\text{-}4.40)$$

$$W_1^e/(1 + N_1/N_2) \xrightarrow{\text{dist}} \chi^2(d_e) \qquad\qquad (13\text{-}4.41)$$

(b) Assume that the filters $G(q^{-1}; \Theta)$ and $H(q^{-1}; \Theta)$ may contain some joint parameters. Let $d = \dim(\Theta)$. We have as N_1, N_2 tend to infinity that:

$$\frac{W_2^e/\det\Lambda_0}{1+N_1/N_2} \xrightarrow{\text{dist}} \chi^2(d) \qquad\qquad (13\text{-}4.42)$$

$$W_3^e/(1+N_1/N_2) \xrightarrow{\text{dist}} \chi^2(d) \qquad\qquad (13\text{-}4.43)$$

We will show the outline of the derivations of the results (13-4.40) and (13-4.41), while the detailed derivations for the others are found in Söderström and Kumamaru (1985a). From (13-4.5) the prediction error $\epsilon(t,\theta)$ can be written as:

$$\epsilon(t,\theta) = H^{-1}(q^{-1};\theta_e)y(t) - H^{-1}(q^{-1};\theta_e)G(q^{-1};\theta_u)u(t) \quad (13\text{-}4.44)$$

with $\theta = [\theta_u^T \ \theta_e^T]^T$

Thus we have:

$$\Psi(t) = [\ \frac{\partial\epsilon(t,\theta)}{\partial\theta_u} \ \frac{\partial\epsilon(t,\theta)}{\partial\theta_e} \ ]_{\theta=\theta^*} = [\Psi_u(t) \ \Psi_e(t)] \qquad (13\text{-}4.45)$$

where:

$$\Psi_u^T(t) = -H_0^{-1}(q^{-1}) \frac{\partial G(q^{-1};\theta_u)}{\partial\theta_u}\bigg|_{\theta=\theta^*} u(t) \qquad\qquad (13\text{-}4.46)$$

$$\Psi_e^T(t) = -H_0^{-1}(q^{-1}) \frac{\partial H(q^{-1};\theta_e)}{\partial\theta_e}\bigg|_{\theta=\theta^*} e(t) \qquad\qquad (13\text{-}4.47)$$

Assume the open-loop operation (i.e., $u(t)$ and $e(t)$ are independent), then P_i in (13-4.26) becomes:

$$P_i = [\ E_i\Psi^T(t)\Lambda_0^{-1}\Psi(t)\ ]^{-1}$$

$$= \begin{bmatrix} [E_i\Psi_u^T(t)\Lambda_0^{-1}\Psi_u(t)]^{-1} & 0 \\ 0 & [E_i\Psi_e^T(t)\Lambda_0^{-1}\Psi_e(t)]^{-1} \end{bmatrix}$$

$$= \begin{bmatrix} P_{iu} & 0 \\ 0 & P_{ie} \end{bmatrix} \qquad\qquad (13\text{-}4.48)$$

On the other hand, from (13-4.28) and (13-4.34) we have:

$$W_1^o \simeq N_1(\hat{\theta}_1-\hat{\theta}_2)^T Q(\hat{\theta}_1-\hat{\theta}_2) \qquad (13\text{-}4.49)$$

$$N_1(\hat{\theta}_1-\hat{\theta}_2) \xrightarrow{\text{dist}} N(0,\bar{P}) \qquad (13\text{-}4.50)$$

with $\bar{P} = P_i(1+N_1/N_2)$, $P_1=P_2$ for $i=1,2$

where from equations (13-4.30) and (13-4.46) the matrix Q has the form:

$$Q = \begin{bmatrix} P_{iu}^{-1} & 0 \\ 0 & 0 \end{bmatrix} \qquad (13\text{-}4.51)$$

Considering the fact that a quadratic form of the Gaussian distributed variable with zero mean has the χ^2-distribution, we can conclude the result of equation (13-4.40) with a normalising factor of $1/(1+N_1/N_2)$. In a similar way, for the criterion W_1^e we have the corresponding weighting matrix Q as:

$$Q = \begin{bmatrix} 0 & 0 \\ 0 & P_{ie}^{-1} \end{bmatrix} \qquad (13\text{-}4.52)$$

Thus we obtain the result of equation (13-4.41).

Note that the results derived so far for the model structure (13-4.37), (13-4.38) assuming open-loop operation show that $W_2^e/\det\Lambda_o$ and W_3^e have the same statistical properties as the sum $W_1^o + W_1^e$. Normalised with $1/(1+N_1/N_2)$ they will all be χ^2-distributed in the ideal case. Furthermore note that the criteria W_2^e and W_3^e have also the same statistical properties in the ideal case. There might be a noticeable difference, though, in the numerical properties. W_3^e is organised as the difference between two large (almost equal) numbers, while W_2^e is organised as the difference of many small positive numbers. The later approach (i.e., W_2^e) can therefore be expected to have better numerical properties. Rounding errors in the calculation should be much less significant for W_2^e compared to W_3^e. Finally we have shown that all criteria introduced are χ^2-distributed under the normal operation of the system. This is not true for finite N_1 and N_2. However, the asymptotic results obtained can conveniently be used for a reasonable selection of the threshold value in the fault detection application (see equation (13-4.22)).

13.4.4. Numerical examples

As a system for numerical illustrations we consider a sampled damped oscillator. In continuous time the system is given by the transfer function:

$$G(s) = \omega^2/(s^2+2\zeta\omega s+\omega^2) \qquad (13-4.53)$$

In the sampled system we get the form:

$$G(z) = (b_1^o z+b_2^o)/(z^2+a_1^o z+a_2^o) \qquad (13-4.54)$$

where ω and sampling interval h are taken to be $\omega=1$, h=0.5. Several different values are considered for the relative damping ζ.

ζ	a_1^o	a_2^o	b_1^o	b_2^o
0.2	-1.597	0.819	0.115	0.107
0.25	-1.567	0.779	0.113	0.104
0.35	-1.498	0.705	0.109	0.097
0.5	-1.414	0.606	0.104	0.088

Table 13-2: The parameter of G(z) for some values of ζ

The corresponding parameter values are given in Table 13-2. We add white noise of unit variance to the output so the data are given by:

$$y(t) = G(q^{-1})u(t)+e(t) \qquad (13-4.55)$$

All realisations contain 500 data points. The input sequence is a PRBS of amplitude one and with the clock period equal to one. This input can well be described as white noise which is independent of e(t).

Two different model structures are applied, namely an ARMAX model M_1, and an output error (OE) model M_2. They are described by:

$$M_1 : A(q^{-1}:\theta)y(t)=B(q^{-1}:\theta)u(t)+C(q^{-1}:\theta)e(t) \qquad (13\text{-}4.56)$$

$$A(q^{-1}:\theta)=1+a_1q^{-1} + a_2q^{-2}, \ B(q^{-1}:\theta) = b_1q^{-1} + b_2q^{-2} ,$$

$$C(q^{-1}:\theta)=1+c_1q^{-1} + c_2q^{-2} ,$$

$$M_2 : y(t) = [B(q^{-1}:\theta)/F(q^{-1}:\theta)]u(t) + e(t) \qquad (13\text{-}4.57)$$

$$F(q^{-1}:\theta)=1+ f_1q^{-1} + f_2q^{-2}$$

When the model M_2 is applied, there are explicit relations among the criteria as given by $W_1^o = W_2^e$ and $W_1^e = 0$.

θ	θ^*	Realisation 1		Realisation 2	
		ARMAX	OE	ARMAX	OE
a_1	-1.597	-1.624	-	-1.578	-
a_2	0.819	0.849	-	0.822	-
b_1	0.115	0.107	0.098	0.076	0.080
b_2	0.107	0.105	0.118	0.162	0.158
c_1	-1.597	-1.654	-	-1.585	-
c_2	0.819	0.871	-	0.836	-
f_1	-1.597	-	-1.618	-	-1.579
f_2	0.819	-	0.844	-	0.823
Λ	1.0	1.012	1.014	1.042	1.044

Table 13-3: Parameter estimates obtained by using realisations corresponding to $\zeta = 0.2$.

Criterion	ARMAX model		OE model	
	Simulated	Theoretical	Simulated	Theoretical
$W_1^\wedge$	0.12	$2\chi^2(1)$	6.10	$2\chi^2(1)$
W_1^o	9.7	-	6.44	$2\chi^2(4)$
W_1^e	2.17	-	0	-
W_2^e	12.21	$2\chi^2(6)$	6.44	$2\chi^2(4)$
W_3^e	7.63	$2\chi^2(6)$	5.13	$2\chi^2(4)$

Table 13-4: Values of the criteria for two realisations corresponding
             to $\zeta = 0.2$.

First we consider two realisations with $\zeta=0.2$. Second order ARMAX and
OE models were used for the parameter estimation using the prediction error method. The obtained parameter estimates are given in
Table 13-3. The estimates are well in accordance with the true
values. The values of the different criteria are given in Table 13-4.
They all indicate that two models identified from the intervals I_1
and I_2 are not distinguishable for both of ARMAX and OE models.

    Next, we considered realisation of data sets with other values of
the damping ζ. These realisations were compared, using the *five*
criteria to one of the realisation with $\zeta=0.2$. The parameter
estimates, given in Table 13-5, are well in accordance with the
theoretical results, see Table 13-2. The obtained values of the
criteria are given in Table 13-6. Some comments on the results given
in Table 13-5 and 13-6 can be made as follows:

(a) *The parameter estimates of* b_1 *and* b_2 *are quite bad. Note however that*
    $b_1 + b_2$ *has a reasonable accuracy*. This means that the static gain
    is well estimated but that the zero location is very badly
    estimated.

(b) *The values of* W_1^e *are small for the ARMAX case*. This is natural,
    since W_1^e expresses the model difference only with respect to
    the noise filter $H(q^{-1}:\theta)$. This filter is the same, namely 1,
    for all cases.

θ	$\zeta = 0.25$		$\zeta = 0.35$		$\zeta = 0.5$	
	ARMAX	OE	ARMAX	OE	ARMAX	OE
a_1	-1.535	-	-1.463	-	-1.382	-
a_2	0.760	-	0.679	-	0.578	-
b_1	0.021	0.019	0.008	0.007	-0.0003	-0.0009
b_2	0.211	0.213	0.215	0.216	0.207	0.208
c_1	-1.542	-	-1.469	-	-1.386	-
c_2	0.777	-	0.695	-	0.592	-
f_1	-	-1.534	-	-1.461	-	-1.378
f_2	-	0.760	-	0.677	-	0.573
Λ	0.982	0.983	0.981	0.982	0.981	0.982

Table 13-5: Parameter obtained for various values of ζ.

Criterion	ARMAX model			OE model		
	$\zeta=0.25$	$\zeta=0.35$	$\zeta=0.5$	$\zeta=0.25$	$\zeta=0.35$	$\zeta=0.5$
W_1^{Λ}	0.11	0.12	0.12	0.12	0.13	0.13
W_1^{o}	28.48	69.49	122.0	25.2	64.1	115.8
W_1^{e}	2.68	2.53	2.30	0	0	0
W_2^{e}	29.91	70.27	122.3	25.2	64.1	115.8
W_3^{e}	27.63	69.86	124.1	25.4	65.8	119.1

Table 13-6: Values of the criteria with various values of ζ.   In the
reference data, $\zeta = 0.2$.

(c) *The difference between the models is more pronounced when $\zeta=0.5$ than when $\zeta=0.25$.* This is fairly natural since $\zeta=0.25$ is closer to the value $\zeta=0.2$ used in the reference data set.

(d) *All the criteria, except $W_1^{\wedge}$ and W_1^e, have almost the same values.*

(e) *Assume for a moment that the models come from the same system.* Then the criteria W_2^e and W_3^e should be distributed as $2\chi^2(6)$ for the ARMAX model. The models for $\zeta=0.2$ and $\zeta=0.25$ can then be discriminated at a significance level of about 95%. For the output error model the distribution is $2\chi^2(4)$ for W_2^e and W_3^e. The significance level becomes in these cases 96%. Needless to say, the significance level for discriminating the models becomes considerably larger when the second model corresponds to $\zeta=0.35$ or $\zeta=0.5$.

From the results, we can conclude that the criteria introduced here can be used effectively as indexes for fault detection.

13.5 CONCLUSIONS

We have proposed two statistical methods for fault and failure diagnosis based on state-space and input-output models, respectively. First, a technique of hierarchical diagnosis for linear dynamic systems with state-space representations has been developed for systematic fault detection and identification. The diagnosis system has two levels. In the first level of diagnosis, a parameter monitoring scheme is used. In the second level, a GLR test is applied based on oriented fault modelling. In particular the parameter monitoring scheme seems to be a new contribution and can be implemented by an on-line procedure for fault detection. These two decision structures are combined in a hierarchical way with exchanging information between them. The effectiveness of the method has been confirmed through simulation studies for a second order system.

Next for black-box type systems, a batch method for fault detection has been proposed by using the detection index of the Kullback discrimination information (KDI). Note that such a batch procedure can easily be extended to an on-line scheme, by replacing $\hat{\theta}_2$ obtained from the interval I_2 with a current estimate $\hat{\theta}_2(k)$. Several new criteria for fault detection have been derived from the

KDI analysis. The criteria have meaningful information about the model structure and the noise statistics. Thus they can effectively be used as indexes for model discrimination. Furthermore, we have shown that the criteria are asymptotically χ^2-distributed in the ideal case. The results obtained can hence be used for a choice of a reasonable threshold value under the consideration of some significance level specifications. Finally several numerical examples have been shown to demonstrate the effectiveness of the method.

Chapter Fourteen

FAULT DETECTION THRESHOLD DETERMINATION
USING MARKOV THEORY

Bruce K. Walker

In designing a fault-tolerant system, it is very useful to have a quantitative means of evaluating the performance obtained from a particular set of design choices. This chapter presents a means for computing performance measures such as the overall reliability for fault-tolerant systems subject to random faults and to random diagnosis system errors. The systems treated by the technique are limited to those which use memoryless diagnostic tests. The methodology is based on the theory of finite state Markov processes, which is reviewed briefly. A means for optimal selection of test thresholds is described and applied to a real example, namely the redundant inertial measurement system of the Space Shuttle.

14.1 INTRODUCTION

The other chapters of this book have shown that a wide variety of techniques are available for diagnosing faults in dynamic systems. After considering the assumptions and restrictions imposed by each of these strategies, the designer of a fault diagnosis system must select the types of diagnostic tests to be used by the system he or she is designing. Then, he or she must design the logic system that combines the outcomes of the various diagnostic tests into decisions regarding the status of the system components. This design process is a formidable challenge for even the most experienced fault-tolerant system designer.

The design problem is complicated by the fact that many of the parameters of the diagnostic tests must be selected. These parameters include the rate at which each test is executed, the number of consecutive indications required of each test before the indication is accepted as correct, and other timing parameters.

The most important of the parameters to be selected are the thresholds against which the numerical values of the decision functions of the tests are compared to generate the test outcomes. The thresholds determine such crucial test properties as the rate of occurrence of false alarms, the likelihood of missed detections and the mean time to detection of faults. As a result, the selection of test thresholds is a significant design problem all by itself. It becomes more difficult when it is part of a larger overall fault diagnosis system design problem where the overall system of which it is a part must have the highest possible reliability over a fixed length operating time.

Assume that the designer has already selected the tests to be used and the reconfiguration logic that will be used in response to the test outcomes. Suppose the rates of execution of the tests are already fixed so that the only remaining design decision is the selection of the threshold values. In order to make this selection, the designer would like to know the overall system reliability as a function of the thresholds. Small values for the thresholds typically result in a high likelihood of *false alarms* which reduces the system reliability because the number of available instruments is likely to be reduced unnecessarily. Large values of the thresholds result in an increased likelihood of *missed alarms* which increases the likelihood that failed components will be used by the system with the result of degraded or unacceptable performance. Therefore, large threshold values can also decrease the system reliability. How can these relationships be quantified so that the designer can make the optimum choice of threshold values?

One method for evaluating quantities such as the overall system reliability is Monte Carlo simulation. The complete system design, including the fault diagnosis tests and the fault identification logic, can be simulated on a computer with the random effects (such as fault events and sensor noises) generated by random number generators. The simulated system can then simulate the mission in question numerous times for each selection of the threshold values until statistically significant reliability results are obtained. The primary problem with the Monte Carlo method is that *numerous times* in this case may mean millions or billions because the probability of fault occurrence for highly reliable components is so low. Furthermore, even if this difficulty could be overcome, a problem still remains in generating results for many different sets of threshold selections. If any of the threshold values are changed, the entire simulation process must be repeated to generate new results. The process must also be repeated if any other system property (such as the failure rate of a component) is changed.

Combinatorial analysis is another option which can be used for evaluating quantities related to overall system performance when random events are involved. It is a well-established method for engineering systems, Shooman (1968). However, combinatorial methods have difficulty accounting for the time order of events, and the order can be crucial to the effect on the system, Luppold et al. (1984). For example, combinatorial analysis will typically treat the combination of the fault of an instrument and a fault alarm for that instrument the same regardless of the order in which they occur. Note, however, that if the fault occurs first, the end result is a detected fault, which should result in a satisfactory operating system, whereas if the fault occurs after the alarm, the reconfiguration logic may have already eliminated the component from use soon enough that the remaining components have a higher probability of developing a fault in the remaining time. Combinatorial analysis can be altered to take the time order of events into account, but only at the expense of very complicated integral formulas to be solved for the quantities of interest, Luppold (1983).

This chapter describes an analysis technique that can be used to select test thresholds such that a measure of the overall performance of the fault diagnosis system is optimised. Furthermore, the technique generates a model for the behaviour of the system that can easily be modified to reflect changes in the threshold selection or in other system parameters such as the fault rates. The technique is based upon the well-known theory of Markov processes. After the fundamentals of the technique are discussed, it is applied to a real-world example - the fault diagnosis system for the inertial measurement units (IMUs) on the Space Shuttle during its reentry and landing phase. The material on the example is taken primarily from Walker and Gai (1979).

In the next section, the modelling assumptions inherent to the technique are discussed and the concept of a single measure of the system performance are explained. The subsequent section reviews the basic concepts of finite state Markov chain models which will be used in the analysis. The steps that are necessary to construct a Markov model of the behaviour of a fault diagnosis system are described in section 14.4 and the optimisation method for selecting test thresholds is described. Finally, the method is applied to the fault diagnosis system for the Space Shuttle IMUs during reentry and landing in the last section.

14.2 ASSUMPTIONS AND RESTRICTIONS OF THE TECHNIQUE

Any actuating, sensing or control system is designed to satisfy given specifications. Very frequently, a single *critical* specification exists that must be satisfied by the system design. For fault-tolerant systems, this critical specification is usually the reliability of the system after a fixed time period. This is not the only possibility, however. In the example to be presented later in this chapter, the critical specification is the probability that, due to combinations of component faults and incorrect fault diagnosis system actions, the navigation system produces errors that are large enough to result in a *vehicle loss*. Other possibilities include the probability that a given contingency status is achieved or the probability that a particular combination of operating components remains available for use at all times. Many other examples of critical specifications exist.

We shall assume here that a critical specification has been selected. The technique to be presented below applies to any scalar-valued critical specification that is an instantaneous function of the status of the system at a specified point in time. This includes, as one possibility, the most frequently used specification, namely the reliability of the system at a given point in time. It will be the value of the critical specification that will be optimised by the choice of the test thresholds.

For the purposes of this chapter, it is assumed that the fault diagnosis tests have a *no memory* property. Specifically, we assume that the outcomes of the fault diagnosis test at a particular sample instant are statistically independent of the outcomes at other sample instants given the failure status of the components involved in the testing. Unfortunately, this assumption rules out the use of the technique on systems that employ sequential fault diagnosis tests (such as the SPRT of Wald (1973)) or tests based on a finite length of data (such as the GLR of Willsky and Jones (1976)). The method can be extended to apply to the latter case, but the resulting system model becomes a dimensional burden unless the finite memory of the test is very short. An extension of the method to systems with sequential tests has also been accomplished, Walker (1980) and Walker (1981), at the expense of a very complicated model to be solved. Walker et al. (1985) and Walker et al. (1986) discuss the issues associated with such models in further detail.

It is also assumed that the probabilities of all the various fault diagnosis test outcomes are known as functions of the test thresholds given the fault/no-fault status of the components. It is sometimes a difficult problem to find these probabilities. However, the reader should note that the methods for finding test thresholds in classical detection theory use these same probabilities, Van Trees (1968). In particular, it is easily shown in classical binary hypothesis testing problems that the optimal test for most of the standard cost functions is a likelihood ratio test. The threshold for the test then depends upon the *false alarm* probability and the *missed detection* probability. These are exactly the probabilities that we assume are known in the analysis that follows. They can be determined in closed form if the test statistic has a simple distribution (e.g. exponential) or they can be determined numerically if the test statistic distribution is more complicated.

To illustrate this assumption, consider the following example. A sensor bias fault is to be detected by comparing the outputs of two sensors where each output is corrupted by Gaussian white noise. Let σ^2 be the known variance of the noise on each sensor and let the fault to be detected have known magnitude B. It can be shown that the optimal test in such a case is a likelihood ratio test, Van Trees (1968). For this particular example, the test can be reduced to the comparison of the test statistic formed from the difference of the sensor outputs to a threshold. In particular, if y_1 is the output of sensor 1 and y_2 the output of sensor 2, then the test takes the form:

    Declare a fault present if $|y_1 - y_2| > T$

where T is the threshold. The difference $y_1 - y_2$ is obviously Gaussian with variance $2\sigma^2$ and mean value zero in the absence of a fault or mean value ±B if a fault is present on one of the sensors. The sign of the mean value in the fault case depends upon the polarity of the fault and upon which instrument is affected. If we assume that the polarities are equally likely and that the fault occurrences of each instrument are equally likely, then positive and negative mean values occur with equal likelihood.

    With this information available, we are now ready to evaluate the probabilities we need. In the absence of a fault, the probability that the test yields a fault indication (or a *false alarm*) is the probability that the value of the test statistic exceeds T given that no faults are present. From symmetry, we have that this probability is given by:

$$P\left\{ |y_1 - y_2| > T \mid \text{no fault} \right\} = 2\left[1 - \Phi\left(\frac{T}{\sigma\sqrt{2}}\right)\right]$$

where $\Phi(x)$ is the probability that the value of a zero-mean, unit variance Gaussian random variable is less than or equal to x. The values of $\Phi(x)$ are tabulated in many standard mathematical references, such as Abramowitz and Stegun (1965). Note that the false alarm probability decreases as T increases because $\Phi(x)$ is a monotonically increasing function of x.

    When a fault is present in one of the instruments, the mean value of the difference $y_1 - y_2$ is no longer zero but rather is +B or −B depending upon the instrument that is affected and upon the polarity of the fault. Once again taking symmetry into account, we find that the probability of a fault indication in this case (i.e. the detection probability) is given by:

$$\Pr\left\{|y_2 - y_1| > T \mid \text{fault present}\right\} = 1 - \Phi\left[\frac{T - B}{\sigma\sqrt{2}}\right] + 1 - \Phi\left[\frac{T + B}{\sigma\sqrt{2}}\right]$$

Note that the detection probability decreases as T increases.

    If the system designer is interested only in optimising the performance of this particular detection test, he or she might choose a threshold that optimises some combination of the false alarm probability and the detection probability. For instance, the minimum probability of error test will result from choosing the threshold to minimise:

$$\Pr\{\text{error}\} = P_{FA} \cdot \Pr\{\text{no fault present}\} + (1 - P_D) \cdot \Pr\{\text{fault present}\}$$

where P_{FA} is the probability of false alarm and P_D is the probability of detection, both of which were expressed above. This is a relatively easy optimisation problem and can usually be solved quickly by trial and error. However, if the goal of the system designer is to optimise a quantity such as the overall system reliability at a specified point in time, then a more sophisticated analysis is required to find this quantity as a function of the threshold T and of any other thresholds that may be involved in the diagnostic strategy.

The probabilities that were calculated above will be referred to as the *performance probabilities* of the test. They will play a key role in the analysis technique to be described below.

When the performance probabilities cannot be obtained in closed form, as was the case for the example just described, we shall assume that they can be obtained as a function of the test threshold in some other manner. This may involve Monte Carlo simulation of the test. The reader might react with suspicion upon encountering a suggestion here that Monte Carlo simulation be used as part of the analysis when the point was made in the introduction that simulation is impractical for the purpose of evaluating system designs. Note, however, that in the previous discussion, the generation of results involved simulating the *entire* fault diagnosis system and all of its random effects. In order to generate the performance probabilities for an individual test, one needs only to simulate the *particular test* with its random effects, which in this case amount to just the noise on the signal being tested. This is almost never impractical to accomplish, although closed form results are still preferable to simulation results whenever a choice is available.

With the assumptions made that a critical specification meeting the restrictions stated above has been selected and that the performance probabilities of each test involved in the fault diagnosis strategy can be found as functions of their respective thresholds, we are ready to describe the method for generating the system performance results. The technique for generating these results is based on the theory of finite state Markov chains. The basics of this theory will be described in the next section.

14.3 BASIC THEORY OF MARKOV CHAINS

In Section 14.4, the method for constructing a Markov model of the behaviour of a fault diagnosis system will be discussed. The result of the modelling process will be a finite state Markov process. The properties of these processes that are of primary interest in analysing these models will be presented in this section. Proofs of the properties are omitted. The interested reader is referred to Howard (1971) for further details.

A finite state Markov process is a random process that possesses the following two properties:

(a) The process can take on only a finite number N of discrete values. These values can be associated with the integers 1, 2,..., N. The process is then said to *occupy state i* when it takes on the value associated with the integer i.

(b) The probability that the next state to be occupied by the process is j conditioned upon any state history reduces to the probability that the next state to be occupied is j given only the current state that is occupied, and this occurs for all j. In other words, the future transitions of the process are independent of all past transitions except for the currently occupied state. This is the so-called *Markov property*.

As a result of these two properties, the probabilistic behaviour of a finite state Markov process is completely characterised by the initial distribution of its value and the history of the state transition probability matrix that describes the probabilities of occupying each of the states at the next transition given the current state that is occupied.

For a discrete time Markov process, let $\pi(k)$ be the vector of N probabilities of occupying the states at time sample k. Thus, $\pi(k)$ is the state probability vector. Let P(k) be the matrix of transition probabilities at time sample k, i.e. $p_{ij}(k)$ is the probability that state j is occupied at time step k+1 given that state i is occupied at time step k. P(k) completely characterises the behaviour of the process because of Property 2 above. P(k) and π(k) have some properties that are useful later in verifying numerical results. Among these properties are the following:

(a) The elements of π(k) sum to unity for all k.

(b) The elements or each row of P(k) sum to unity for all k.

(c) The eigenvalues of P(k) for any k are all within the closed unit disc. Furthermore, P(k) has no repeated unit magnitude eigenvalues that do not possess a linearly independent set of eigenvectors (i.e. geometric multiplicity of *zero*).

Given the initial distribution of the value of the process, i.e. given π(0), the value of π(k) for any k $\geqslant$ 0 can be found from the difference equation:

$$\pi(k+1) = \pi(k)\ P(k) \qquad k \geqslant 0 \qquad\qquad (14\text{-}3.1)$$

The solution to this equation is:

$$\pi(k) = \pi(0) \cdot P(0)...P(k-1) \qquad k > 0 \qquad\qquad (14\text{-}3.2)$$

For the special case where P(k) is a constant matrix P, i.e. for the case of a time-invariant process, the solution for the state probabilities reduces to:

$$\pi(k) = \pi(0)\ P^k \qquad k \geqslant 0 \qquad\qquad (14\text{-}3.3)$$

If P is diagonalisable, this solution can be expressed in modal form as:

$$\pi(k) = \sum_{i=1}^{N} \lambda_i^k\ \pi(0) v_i w_i^T \qquad k \geqslant 0 \qquad\qquad (14\text{-}3.4)$$

where the λ_i are the eigenvalues of P, the v_i are the corresponding right eigenvectors, and the w_i are the corresponding left eigenvectors. Because of the properties of P above, the solution for π(k) is guaranteed not to grow unbounded.

If P has a single eigenvalue of unity, then a unique steady-state solution exists for π(k), namely:

$$\lim_{k \to \infty} \pi(k) = C\ w_1^T \qquad\qquad (14\text{-}3.5)$$

where w_1 is the left eigenvector associated with the eigenvalue of unity and C is a scaling constant chosen such that C w_1^T has elements that sum to unity.

In many instances, the first Markov model that a designer produces for a fault diagnosis system of interest is so large that it is unwieldy. In such cases, it is sometimes possible to reduce the size of the model by formally merging some of the states. States i and j of a Markov model can be merged if $p_{ik} = p_{jk}$ for all $k \neq i,j$. The states can then be aggregated as one state. Let us denote this state as state i, thus state j is eliminated. The merging is accomplished by deleting the jth row of P, replacing the ith column of P with the sum of the ith and j^{th} column, and eliminating the jth column. This reduces the dimension of the model, but it also eliminates the possibility of distinguishing between the conditions that were represented by states i and j before they were merged. Thus, it is not wise to merge states that have different performance or operating characteristics in a fault diagnosis system model even if the mathematical conditions for merging are satisfied.

Because the sequence of P(k) and the value of $\pi(0)$ provide a complete characterisation of the behaviour of a Markov process, a number of other statistical properties of the process can be derived from knowledge of their values. These include the statistics of the time to first passage of particular states, the statistics of the occupancy times of the states over a fixed time period, and the probabilities of various trajectories through the states. (For details on computing such quantities, see Howard (1971)). Many of these properties have meaning in a fault diagnosis system analysis setting. For example, the first passage to a particular state of the model may represent the first occurrence of a particular operating condition for the fault diagnosis system. If the goal of the system design is to maximise the time to this first occurrence, then the corresponding first passage time statistics are of prime interest to the system designer in evaluating a particular set of design parameters.

This section has described the powerful results that can be generated from a Markov model of fault diagnosis system behaviour. In the next section, the steps which are necessary to construct these models will be discussed.

14.4 CONSTRUCTING MARKOV MODELS OF FAULT DIAGNOSIS SYSTEM BEHAVIOUR

The preceding section described the basic theory of Markov processes. The purpose of this section is to describe the steps that are necessary to construct a Markov model of a fault diagnosis system that satisfies the restrictions discussed previously.

Fault-tolerant systems are subject to many different sequences of random events that can result in changes in their performance or in the loss of their effectiveness. The random events comprising these sequences are component faults and the outcomes of the various fault diagnosis tests employed by the system. Combinatorial analysis does not capture the nature of all of the various sequences of events because the order in which they occur can be significant, as has been discussed already. Markov models capture these effects accurately because, as we saw above, a Markov model completely characterises the behaviour of the process.

A Markov model description of the behaviour of a fault diagnosis system consists of three entities: the definitions of the states, the definition of the initial state probabilities, and the definition of the transition probabilities. Let us examine these three entities one at a time.

(a) *State definitions*. Each of the states of a Markov model of fault diagnosis system behaviour represents a particular operating condition, where the operating condition is defined by the number of instruments that are operating, how many (if any) of them are faulty, the nature of these faults, the status of each of the fault diagnosis tests, and the logic that is in use for selecting components to be used by the system. Thus, knowledge of the operating condition or state of the system allows us to specify exactly what the performance of the system is at that time. Among other things, this knowledge allows us to determine whether the system is operating satisfactorily or not. If not, we say that the system is *lost*.

For example, the system is normally started with all of its components in working order and no fault indications from any of the diagnostic tests. This condition is a state of the system, and in fact it is the state that ordinarily yields the best performance for the system. If an instrument is faulty but its malfunction is not immediately detected, this condition becomes another state of the system. Obviously, the performance of the system in this state is degraded relative to the initial performance. If the degradation is unacceptable, then the system is lost when it reaches this state. If a component develops a fault or fails and is immediately diagnosed properly so that corrective action takes place immediately, this yields a condition for the system whereby fewer than the maximum complement of components are in use. This is yet another state for the system, and in most cases it is an acceptable state.

As we shall see shortly, it is not necessary to specify all of the states for the system at the beginning of the modelling process. However, it is important to specify the states that are the most likely to be the initial states of the system operation.

(b) *Initial state probabilities.* In most cases, the initial condition of the system will be known and it will typically be the condition where all instruments are working and no failure indications have been produced by the fault diagnosis tests. In this case, the corresponding state of the Markov model has *probability one* associated with it and the other states all have *probability zero* associated with them.

This need not always be the case, however. If the designer knows that under some circumstances the system does not begin its mission with all good instruments and no test indications, then he or she must specify what the other possible initial system conditions are and specify probabilities for each such condition such that the sum of these probabilities is one. The probabilities associated with the various conditions are used to construct $\pi(0)$.

(c) *Transition probabilities.* The construction of transition probabilities is the crucial step in developing a Markov model of fault diagnosis system behaviour. The most straightforward procedure to follow, though not necessarily the fastest, is to construct event trees for each of the states that were initially specified. An event tree is a graphical representation of the event sequences that can occur in a single time step. From it, the transition probabilities can easily be found and any other states of the system that need to be considered will become apparent.

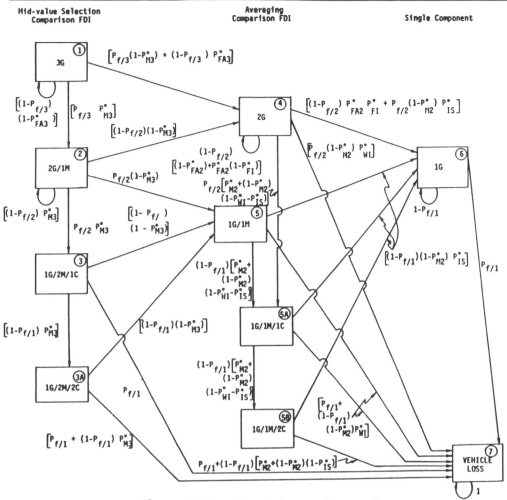

Figure 14-1: Event tree of example

Let us explain the construction of event trees with an example. Suppose the system is known to experience only one instrument fault at a time. Furthermore, suppose that the diagnosis logic for this system consists of a no-memory fault detection test which, if it indicates a fault, is instantaneously followed by a no-memory isolation test to isolate the faulty instrument and also to reject false detection alarms. The event tree for this situation is shown in Figure 14-1.

In a single time step, a number of event sequences are posssible. First, a component can malfunction or no components can malfunction, as indicated by the first branch point on the tree. The symbols on the branches indicate the probabilities with which each of these events occurs (P_f and $1-P_f$, respectively, in this case).

Regardless of whether a fault has occurred or not, at the end of the time step the detection test will reach a decision. This is the second branch point in the tree in Figure 14-1, and the possible decisions by the test are that a fault has occurred or that no fault has occurred. The probabilities associated with these decisions are *conditioned* on the fault status of the components being tested, which is indicated by the events on the preceding branches of the tree and the state of the system at the previous time step. The probabilities that appear on these branches are the false alarm probability P_{FA}, which is conditioned on the absence of a fault, and the detection probability P_D, which is conditioned on the presence of a fault. Note that these are precisely the performance probabilities for the detection test, which we assumed were known in the discussion of the restrictions and assumptions associated with the technique.

If the detection test indicates that a fault is present (even if one is not), then an immediate isolation test is executed, and it has three possible outcomes, as shown by the third branch point in Figure 14-1. Once again, the probabilities that appear on the branches following the third branch points are the *conditional* probabilities of the various isolation test decisions given the system condition indicated by the state at the preceding time step and by the branches preceding this branch point. For the isolation test outcomes, this includes conditioning on the outcomes of the detection test. If the same data are used by both tests, this may present a problem because the correlation of the test outcomes must be accounted for. However, it is often the case that the outcomes of the tests are sufficiently uncorrelated that the performance probabilities for the isolation test without conditioning on the detection test outcome provide adequate accuracy. In any case, once again the probabilities to be used here are exactly the performance probabilities that were assumed to be known. In this example, these probabilities are the probability that the isolation test will falsely isolate an instrument following a false detection alarm (P_{FI}), the probability that it will correctly isolate the faulty component following a correctly detected fault (P_{IS}), and the probability that the test will isolate the wrong component following a proper detection alarm (P_{WI}).

The outcomes that are listed at the ends of the tree branches are the various operating conditions for the system that result from the condition of the system at the preceding time step and from the events during the current time step that are indicated by the path through the tree leading to each outcome. For instance, one of the outcomes in the example in Figure 14-1 is *correct isolation*. This results from the condition of no faults and no alarms at the previous time step when, during the current time step, a fault occurs, is detected, and is properly isolated. The result is that the system is now operating with a reduced number of components (one has been eliminated), all of which are good (because the faulty one was properly isolated), and with no outcomes from any diagnostic tests that may be in place for this new operating condition. Clearly, this new status for the system represents another state in the model. Another example of an outcome indicated in Figure 14-1 is *missed isolation*. This results from the conditions of the previous time step if a fault occurs and is properly detected but the isolation test fails to isolate any instrument and rejects the detection alarm. The result is a system that continues to use the faulty component that is present and also continues executing the detection test to monitor the components for further malfunctions. This is another system condition that should be represented by a state in the model.

Thus, we see that the construction of the event trees leads to the definitions of the other states that must be included in the model. For each of these additional states, an event tree similar to the one just described must be constructed. This may, in turn, lead to additional states. Eventually, all of the conditions under which the system continues to operate will appear as outcomes of the various event trees. At that point, the model is complete.

The key to the successful construction of event trees is to be certain that all of the events that can occur in a single time step are included in the tree. This automatically leads to a model that possesses all of the properties that characterise Markov models.

Once all of the event trees for the states have been constructed, it is a relatively simple matter to generate the transition probabilities. The transition probability associated with transitions from some state i to some other state j of the model is found from the event tree that is constructed with state i as its initial condition. The desired probability is just the product of the conditional probabilities along the branches of the tree that lead to state j. If more than one branch leads to state j, then the probabilities from each branch are summed to yield the final transition probability. In the example, then, the transition probability leading from the state representing a system with all good components and no test outcomes to the state where a fault has occurred and been correctly isolated is $P_f \cdot P_D \cdot P_{IS}$. Similarly, the transition probability from the *all good/no outcome* state to the state where a component has been eliminated by a false isolation is $(1-P_f) \cdot P_{FA} \cdot P_{FI}$. If there is no distinction between the state representing the loss of one component due to a false isolation and the state where a fault has been correctly isolated, then they can be combined and the appropriate transition probability to this combined state from the *all good/no outcome* state is the sum of these two probabilities, namely:

$$P_f\, P_D\, P_{IS} + (1 - P_f)\, P_{FA}\, P_{FI} \qquad\qquad (14\text{-}4.1)$$

Finally, the transition probabilities derived from the event trees can be collected into the single step transition probability matrix P. This matrix can then be used in the manner described previously to generate the desired statistical properties of the fault diagnosis system behaviour. For instance, if the probability that a particular operating condition represented by one of the states is in force at a particular time step k is the quantity of interest, then it can be found as the appropriate element of the state probability vector $\pi(k)$, which is obtained from the formula:

$$\pi(k) = \pi(0)\, P^k \qquad\qquad (14\text{-}4.2)$$

where we assume that P is constant (if P varies with time, we use the product formula given earlier).

Most fault-tolerant-system behaviour models include one state that represents the condition of unacceptable system performance, or *system loss*. This state is almost always a trapping state in the sense that there is no nonzero probability associated with any transitions out of this state. When this is the case, the state probability associated with this state is the system unreliability, i.e. the probability that the system function has been lost. As we have already discussed, this quantity is often the primary measure of the performance of a fault diagnosis system. Thus, having a means for its calculation is very valuable to the system designer. The use to which such performance measures can be put will be made clearer by the example to be discussed later in this chapter.

14.5 OPTIMAL SELECTION OF FAULT DETECTION THRESHOLDS

As we have mentioned, the ability to calculate various statistical system properties is an important capability available to the fault diagnosis system designer. It allows him or her to examine the impact that different sets of design parameters have on such quantities as the reliability and availability of the overall system. However, an even more valuable capability is a means to choose the design parameters. With this additional capability, the designer can optimise the choice of parameters such as the thresholds for the diagnosis tests that are used by the fault diagnosis system. We shall now describe a method which uses the Markov modelling technique just presented to generate optimal diagnosis test thresholds.

Suppose that the critical specification has been chosen for the system design and that the Markov model has been constructed such that one of the elements of $\pi(k)$ for some fixed k is the probability that the critical specification is violated. The system designer wishes to select the thresholds for the diagnostic tests that are used by the system such that this probability is minimised. It is assumed again here that the performance probabilities of the diagnosis tests are known as functions of the thresholds.

Since the Markov model is available, one means for selecting the optimal thresholds is to use the model to calculate the appropriate state probability for many different choices for the threshold values. Then, the set of values yielding the smallest value of the probability of violating the critical specification is selected as the optimal set. The primary difficulty with this method is that many trial sets of values may be required before the optimal set is established. This may require many evaluations of the state probabilities using the Markov model, which may require a large investment of computer time and of the designer's time (although still not nearly as much time as Monte Carlo simulation of all these cases would require). Furthermore, this type of trial and error search yields very little information regarding the sensitivity of the system loss probability to the various thresholds that are selected.

A more elegant and efficient approach to determining the optimal thresholds is by a relatively simple gradient optimisation technique. Suppose for the moment that the sensitivities of the system loss probability to each of the elements of P were available. Denote these sensitivities as $\partial P_{SL}/\partial p_{ij}$ where $1 \leqslant i$, $j \leqslant N$ where N is the number of states in the model. The partial derivatives of the p_{ij} with respect to each of the test performance probabilities can be found analytically because the p_{ij} are constructed from the performance probabilities (among other things) in the modelling process. Denote these partials as $\partial p_{ij}/\partial \gamma_m$ where the γ_m are the performance probabilities. Finally, since the performance probabilities were assumed to be known as functions of the thresholds, we can find the partial derivatives of the performance probabilities with respect to the threshold from the known forms. If the forms are tabular, then the partial derivatives must usually be found numerically by differencing consecutive entries in the table and dividing by the difference in threshold values. However, it is interesting to note that some distributions of the test statistic yield closed forms for these partial derivatives even though they do not yield closed forms for the performance probabilities as functions of the threshold. Two examples of such distributions are the Gaussian distribution and the chi-squared (χ^2) distribution because they possess closed form density functions. In any case, let these partial derivatives be denoted $\partial \gamma_m/\partial T_n$ where the T_n are the thresholds to be selected.

With these partial derivatives available, it is now possible to construct a simple gradient technique for selecting the optimal thresholds. We begin the algorithm with an initial guess at the thresholds T_n^0. For this set of thresholds, we compute the system loss probability by using the Markov model and the $\partial \gamma_m/\partial T_n$ using the known forms of the performance probabilities as functions of the thresholds. We also compute the $\partial p_{ij}/\partial \gamma_m$ from the closed forms of the p_{ij}. Assuming the $\partial P_{SL}/\partial p_{ij}$ are available, we have that the nth element of the gradient of P_{SL} with respect to the vector $\underline{T}$ of thresholds is given by:

$$\frac{\partial P_{SL}}{\partial T_n} = \sum_{i=1}^{N} \sum_{j=1}^{N} \frac{\partial P_{SL}}{\partial p_{ij}} \sum_{m} \frac{\partial p_{ij}}{\partial \gamma_m} \cdot \frac{\partial \gamma_m}{\partial T_n}$$

The algorithm proceeds by adjusting the threshold vector $\underline{T}$ according to:

$$\underline{T}^{i+1} = \underline{T}^i - \alpha \frac{\partial P_{SL}}{\partial \underline{T}}$$

where $\alpha > 0$ is a step size control parameter that can be chosen by any one of a number of methods. In the above equation, $i=0$ on the first step of the algorithm. The new threshold values are then used with the Markov model to recalculate the system loss probability. If it has been reduced, another step of the algorithm is executed beginning with the evaluation of the partials. If the system loss probability is unchanged to within a pre-specified tolerance, the last set of thresholds are selected as the optimal set. If the system loss probability is increased, the step size control parameter is reduced and the step is re-executed.

This method has been applied to the problem of selecting the optimal thresholds for the redundant Inertial Measurement Units (IMUs) on the Space Shuttle. The next section describes the results of this application.

14.6 APPLICATION TO SPACE SHUTTLE IMUs

In this section, we shall consider the application of the threshold determination technique to the problem of choosing the thresholds for the redundancy management system of the Space Shuttle inertial measurement system. The Shuttle is equipped with a triplex array of identical inertial measurement units (IMUs). Each provides an independent gimballed reference with respect to which the three-dimensional measurements of angular velocity and linear acceleration are made. The redundancy management logic is prespecified, and it calls for the termination of use of a complete IMU upon the identification of a fault in either a gyroscope or an accelerometer within it. As will be explained below, the RM logic consists of a detection test for each of the two types of instruments which, if it indicates a fault is present, triggers an isolation test to determine which IMU contains the faulty component. When all three IMUs are operating, this process is essentially equivalent to a majority vote. However, when only two units remain operational, the identification process becomes more complex. The system is intended to remain operational despite the loss of as many as two IMUs.

The discussion below of the application of the threshold determination method described in the previous sections to this example follows closely the presentation in Walker and Gai (1979). The interested reader should turn to the reference for further explanation.

The mission to be considered is the first Orbital Flight Test (OFT) mission of the Shuttle. The mission is of approximately 45 min duration beginning with a deorbit burn to re-enter from a 120-mile-high orbit and ending with touchdown on a runway at Edwards AFB. For the last 7 min of the mission, the navigation system has radio and air data navigation aids available in the form of TACAN and a barometric altimeter and, for the last 2 min, the Microwave Landing System (MLS) at Edwards AFB. Otherwise, all navigation data are derived from the redundant IMUs.

The first step in applying the method is to choose an appropriate critical specification. An appropriate choice for a cost function reflecting the performance of the FDI system in this case is the probability of vehicle loss, denoted P_{vl}. A vehicle loss is said to occur when, due to some sequence of sensor malfunctions and FDI decision errors, the navigation system is forced to utilise data from faulty sensors. A sensor is defined to be faulty (in this context) when its measurements are corrupted by a bias error sufficiently large to cause violation of the mission requirements if its output is used in the navigation computations. Otherwise, the sensor is considered good. The fault probabilities of the instruments of interest over each time step are denoted P_f. The Shuttle IMUs have a mean time to fault of 1000 hours, therefore, P for each IMU over the 16-second FDI period is approximately 4.44×10^{-6}.

The probabilities of the various possible FDI decision outcomes will be referred to as the FDI performance probabilities. We shall now determine the values of the performance probabilities as functions of the thresholds. The reader will recall that this is a necessary step in applying the threshold optimisation procedure.

It is assumed that the two stages of FDI (fault detection and identification) are independent. Let *three-level* represent all operating conditions where the FDI system indicates that three instruments are operating normally, and similarly for *two-level* and *one-level*. The navigation system selects a single measurement for use at each time point. At the three-level, a single measurement is selected by mid-value selection. At the two-level, the two available measurements are averaged.

The Shuttle IMU FDI system operates at a 16 Hz. rate and consists of two independently operating sections, one for accelerometer FDI and one for gyro FDI. For both cases, and for both three-level and two-level, the detection decision function is the squared magnitude of a three-dimensional vector representing the vector difference of the measurements from two of the IMU's. This magnitude is compared to the detection threshold to detect failures. The three-level and two-level detection thresholds are different. For three-level isolation, the IMU that is common to the two largest difference magnitudes is indicated as containing the faulty component. This amounts to a majority vote on the sensed quantity. For two-level isolation, the skew geometry of the platforms is exploited, Solov and Thibodeau (1973). The vector difference used for detection is transformed into the coordinate frame of each of the two operating platforms. In each of these frames, the two components lying in the input plane of the instrument which is used as a *two-degree-of-freedom* (2 DOF) sensor are squared and added. The remaining single-axis components are also squared. This procedure results in four decision functions, two of the 2 DOF type and two of the single DOF type. When a single bias fault is present, the component(s) of the measurement difference which is (are) orthogonal to the faulty axis (axes) are unaffected by the fault. Thus, the two-level isolation logic identifies a fault when exactly three of the four decision functions exceed their thresholds. The identified platform is the one which produces the single decision function that does not exceed the threshold.

Accelerometer bias	50 µg
Accelerometer scale factor	141 ppm
Accelerometer IA misalignment	15 arc-s
Accelerometer quantisation (1 cm/s)	1/12 cm/s
Gyro bias drift	0.035 deg/h
Gyro g-sensitive drift	
Input axis	0.025 deg/h/g
Spin axis	0.025 deg/h/g
Gyro g^2-sensitive drift	
Input/spin plane	0.025 deg/h/g^2
Spin/output plane	0.025 deg/h/g^2
Input/output plane	0.005 deg/h/g^2
Random gyro drift	0.003 deg/h
Effective resolver error (rss)	40.1 arc-s
Case to nav. base misalignment	100 arc-s
Gimbal nonorthogonalities	50 arc-s

Table 14-1: IMU error model (3σ values)

Assume that the components of the decision vectors are uncorrelated and jointly Gaussian with identical variances, approximated by the largest of the three. The decision functions can then be approximated by chi-squared (χ^2) random variables with the same number of degrees of freedom as the number of components making up the decision function (3, 2, or 1). The noncentrality parameter for this distribution is directly related to the magnitude of any fault that might be present.

By performing a covariance analysis for a normal Shuttle IMU, the required statistics are obtained of the vectors from which the decision functions are formed. The Shuttle IMU error model is presented in Table 14-1. This error model, combined with the characteristics of the mission described earlier, provides the basis for generating the variances of the decision vector. These variances will be time varying. Let $\sigma(t)$ represent the standard deviation of each component of a vector that is to be used by a FDI test. We shall approximate $\sigma(t)$ as piecewise constant for reasons explained below.

Let the performance probabilities of interest be denoted as follows:

P_{FA3} = Pr [false alarm (or Type I) error by the three-level detection logic]

P_{M3} = Pr [missed alarm (or Type II) error by the three-level detection logic]

P_{FA2} = Pr [false alarm error by the two-level detection logic]

P_{M2} = Pr [missed alarm error by the two-level detection logic]

P_{FI} = Pr [isolation decision by the two-level isolation logic given the occurrence of a false alarm]

P_{IS} = Pr [correct isolation decision by the two-level isolation logic given detection]

P_{WI} = Pr [incorrect isolation decision by the two-level isolation logic given detection]

No three-level isolation error probability is included because it is assumed that the majority voting technique used for three-level isolation has negligible probability of error. Since this isolation technique employs no threshold, an isolation decision always follows a detection, therefore missed isolation errors cannot occur. Furthermore, the probability of an incorrect isolation at this level is negligible because it is smaller than the product $P_f P_{FA3}$.

The probability of occurrence of a false alarm on a single FDI test can now be calculated directly as:

$$P_I(\eta, t) = \int_{\eta}^{\infty} p_N(u, t) \, du \qquad (14\text{-}6.1)$$

where η is the threshold and $p_N(u, t)$ is the appropriate central chi-squared (χ^2) pdf. P_{FA3} and P_{FA2} are evaluated from this equation by using the 3 DOF chi-squared (χ^2) Pdf and the $\sigma(t)$ for detection tests. A false isolation at the two-level requires three of the four

isolation decision functions to exceed the single isolation decision
threshold. Let α_1 and α_2 be the values of P_I for 1DOF and 2DOF tests,
respectively, using the isolation value for $\sigma(t)$. Then:

$$P_{FI} = 1/2 \ \alpha_2^2 \ \alpha_1 \ (1-\alpha_1) + 1/2 \ \alpha_1^2 \ \alpha_2 \ (1-\alpha_2) \qquad (14\text{-}6.2)$$

Notice that P_{FA3}, P_{FA2} and P_{FI} will all be piecewise constant functions of time because $\sigma(t)$ is.

In order to compute the performance probabilities for fault
conditions, it is necessary to know the magnitudes of the instrument
biases that represent faults. For this application, a sensitivity
analysis, Gai et al (1976) and Gai et al (1978), provided a means of
determining the probability of violating the mission requirements as
a function of the bias magnitude. These results were stored as
numerical tables, one table each for gyros and for accelerometers.

With these data in hand, the missed alarm error probability on
each FDI test is given by:

$$P_{II}(\eta,t) = \int_0^\eta \left[\int_\infty^\infty p(u,t|B) p_{MF}(B) dB \right] du \qquad (14\text{-}6.3)$$

where $p_{MF}(B)$ is the pdf derived from the distribution of violating
the mission requirements as a function of the failure magnitude B,
which is obtained from the sensitivity analysis discussed above, and
$p(u,t|B)$ is the noncentral chi-squared (χ^2) pdf given the failure
magnitude B. P_{M3} and P_{M2} are evaluated using this equation and the
detection value for $\sigma(t)$. An incorrect isolation at the two-level
occurs when exactly three isolation decision functions exceed the
isolation threshold with one threshold excursion due to a false alarm
error. A correct isolation occurs when the three appropriate
isolation decision functions exceed the isolation threshold. Let β_1
and β_2 be the values of P_{II} for 1DOF and 2DOF tests, respectively,
using the isolation value for $\sigma(t)$, and assume that failures are
equally likely affecting each of the three orthogonal input axes of
the instruments. Then:

$$P_{IS} = 1/3 \ (1-\beta_2)(1-\beta_1)[2(1-\beta_2)(1-\alpha_1) + (1-\beta_1)(1-\alpha_2)] \qquad (14\text{-}6.4)$$

$$P_{WI} = 1/3 \ (\beta_2+\beta_1-2\beta_2\beta_1) \ [2(1-\beta_2)\beta_1 + \beta_2(1-\beta_1)] \qquad (14\text{-}6.5)$$

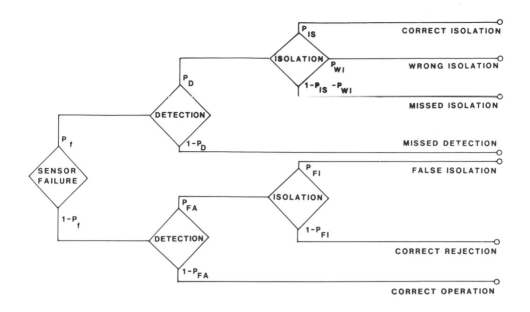

Figure 14-2: Markov model for Shuttle IMU RM system

With all of the elements in place, it is now possible to apply the Markov modelling procedure described earlier in this chapter to this system. The Markov model describing the behaviour of the Shuttle IMU RM system is shown in Figure 14-2. Each state of the model is characterised by (a) the number of good components still available for use in navigation (denoted by G), (b) the number of faulty components not yet identified by the FDI system as faulty (i.e. missed) and therefore still in use by the navigation system (denoted M), and (c) the number of components eliminated from use by the FDI system. Recall that once a sensor has been declared faulty by the FDI logic, it is removed from future consideration. It is assumed in constructing the model that simultaneous faults occur with negligible probability (order of P_f^2). In Figure 14-2, the probability of the occurrence of a sensor fault during a single FDI time interval when i good sensors are operating is denoted $P_{f/i}$.

The three transient substates labelled 3A, 5A, and 5B are included in the diagram. These are included because it is not desirable for the system to remain in states 3 or 5 for more than a few consecutive FDI tests. In the case of state 3, the presence of two soft faults might result in the transmission of bad data to the navigation system when a midvalue measurement selection scheme is employed because the two failures might be similar in the affected axes and the polarity. This would also tend to invalidate the assumption made above on the likelihood of three-level isolation errors. In the case of state 5, averaging of the two available measurements for transmission to the navigation system would produce navigation data corrupted by the presence of the bias in the output of the faulty instrument. The persistence of these states for three or more consecutive FDI tests is therefore considered a vehicle loss. The number of consecutive tests with the same system state is represented by C in the transition diagram.

Recall that the assumption was made that $\sigma(t)$ was piecewise constant and that this resulted in piecewise constant values for the FDI performance probabilities. We now make the further assumption that the thresholds are also approximately piecewise constant over the same subintervals. This modifies the formula given earlier in this chapter for propagating the Markov model to:

$$\pi(N) = \pi(0)\left[\prod_{k=1}^{K} P_k^{N_k}\right] \qquad (14\text{-}6.6)$$

where the thresholds and performance probabilities are constant over each of K subintervals, P_k is the constant one-step probability transition matrix for the kth subinterval and N_k is the number of FDI tests during the kth subinterval. The form of this equation lends itself to a concise calculation of the cost gradient for use in the steepest descent procedure to optimise the thresholds. This gradient is given by:

$$\frac{\partial P_{vl}}{\partial P} = \sum_{k=1}^{K} \frac{\partial P_{vl}}{\partial PR_k} \frac{\partial PR_k}{\partial T_k} \frac{\partial T_k}{\partial P} \qquad (14\text{-}6.7)$$

where:

    P   = threshold parameter vector (to be determined)
    PR_k = vector of performance probabilities during the kth subinterval (constant)
    T_k = vector of thresholds during the kth subinterval (constant)

The time-varying nature of the decision function statistics leads one to expect that the FDI performance would be enhanced by the use of time-varying thresholds. For the Shuttle IMU's, the specified threshold functional forms are (from Rockwell International (1977))

$$T_3 = T_{2D} = (a+bt)^2$$
$$T_{2I} = (c+dt)^2$$

(14-6.8)

where T_3 is the three-level detection threshold, T_{2D} is the two-level detection threshold, and T_{2I} is the two-level isolation threshold. These functional forms are used for both gyros and accelerometers. The parameters to be determined for each sensor type, therefore, are a, b, c and d.

Application of the optimisation procedure of the previous section to the accelerometers produces the results shown in Table 14-2. As is indicated in the table, occasionally it was necessary to use manual intervention in the iterative gradient procedure to aid the convergence of the algorithm. In nearly every case, this intervention merely overcame some convergence difficulty related to the fact that the performance probability functions, since they are not expressible in closed forms, were tabularised for storage in the computer. It should be noted that different initial guesses from the one indicated by Table 14-2 were tried and the results (not shown here) indicated convergence to the same optimal parameter values.

Pass	a, m/s	b, 10^{-4}m/s^2	c, m/s	d, 10^{-4}m/s^2	P_{VL}
1	0	5.00	0.050	4.000	1.187×10^{-3}
2	0	4.00	0.070	3.000	1.023×10^{-3}
3[a]	0.280	2.00	0.280	1.300	7.057×10^{-5}
4	0.280	2.42	0.280	1.216	2.651×10^{-5}
5	0.280	2.59	0.280	1.191	1.673×10^{-5}
6	0.280	2.71	0.280	1.178	9.633×10^{-6}
7[a]	0.300	3.00	0.280	0.800	4.784×10^{-6}
8	0.300	3.78	0.280	0.803	1.126×10^{-6}
9[a]	0.310	4.50	0.280	0.900	0.054×10^{-7}
10	0.310	4.48	0.280	1.205	8.967×10^{-7}
11	0.310	4.54	0.280	1.179	8.889×10^{-7}
12	0.310	4.55	0.280	1.087	8.889×10^{-7}
13	0.310	4.55	0.280	0.993	8.888×10^{-7}
14	0.310	4.55	0.280	0.969	8.888×10^{-7}
15[a]	0.350	4.35	0.306	0.862	8.887×10^{-7}

[a]Interactive pass

Table 14-2: Optimisation results of acceleration threshold

The final accelerometer results are:

$a = 0.350$ m/s $\qquad b = 4.35 \times 10^{-4}$ m/s^2

$c = 0.306$ m/s $\qquad d = 2.83 \times 10^{-4}$ m/s^2

$\qquad P_{vl} = 8.89 \times 10^{-7}$

Table 14-3 lists the complete results including the performance probabilities.

i	Time, s	N_i	Detection threshold, m^2/s^2	P_{FA}	P_M	Isolation threshold, m^2/s^2	P_{IS}	P_{WI}	P_{FI}
1	0	66	0.132	1.2×10^{-29}	0.051	0.085	0.996	6.7×10^{-23}	4.4×10^{-59}
2	1060	12	0.661[a]	0	0.999	0.164	0.605	5.0×10^{-31}	0
3	1252	8	0.784[a]	2.1×10^{-20}	0.999	0.185	0.399	1.6×10^{-6}	8.0×10^{-6}
4	1380	20	0.822	1.1×10^{-9}	0.990	0.176	0.385	1.2×10^{-3}	4.3×10^{-7}
5	1700	18	1.11	2.3×10^{-6}	0.990	0.219	0.227	8.7×10^{-3}	1.4×10^{-4}
6	1988	16	1.65[a]	1.1×10^{-17}	0.999	0.221	0.221	0.015	6.9×10^{-4}
7	2244	30	1.63	2.1×10^{-7}	0.251	0.359	0.954	2.8×10^{-4}	9.4×10^{-6}

$P_{VL} = 8.89\times10^{-7}$

Prob [one or more false alarms during mission] = 4.9×10^{-5}

Prob [one or more undetected three-level faults during mission] = 1.1×10^{-3}

[a] At extreme of data.

Table 14-3: Complete accelerometer threshold results

The procedure is completely analogous for gyro FDI. The final results are:

$a = 5000$ arc-s $b = 0.2$ arc-s/s
$c = 91.7$ arc-s $d = 0.527$ arc-s/s
          $P_{v1} = 1.89 \times 10^{-5}$

Table 14-4 lists the complete gyro results.

i	Time, s	N_i	Detection threshold, arc-s^2	P_{FA}	P_M	Isolation threshold, arc-s^2	P_{IS}	P_{WI}	P_{FI}
1	0	3	1.0×10^{7a}	9.3×10^{-4}	0.999	9100	0.227	0.072	0.111
2	48	4	1.1×10^{7a}	8.9×10^{-4}	0.999	17,100	0.270	0.090	0.105
3	112	4	1.1×10^{7a}	7.3×10^{-4}	0.999	29,700	0.344	0.076	0.085
4	176	9	1.1×10^{7a}	5.2×10^{-4}	0.999	28,000	0.220	0.039	0.109
5	320	17	1.3×10^{7a}	1.7×10^{-4}	0.999	68,700	0.345	0.030	0.103
6	592	13	2.3×10^{7a}	1.5×10^{-7}	0.999	168,000	0.562	0.004	0.005
7	800	50	3.0×10^{7}	2.6×10^{-8}	0.999	425,000	0.715	0.003	0.019
8	1600	25	2.6×10^{7}	8.2×10^{-6}	0.900	1.10×10^{6}	0.873	1.3×10^{-5}	0.002
9	2000	30	2.8×10^{7}	8.2×10^{-5}	0.702	1.61×10^{6}	0.847	1.7×10^{-5}	0.004
10	2480	16	2.9×10^{7}	3.7×10^{-4}	0.400	3.50×10^{6a}	0.951	5.2×10^{-6}	1.4×10^{-4}

$P_{VL} = 1.89 \times 10^{-5}$

Prob [one or more false alarms during mission] = 0.0251

Prob [one or more undetected three-level faults during mission] = 2.0×10^{-3}

[a]At extreme of data.

Table 14-4: Complete gyro threshold results

It is interesting to note that the false alarm probability is several orders of magnitude smaller than the missed alarm probability for all the subintervals in both Table 14-3 and Table 14-4. This is a reflection of the relatively small failure probability (1.33×10^{-5} for each time step when all three IMUs are operating). In fact, P_{FA} is usually within an order of magnitude of the product of the failure probability and P_M.

It is common practice to set the thresholds for fault diagnosis tests at a small multiple, commonly three, of the standard deviation of the decision function. This yields a probability of false alarm for each test of 0.0026 when the decision function is approximately Gaussian. Note that the false alarm probability for the optimised Shuttle IMU FDI thresholds is always smaller than 0.0026. By using the Markov model of the Shuttle IMU RM system, it can be shown that the use of the common "3σ" thresholds yields a vehicle loss probability of 1.4×10^{-4} for either accelerometers or gyros. This is considerably degraded relative to the results obtained with the optimised thresholds.

14.7 CONCLUSIONS

In this chapter, we have described a technique for selecting such crucial fault diagnosis system parameters as the thresholds for the diagnostic tests. The method involves the construction of a Markov model to describe the random behaviour of a fault-tolerant system, where this behaviour consists of random fault diagnosis events as well as random faults. This model allows us to relate the performance of the system directly to the parameters to be chosen. In particular, it allows us to evaluate the probability of meeting a given system specification as a function of the parameters to be selected. This is a very powerful design tool.

We then demonstrated the technique by applying it to the problem of selecting the thresholds for the fault diagnosis system of the Space Shuttle inertial measurement system during a reentry-to-touchdown flight. The application showed that the major tasks to be accomplished in applying the technique are the calculation of the probabilities of fault diagnosis events as functions of the thresholds and the construction of the Markov model. Once these tasks are accomplished, the problem of optimising the thresholds becomes a relatively straightforward one.

Despite its great power, the technique is not globally applicable. Fault diagnosis systems that use tests which do not possess the no-memory property cannot be treated by this method. Research continues on modelling these systems, Walker (1981), Walker et al. (1985), Walker et al. (1986). The future may see a method for evaluating the performance of any fault diagnosis system quickly and easily for many choices of the design parameters. Such a method would be absolutely indispensable to a fault diagnosis system designer.

Chapter Fifteen

SYSTEM FAULT DIAGNOSIS

USING THE KNOWLEDGE-BASED METHODOLOGY

Spyros G. Tzafestas

Fault diagnosis of man-made systems lies in the centre of modern technology and is continuously receiving increasing attention both theoretically and practically. In recent years attempts have been made to apply artificial intelligence and knowledge-based techniques which provide the means of combining numeric and symbolic models for performing the fault diagnosis task. The aim of this chapter is to discuss the knowledge-based approach to fault diagnosis and provide an overview of the underlying concepts, issues and tools. The emphasis is given to the diagnosis methods which are based on *deep modelling* and are appropriate for diagnosing technological systems. In addition to these techniques, a number of existing fault diagnosis expert systems are described briefly.

15.1 INTRODUCTION

Knowledge-based expert systems are designed using artificial intelligence (AI) techniques and emulate human performance presenting a human-like action to the user. Artificial intelligence is the branch of computer science which studies how to enable computers to learn, reason and make judgements or, according to Elaine Rich, how to make computers do things at which, at the moment, people are better. Expert systems are currently finding application to an increasing repertory of human-life domains, in the centre of which lies the fault diagnosis and repair domain of technological processes.

Fault diagnosis and repair are knowledge-intensive and experiential tasks, which in actual processes can sometimes go beyond the capabilities of skilled technicians, operators or engineers.

Expert systems can perform at least at the level of a highly experienced human trouble shooting/repair expert whose knowledge greatly exceeds the contents of service manuals. Thus expert systems provide the critically required assistance for prompt detection, location and repair of process faults leading to an improved overall field service efficiency, effectiveness and performance. The field of system diagnosis/repair is presently at the heart of industrial automation and has all the required characteristics (closed domain, rich expertise available, good underlying models, heuristic methods, readily performed test/validation procedures) that make expert systems very likely to succeed in real industrial environments.

The main functions of the fault diagnosis and repair task are:

(a) *Symptom interpretation*
(b) *Fault diagnosis*
(c) *Trouble-shooting*
(d) *Repair/replacement*
(e) *Test design/generation*
(f) *Test/repair scheduling*
(g) *Monitoring*

In general the domain model involves all functional relationships brought down to the level of *field replaceable* (*or repairable*) *units* (FRUs). Following the hierarchy of the model, a heuristic *best-first* strategy is used to locate and replace defective FRUs, restoring operation gradually until total operation is gained.

The importance of fault diagnosis/repair in technological systems (chemical plants, nuclear reactors, manufacturing systems, aircraft, etc.) is obvious. For example, in an operating chemical plant product quality is maintained by assuring that process variables fluctuate within permissible ranges. If operating conditions go beyond these limits the product quality is in jeopardy but, more critically, these undesired fluctuations could lead to some catastrophic event (explosion, fire, release of toxic chemicals, etc.). Diagnosis of process faults/malfunctions is a very difficult task for the human operator. Especially, even well trained operators who behave efficiently in the case of standard operating fluctuations, have difficulties in handling unanticipated events and faults or failures with low probability of occurrence. Clearly, time is critical in these situations, and hesitation, or improper action could lead to disaster. Here is exactly where automated fault diagnosis can offer a unique aid to operators in detecting, locating and identifying process malfunctions. Throughout the years a large variety of automated fault diagnosis techniques have been proposed and developed. These techniques fall in two main categories, namely: *quantitative* and *qualitative* techniques. Of course several techniques of a combined nature also exist or are under development. Quantitative techniques are usually based on the concept of dynamic redundancy and make use of state filters/estimators and parameter identification procedures. These techniques are studied extensively in the other chapters of this book and will not be discussed further in this chapter. Here we are concerned with *qualitative* and *quantitative* techniques based on *shallow* and/or *deep knowledge* about the technological systems at hand, and implemented using knowledge-based AI tools and methods.

Firstly, a short survey of some existing fault diagnosis expert systems is made followed by a general discussion of the architectural and design issues of expert systems. Then the fundamental techniques for knowledge-based fault diagnosis are presented in some detail with emphasis on the deep knowledge diagnosis techniques (causal search, constraint suspension and governing equations techniques). Included in the chapter are samples of LISP-based and Prolog-based implementations, as well as numerical examples whenever appropriate. Some special attention is given to non-Boolean inference techniques which are suitable for diagnosis under uncertainty. The final section of the chapter presents a more elaborate description of three fault diagnosis expert systems, namely those developed at Vanderbilt University, Lockheed, and at the National Technical University of Athens.

15.2 SURVEY OF FAULT DIAGNOSIS EXPERT SYSTEMS

Some recent surveys of expert systems for fault detection/diagnosis of technological devices and processes are provided in (Tzafestas (1986), Pau (1986) and Georgeff and Firschein (1985)). A survey of analogous systems for controller design and industrial process supervision can be found in (Tzafestas (1987)). General material about expert systems architecture, design and implementation is presented in (Nau (1983), B.A. Thompson and W.A. Thompson (1985), Hayes-Roth (1984), Murphy (1985) and Sowizral (1985)). Two collective works which contain up-to-date contributions and results on various aspects, techniques and applications of the knowledge-based approach to system fault diagnosis and reliability are (Tzafestas et al. (1987) and Tzafestas (1988)). In this section a survey of some recent fault diagnosis expert systems will be made, which although unavoidably is not intended to be exhaustive, will play a preparatory role for the material of the chapter.

Diagnostic expert systems were first used in the medical domain, and were naturally designed following the *shallow reasoning approach* since the human disease mechanisms are difficult to describe or completely unknown (Shortliffe (1976) and Pople (1982)). One of the earliest medical expert systems for the diagnosis of bacterial infections in blood is MYCIN, which uses production rules and backward chained inference. It has a modest size knowledge base and contains about 700 rules (Shortliffe (1976)). The NEOMYCIN is a new version of MYCIN implemented using a distributed problem-solving methodology (Clansey and Letsinger (1981)). EMYCIN (Essential MYCIN) is the expert system shell (core) resulting from MYCIN when removing the domain knowledge base (van Melle et al. (1984)). Presently there are many expert systems for medical diagnostic applications (e.g. AI/RHEUM for rheumatology, ANNA for heart arrythmias, CASNET for the glaucoma diseases, CENTAUR for pulmonary function test interpretation, EEG ANALYSIS SYSTEM for electroencephalogram analysis, etc.).

Some of the available expert systems for diagnosing malfunctions of technological devices and systems are the following.

Dart: A device-independent diagnostic system that works directly from design descriptions rather than MYCIN-like symptom-fault rules (Genesereth (1982), Genesereth (1984), and Basden and Kelly (1981)). DART is intended for use in conjunction with a tester that can manipulate and observe a malfunctioning device. The diagnostician receives from the tester a description of an observed malfunction, prescribes tests and accepts the results, and finally locates the faulty components responsible for the fault. The DART system does not follow the MYCIN-like *rules* that associate symptoms with possible diseases (shallow approach), but it works directly using *deep* information about *intended structure* (a device's components and their interconnections) and *expected behaviour* (equations, rules, and procedures relating the inputs, states and outputs of the device at hand). DART has been implemented in MRS (Geneserth et al. (1983)) and tested in circuit troubleshooting in the teleprocessing system of the IBM 4331, and also in non-electronic devices such as the cooling system of a nuclear reactor. The time to diagnose each case was on the order of seconds or minutes. DART stands for *Diagnostic Assistance Reference Tool*.

Other works that have made substantial contributions on the *deep approach* to fault diagnosis in terms of system structure and behaviour (with emphasis on digital systems and computers) are described in (Davis and Shrobe (1983), Hamscher and Davis (1984), and Davis (1984)). An application of these reasoning techniques (based on structure and behaviour) to the fault diagnosis of power systems has been presented by Matsumoto et al. (1985). Use is made of information on relays and circuit breakers, and the relay sequence is represented by a logical circuit, the describing equations of which are converted in production-rule form.

Les: This is the *Lockheed Expert System* shell which is similar to EMYCIN but is more powerful because it uses a deep structural description of the device in its troubleshooting, and allows the knowledge engineer to explicitly control the reasoning process through WHEN rules. LES is presently being applied to the so-called *Baseband distribution subsystem* (BDS), a large signal switching network. The various FRUs of the system are described by using the frame representation (Perkins and Laffey (1984) and Laffey et al. (1986)) (See Section 15.5.2).

Fis: This is a *Fault Isolation System*, developed at the US Naval Research Laboratory, the purpose of which is to assist a technician in diagnosing faults in a piece of electronic equipment. It is written in Franz Lisp, runs on a VAX 11/780 computer, and was designed primarily to diagnose analogue systems, isolating faults to the level of amplifiers, power supplies and other larger components. The techniques of FIS are also applicable to the automatic generation of the programs that drive conventional *automatic test equipment* (ATE), to the real-time control of ATE, and to fault diagnosis in systems containing mechanical, hydraulic, optical, and other types of components. The diagnosis is based on the computation of the fault probabilities of the individual modules after each test (Pipitone (1986)).

Forest: This system supplements the fault detection and isolation capabilities of current ATE diagnostic software. It involves experiential rules of thumb from expert engineers, deep knowledge on circuit diagrams, and general troubleshooting principles. FOREST is encoded with rules using certainty factors and a MYCIN-like explanation facility, and is implemented in PROLOG (Finin et al. (1984)).

Nds: This is an expert system that detects and locates multiple faults in a nationwide communications network (COMNET) through expert diagnostic strategies based on knowledge about the structure and composition of the network. The components that can be diagnosed include telecommunication processors, modems, computer terminals, and telephone circuits. It is a rule-based system developed by Shell Company in cooperation with Smart Systems Technology (Williams et al. (1983)).

Vmes: This is the *versatile maintenance expert system* for trouble-shooting circuits that uses structural and functional descriptions. VMES was embedded in the Semantic Network Processing System (SNePS) and used as a kind of expert system shell. While troubleshooting, it displays the device at hand and indicates dynamically the state of the reasoning process (Shapiro et al. (1986)).

Diamon: This is a knowledge-based diagnosis and maintenance planning system aimed at increasing the efficiency and user-friend-liness of systems for condition monitoring of rotating machinery. It is based on the concepts used in a traditional system for turbomachinery condition monitoring and health control (Skatteboe et al. (1986)).

Relshell: This is an expert system shell for fault diagnosis of multiparametric, gradually deteriorating technological systems. It uses a hybrid frame/rule based representation of knowledge, fuzzy arithmetic and logical operations together with a specially defined operator (Gazdik (1987)).

Many other expert systems and expert system shells (tools) exist for diagnosis/maintenance of technological systems. Information for most of them is readily available in the literature, while some of them can be found commercially. Some examples in the chemical process field are given in (Chester et al. (1984), Davis et al. (1985), Andow and Lees (1975), Andow (1985), Beazley (1986), and Kumamoto et al. (1984)).

15.3 A LOOK AT THE ARCHITECTURE AND DESIGN OF EXPERT SYSTEMS

The key issues of expert system building are:

(a) *symbolic representation* of the knowledge (predicate logic, semantic networks, frames, production rules, etc.).
(b) *symbolic reasoning* using rules and methods suitable for deducing, examining, judging, determining, selecting, etc.
(c) *graphic explanations* using developer-oriented graphic means (knowledge-based graphs and rule graphs, etc.) and end-user-oriented graphic means to provide (developers and users) the representation and reasoning.

Some of the benefits resulting from the above AI issues are:

(a) *Explicit representation* of symbolic structures, performance and reasoning, and easy integration of various representation types.
(b) *Visibility, flexibility* and *adaptibility* obtained from the declarative nature of the representation.
(c) *Reasoning in radically different ways* from those tried by non-AI systems.
(d) *Intelligent interaction* based on the representation.
(e) *AI/expert systems attempt to remove the programmers* as problem translators, and allow the expert to interact directly with the system to explore the problem space.

The expert system field is a branch of AI which deals with developing computer programs that embody the knowledge of experts enriched with some common sense in an effort to solve difficult problems usually requiring human expertise.

Knowledge engineering (i.e. knowledge extraction from human experts) is profitable if the following are true:

(a) At least one human expert exists who can carry out the task very well.

(b) The **expert ability** is mainly due to special knowledge, experience and judgement.

(c) The **task** belongs to a well defined application domain.

(d) The **expert** can combine knowledge, judgement and experience, and can **explain** his path of reasoning.

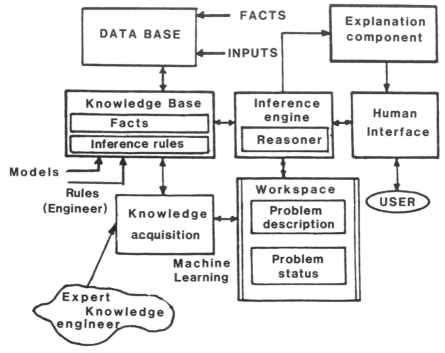

Figure 15-1: Expert system architecture

An **expert system** consists of the following architectural, clearly separated, components (Figure 15-1):

(a) *Knowledge base* (general knowledge about the problem, i.e. facts and rules).

(b) *Data base* (information about the current problem, i.e. input data).

(c) *Inference engine* (methods for applying the general knowledge for the problem).

(d) *Explanation component* (which can inform the user on why and how the conclusions are obtained).

(e) *User interface* and *knowledge acquisition* component.

(f) *Work space* (i.e. an area of memory for storing a description of the problem constructed from facts supplied by the user or inferred from the knowledge-base).

An *expert system* should be clearly distinguished from an *expert workstation* (*expert system shell*). The *expert system* solves a problem in the same way as a human expert, makes decisions (or suggests the decisions) and explains its conclusions to the user who is not an expert and is only familiar with the domain. An *expert workstation*, on the other hand, provides representation, reasoning, and graphics to an expert who uses the workstation to solve his problem(s). The user must be an expert because he (she) is solving the problem, with the system making the expert's job easier by performing extensive book-keeping. A usual development approach is in the first instance to build an expert workstation, and then add the expert system. The workstation provides the representation, graphics and some reasoning and plays the role of a nonrisky testbed for the expert system.

Rule-Based Systems: First generation expert systems use the knowledge representation in the form of IF...THEN... (the so-called *production*) *rules*. The basic form of a production rule is:

RULE R_k

IF $c_1, c_2, \ldots, c_m$

THEN $h_1, h_2, \ldots, h_n$

where c_i (i=1,2,...,m) are predicates known as *conditions* (antecedents, premises, etc.) and h_i (i=1,2,...,n) are referred to as *consequents* (deductions or actions). The fundamental *reasoning* (*syllogism*) which applies here is: *If A implies B and B implies C, Then A implies C.*

When all c_i are true, Rule R_k is said to be *triggered*. The set of triggered rules is called the *conflict set*. A rule is selected from the conflict set using a conflict resolution strategy. A triggered rule is said to be *fired* when its consequences are performed.

Some strategies for selecting the rule for firing from the conflict set are:

(a) *Rule ordering* (rule appearing earliest has highest priority).
(b) *Data ordering* (rule with highest priority data-condition has highest priority).

(c) *Size ordering* (rule with longest list of constraining conditions has highest priority).

(d) *Context limiting* (activate or deactivate groups of rules at any time to reduce the occurrence of conflict).

(e) *Specificity ordering* (arrange rules whose conditions are a super-set of another rule).

The conflict resolution strategy is selected *ad-hoc*. Most popular are the *specificity ordering* and *context limiting* strategies.

The *control* (or *interpretation*) mechanism used by synthesis systems is as follows:

(a) Find rules whose *IF* parts are triggered, and select a rule using a certain conflict resolution strategy.

(b) Fire the rule (i.e. do what the rule's THEN part says).

In analysis systems the antecedents of rules can be either *observed* or *derived* facts and the consequents are new facts that are deduced.

The above mechanism is known as *forward inference* mechanism. In analysis of systems the control mechanism can be either of the *forward* or the *backward* chaining type. In the backward mechanism, a particular hypothesis is selected, and the rules are searched to see if the hypothesis is a consequent. If yes, the antecedents of the rule, constitute the next set of hypotheses. The process is continued until some hypothesis is not true or all hypotheses are true based on the data. Forward and backward inference chaining resemble the *bottom-up* and *top-down* control in general computer algorithms (compilers).

Rule-based systems have many advantages. For example:

(a) They provide a homogeneous representation of knowledge.

(b) They allow incremental growth of knowledge through addition of new rules.

(c) They allow unplanned but useful interactions.

A well known example of rule based systems is XCON, a synthesis expert system for configuring DEC's VAX computers. The main disadvantage of rule-based systems is that the knowledge acquisition from domain expert is time consuming and difficult. All possibilities have to be explicitly enumerated and there is no capability of system generalisation.

Very often in practice we need to associate *levels of confidence* with rules, antecedents and consequents. In these cases the inference mechanism must be able to compute the confidence of the consequent given the confidence values of its antecedents and the confidence of the rule. This is the subject of *fuzzy reasoning systems* which are based on many techniques. Some of these techniques are the *certainty factors approach* developed by Shortliffe and applied to MYCIN (Shortliffe (1976)), the probabilistic (Bayesian) approach used in the mineral exploration expert system PROSPECTOR (Duda et al. (1978)), and the *Rule Value Approach* developed by Naylor (1984), which uses neither forward nor backward chaining, both of which tend to concentrate on the hypotheses, but rather, it concentrates on the evidence. Actually, for each *item* of evidence it assigns a value (the value of this rule in the process of inferencing) and asks that question with the *highest value* first.

Rule-based systems can be efficiently implemented using various programming languages such as FORTRAN, PASCAL, C, LISP, PROLOG or expert system tools (shells) such as EMYCIN, KAS, ROSIE, OPS5, M1 which possess built-in inference mechanisms. EMYCIN was derived from MYCIN by removing the domain knowledge. In the same way KAS was derived from PROSPECTOR. OPS5 provides flexibility and generality in that it is easy to tailor the system to the domain, but unlike EMYCIN and KAS it does not have a sophisticated front end. ROSIE uses English-like syntax but does not possess a sophisticated data base structure. M1 was developed by Teknowledge for the IBM PC and uses English-like syntax. Most first generation production rule systems are written in LISP dialects. For example EMYCIN is written in INTERLISP (Stanford Univ.), OPS5 is written in FRANZLISP (Carnegie Mellon Univ.) and ROSIE is again written in INTERLISP (Rand Corporation).

Developing a rule-based expert system via an expert system tool reduces the need to program it directly in an AI language (LISP or PROLOG), but most expert system tools presently available are not designed for real-time applications, i.e. do not support interrupts, nor do they accept on-line data. But there is no reason why these properties cannot be added, particularly if the host computer allows access to its operating system. One point about employing LISP-based systems for real-time control, is that during execution, a LISP system carries out periodically *garbage collection* of unused memory, and usually garbage collection is an *automatic* and *uninterruptible process*. Thus if interrupting procedures are needed during garbage collection, it may not be possible for the system to respond in time. This point should be carefully examined (i.e. the way of handling garbage collection) whenever one wants to develop a LISP-based expert system. This problem appears always when the system makes use of linked lists and heap storage, no matter what language or expert system tool is used.

Systems Based on Semantic Networks and Frames: *Semantic networks* are based on the very simple and ancient idea that *memory* consists of *associations* between concepts. The associative memory concept has its origin in Aristotle, and entered the computer science work through the use of simple associations to represent word meaning in databases. A semantic network is a method of knowledge representation where *concepts* are represented as *nodes* (circles) and *relations* between pairs of concepts by *directed arcs* (labelled arrows). Such pairs of related concepts may be thought of as representing simple facts. Not all node-and-link nets are necessarily semantic nets, but only when there is a way for associating meaning with the network. One approach to this, the so-called *procedural semantics*, is to associate a set of programs that operate on descriptions in the representation (Minsky (1986)).

The concept or *frame* is the result of organising the properties of some object or event (or a class of objects) to form a *prototype* (structure). A frame, for example, can be used to represent an aircraft or a bank, a manager, a controller or even a class of vehicles, etc. The power of frame representation is that those elements that are conventionally present in a description of an object or event are grouped together and can be accessed and processed as a unit. A frame can be regarded as a *collection* of semantic *net nodes* and *slots* that all together describe a prototyped (stereotyped) object or class of objects, etc. A frame language has special constructs for organising frames that represent classes into hierarchical taxonomies. Associated with this, there are special purpose reasoning procedures that utilise the structural characteristics of frames to carry out particular inferences that extend the explicitly held set of beliefs to a larger (virtual) set of beliefs. Examples of semantic networks and frames are shown in Figure 15-2(a,b).

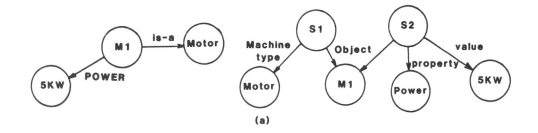

(a)

Figure 15-2(a): Semantic network of Ml: "Ml is a motor working at 5KW", and its equivalent network using SNePS.

```
name CONTROL ENGINEER
       Specialization-of: CONTROL ENGI-
             NEERING
       name: unit (family-name, first-
             name, initial)
       age: unit (years)
       address: ADDRESS
       department: range (production,
             maintenance)
       salary: SALARY
       Special experience: range (con-
             troller design, actuators,
             transducers)
       start-date: unit (month, year)
       to: unit ((month, year) (default:
             now))
                            (b)
```

Figure 15-2(b): The frame of the general concept "CONTROL ENGINEER". Appropriate frames are needed for the "ADDRESS" and "SALARY" fillers of the CONTROL ENGINEER frame.

Many systems combine the frame representation with production rule languages. Such systems are the KEE and LOOPS. KEE is written in INTERLISP and was originally for the Xerox 1100 and Symbolics 3600 machines, but now is available as a general purpose system (Knowledge Engineering Environment). KEE also supports *procedure-oriented* and *object-oriented* representation methods. *LOOPS* was developed at the Xerox Palo Alto Research Center and is appropriate for object-oriented, rule-based, access-oriented and object-oriented representation methods. Its main feature is the integration of the above four programming methods to permit the paradigms to be used together in system building. *LOOPS* is written in INTERLISP-D and works on Xerox 1100 series workstations. Other systems for frame-based representation are KL-ONE which possesses automatic inheritance and support for semantic nets using subsumption, and KMS which consists of a collection of subsystems, each with its own knowledge representation and inference mechanism. KL-ONE operates on DEC VAX's and KMS on UNIVAC 1100/40. SNePS (see Figure 15-2(a)) is a semantic network system permitting to represent as semantic network both relational knowledge and production rules (Shapiro et al. (1986)).

15.4 KNOWLEDGE-BASED TECHNIQUES FOR SYSTEM FAULT DIAGNOSIS

As we have already seen the knowledge-based diagnosis techniques fall in *two* categories:

(a) Shallow diagnostic reasoning techniques,
(b) Deep diagnostic reasoning techniques.

15.4.1 Shallow knowledge approach

Shallow diagnostic expert system building techniques are appropriate for evidentially oriented systems which are typical in medical diagnosis. Shallow knowledge consists of inference rules that can be made given a certain situation and are based on prespecified relationships between fault symptoms and system malfunctions. This means that a direct capture of the relationships between system behaviour irregularities and system faults is required. The production rules simply try to mimic the deductive processes of the skilled expert and do not provide any kind of inherent reason for the cause between the conditions and the conclusions. This is what happens in medicine where the underlying mechanisms cannot be described exactly or are completely unknown. Moreover, very often these rules are shortcuts of a deeper understanding of the problem domain and loosely model the decision making process of how the expert models the domain. The classical example of shallow expert system is MYCIN.

The main characteristics of the shallow reasoning approach are:

(a) Difficult knowledge acquisition
(b) Unstructured knowledge requirements
(c) Diagnosability or knowledge-base completeness non-guaranteed
(d) Excessive number of rules
(e) Knowledge-base highly specialised to the individual process.

These disadvantages can be partly overcome by decomposing the problem into smaller problems either in a hierarchical manner or according to unit operations.

The *two* common ways of carrying out the fault diagnosis are (Lees (1983)):

(a) *The fault dictionary*: A list of causes and effects. Diagnosis is performed by looking up the effects in a cause-effect look-up table to see what the cause was.
(b) *The diagnostic tree*: It is a way of restricting the search to go along different diagnosis paths.

Since both methods use look-up tables and all potential faults have to be precomputed in advance, in practice the look-up table has a large number of entries. Thus the diagnostic procedure becomes difficult and time consuming for large systems. On the other hand, if this *a priori* knowledge is not complete and correct, the diagnosis may be erroneous or impossible.

All these disadvantages of the shallow approach do not exist in the deep approach which makes use of deep models and is particulary suitable for diagnosing technological (man-made) systems.

15.4.2 Deep knowledge approach

The deep approach is based on a structural and functional model of the problem domain (i.e. on the so-called *deep model*). *Deep model* is a model that can derive its own behaviour for a given set of *parameters* and *signals*, and predict what should be the effects of changes in them. Deep knowledge expert systems attempt to compute the underlying principles of the process (or domain) explicitly, and the need to predict every possible fault scenario or to use precomputed rules is eliminated. It is clear that this approach leads to expert system tools that are able to handle a wider range of problem types and larger problem domains.

Three basic representative deep-knowledge diagnosis techniques are:

(a) Causal search/Hypothesis test technique
(b) Constraint suspension technique
(c) Governing equations technique

A brief outline of these techniques follows.

15.4.2.1 Causal search technique

The causal search technique is based on the tracing of process malfunctions to their source. Causality is usually represented by *signed directed graphs* (digraphs), the *nodes* of which represent state variables, alarm conditions or fault origins, and the *branches* (edges) represent the influence between nodes. The digraph can include (besides the positive '+' or negative '-' influences on the branches) the intensity of the effect, the time delays and the probabilities of fault propagation along the branches. Digraph-based qualitative diagnostic techniques are popular since little information is required to construct the digraph and carry on the diagnosis (Shiozaki et al. (1985), Tsuge et al. (1985) and Umeda et al. (1980)).

Here we shall describe the rule-based signed directed graph (SDG) technique developed by Kramer (Kramer and Palowitch) which is much more flexible and requires much less computation time than previous analogous methods (e.g. Shiozaki et al. (1985)). This technique converts the SDG into a concise set of logical rules which lead to increased diagnostic resolution and reduced alarm threshold sensitivity. These rules can be integrated with other rules (e.g. heuristic rules) of fault diagnosis or rules referring to other aspects of the system operation.

The basic limitation of the technique is that: *correct diagnosis can be ensured only if each variable undergoes at maximum one transition between qualitative states during fault propagation.* Also it is assumed that a single fault, affecting a single node in SDG (i.e. the root node), is the source of all disturbances, and that the fault does not change other causal pathways in the digraph.

Fault diagnosis generates a hypothesis about the operating state of the process (plant) at hand on the basis of a set of symptoms. Diagnosis is the *dual* (or *inverse*) of *fault modelling* (*simulation*) that predicts the response of the plant, given the operating state. All diagnostic systems contain a fault simulation part, which in the digraph approach involves the construction of a *directed tree branching* from the root node. The branches represent the actual active pathways of fault propagation. From a given root node one can generate more than one tree. Only one or a small set of these trees (called *interpretations of the fault propagation*) reflect the true behaviour of the plant. It should be emphasised that a branch in the digraph represents a potential pathway of fault propagation, whereas a branch in the simulation tree is the specific active pathway along which the fault is expected to propagate. Not all pathways in the digraph are active for a given fault. Diagnosis involves detecting all root nodes given on-line sensor data classified as *high* (+1), *normal* (0) and *low* (-1).

The technique of converting the SDG into logical rules is illustrated below with an example taken from (Kramer and Palowitch).

Example 1: Conversion of SDG into Rules

Consider the abstract network of interactions between variables shown in Figure 15-3(a). A variable X and its measured value (sensor response X_s) can be lumped as a single node (except when the sensor belongs to a loop), but this lumping is valid only if the measurement is very fast. Let us now assume that the root node (i.e. the fault origin) is A=+1 which means that the fault enters the network at A by perturbing A in the positive direction.

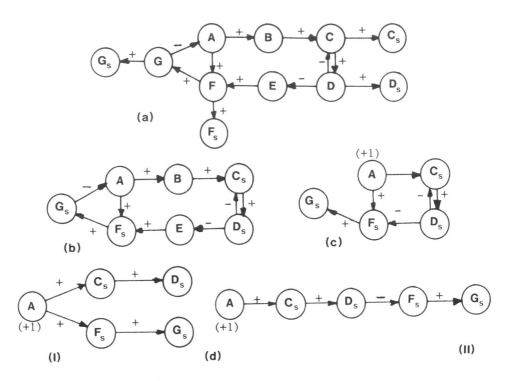

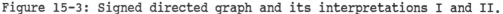

Figure 15-3: Signed directed graph and its interpretations I and II.

Applying the node lumping process to the digraph of Figure 15-3(a)
yields the simplified digraph of Figure 15-3(b), which is further
simplified by removing the nonmeasured variables B and E as shown in
Figure 15-3(c). The unmeasured nodes are not of interest in the
diagnosis except as potential root nodes. The root node A is not
removed (at this stage) even if it is not measured. In Figure 15-3(c)
the feedback branch from G was removed since we assume that the
amplitude of perturbation at A is large enough to override the
feedback effect from G. This is true since when a root node is
selected it is always tacitly assumed that the dominant causal
pathway to that node originates outside of the digraph of the normal
process.

In the present case (see Figure 15-3(c)) one obtain two *directed*
trees rooted at A as shown in Figure 15-3(d). These are called
interpretations I and II, respectively. Due to the assumption of single
transitions, C cannot be returned to normal (by the negative FB from
D) after the initial disturbance, and so in both cases the FB branch
from D to C was removed. These interpretations express the two ways

in which the disturbance can be transmitted to F (directly from A, or through B, C, D and E).

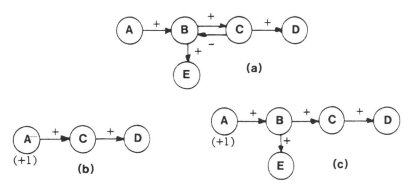

Figure 15-4(a): Digraph for control loop
      15-4(b): Loop working (resulting in disturbance cancellation)
      15-4(c): Loop saturated (passing the disturbance to E).

In constructing simulation trees, any *feedback* (*FB*) loops must be removed. In case of non-controlled variables in negative FB loops, it is assumed that the negative FB cannot completely compensate nor override the initial disturbance. In FB control loops, control action can prevent significant deviation of the controlled variable by transferring disturbances to the manipulated variable and then out of the loop, making it appear as though disturbances are passed through a normal node (see Figure 15-4). Two interpretations are applied when a disturbance enters a control loop; one corresponding to disturbance cancellation on the controlled variable, and one to the case where the disturbance exceeds the ability of the loop to compensate (loop saturation). The sensor patterns which may appear during fault propagation in interpretations I and II are given in Table 15-1. Due to the single fault assumption, all variables not accessible from the root node must be at normal state (0). These patterns involve the true response to a fault originated at A, and also some spurious patterns. Clearly, given m measurements accessible from the root node, only m patterns may arise during the dynamic propagation of the fault. The extra patterns correspond to spurious effects which are consistent with the digraph model of the plant.

Interpretation I

C	D	F	G
0	0	0	0
1	0	0	0
1	1	0	0
0	0	1	0
0	0	1	1
1	0	1	0
1	1	1	0
1	0	1	1
1	1	1	1

Interpretation II

C	D	F	G
0	0	0	0
1	0	0	0
1	1	0	0
1	1	-1	0
1	1	-1	-1

Table 15-1 Sensor patterns corresponding to interpretations I and II

The data in Table 15-1 could be used directly for diagnosis by matching the stored patterns with online sensor data. This is exactly the *fault dictionary* approach, which is somewhat unattractive, since in large systems for each fault there may be a large number of patterns. This difficulty is overcome by converting the simulation tree (or, equivalently, its interpretations I and II) in Boolean logical form and then in standard "IF ... THEN" rule form. This is done in the following way.

Consider a branch $X \overset{+}{\to} Y$ of a simulation tree. This means that if X=+1, Y can be +1 or 0 (due to the existing unspecified time delay of the branch). If X=0 then B=0, and if X=-1, then Y=-1 or Y=0. Analogous results hold for a branch $X \overset{-}{\to} Y$.

Now let us introduce the logical predicates p and m as:

$$(p \ X \ Y) <=> (X = Y) \text{ or } (|X| > |Y|)$$
$$(m \ X \ Y) <=> (X = -Y) \text{ or } (|X| > |Y|)$$

It is easy to verify that the truth tables of $X \overset{+}{\to} Y$ and $X \overset{-}{\to} Y$ are the same as the truth tables of (p X Y) and (m X Y) respectively.

Thus the simulation tree of Figure 15-3(d) (interpretation I) can be represented in logical form as (the subscript s was suppressed for convenience):

Root node (A=+1)

$$A \overset{+}{\to} C \Rightarrow (p \; A \; C), \quad A \overset{+}{\to} F \Rightarrow (p \; A \; F)$$
$$C \overset{+}{\to} D \Rightarrow (p \; C \; D), \quad F \overset{+}{\to} G \Rightarrow (p \; F \; G)$$

If the measured (on-line) data violate any of these logical relationships, then the state of the plant is not consistent with interpretation I. If this happens to both interpretations I and II, then A cannot be considered as a possible fault origin.

The above logical relations contain the unmeasured node A. To avoid this we note that [(X=+1) and (p X Y)] is equivalent to Y ≠ -1 and so the above set of four logical relationships can be written in the following rule form:

IF ((C ≠ -1) and (F ≠ -1) and (p C D) and (p F G))
THEN A=+1 is a possible fault origin.
A similar rule can be written down for interpretation II, i.e.
IF ((C ≠ -1) and (p C D) and (m D F) and (p F G))
THEN A=+1 is a possible fault origin.

These two rules have the same THEN part, and so they can be combined in one rule by ORing their IF parts. The resulting rule after some logic simplification* takes the final form:

IF ((C ≠ -1) and (p C D) and (p F G)
       and ((F ≠ -1) or (m D F)))
THEN A=+1 is a possible fault origin.

The premise of this rule is true when any measurement pattern from Table 15-1 occurs and false otherwise.

The above method of writing rules for each interpretation is not feasible because in practice a large number of interpretations can be derived from a realistic digraph. This difficulty can be overcome by deriving the rules directly from the SDG. To this end, we observe that, in general, converging pathways represent a choice in the construction of the simulation tree where one of the paths is used to form an interpretation. By using all combinations of these choices one can obtain the full set of interpretations. Thus the combined set of interpretations can be represented by making explicit the choices,

instead of enumerating each interpretation. The method is illustrated
by the following example.

Example 2: Direct rule derivation from SDG

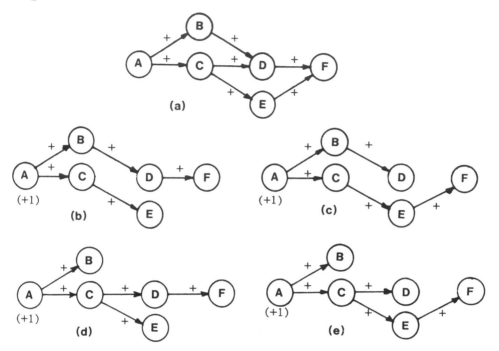

(a)

(b)   (c)

(d)   (e)

Figure 15-5: A partially developed SDG (a), and its four interpretations (b)——(e).

Consider the partially developed simulation tree of Figure 15-5(a)
where all nodes are assumed to be measured. Since we have two nodes
(namely D and F) where two feedforward paths converge, we obtain four
interpretations as shown in Figure 15-5(b-e). The combined set of
interpretations can be represented by the following set of branches
(*eor* denotes *exclusive or*):

$$A \overset{+}{\to} B \text{ and } A \overset{+}{\to} C \text{ and } C \overset{+}{\to} E$$

$$\text{and } (B \overset{+}{\to} D \text{ eor } C \overset{+}{\to} D) \text{ and } (D \overset{+}{\to} F \text{ eor } E \overset{+}{\to} F)$$

---

* Using the logical distributive law (X and Y) or (X and Z) <=> X and
(Y or Z).

This set of branches can be transformed to the following logical form:

IF ((p A B) and (p A C) and (p C E)
   and ((p B D) or (p C D))
   and ((p D F) or (p E F)))
THEN A=+1 is a possible fault origin.

This rule covers all measurement patterns resulting from all interpretations of the digraph.

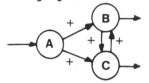

Figure 15-6: A digraph with feedback loop.

Special care is required when feedback loops exist in the digraph. In this case loop isolation may occur which leads to spurious pattern measurements. To avoid this possibility, at least one branch must enter the feedback loop. For example, it is easy to see that in the case of the SDG of Figure 15-6 the condition *(p A B) or (p A C)* must be joined with *and* to the premise of the rule:

IF (((p A B) or (p C B)) and ((p A C) or (p B C)))
THEN A=+1 is a possible fault origin.

Also a special clause is needed to prevent spurious interpretations when removing unmeasured nodes at which multiple branches enter and leave them. Finally to avoid the need of multiple rules corresponding to different interpretations of control loop behaviour, a single clause covering both *normal* (working) and *saturated* loop can be derived by joining the logical clauses for normal and saturated behaviour with *or*.

Hypothesis formulation/testing: The causal search method is also called hypothesis formulation/hypothesis testing method since it follows the usual human diagnostics path, i.e. a cause for a system malfunction (upset) is postulated, the symptoms of the postulated fault are determined and the result is compared with the system observables. Of course the search for the location of a fault can be narrowed by using appropriate heuristics or experiential knowledge and precedence rules in order to be able to resolve competing causal influences on the same process-variable.

Hypothesis testing requires qualitative (non numerical) simulation of the effects of the postulated malfunctions which needs prediction of the deviation of process measured variables as a result of faults.

Forbus (1985) provides a discussion of how a qualitative account of behaviour can be obtained using a structural description of objects and their connectivity. Very useful is the work of Stanfill (1985) where he studies some naive physics aspects of reasoning about simple machines. He developed a program (**MACK**) which deduces the behaviour of simple machines starting with a geometric representation of the shapes of the parts. This is done through the construction of a series of progressively more abstract models.

In simple cases the behaviour representation of the system under diagnosis can be the function, but in general, functional specifications include an account of the intentions for which the system or device is used (**teleology**). Moreover very frequently, behaviour needs to be abstracted to a level higher than that at which the component is specified, e.g. in an electronic circuit transistors and resistors are described in terms of voltages and currents, but a device can be described as an amplifier or an oscillator.

15.4.2.2 Constraint suspension technique

This technique belongs to the approach of diagnostic reasoning from behaviour to structure or, more accurately, from misbehaviour to structural defect. That is, given symptoms of misbehaviour one wishes to determine the structural aberration (Davis and Shrobe (1983), Hamscher and Davis (1984), and Davis (1984)).

System Structure: Here by structure we mean information about the interconnection of the components (or modules) of the physical process or device. There are two ways to look at structure: *functionally* and *physically*. Functional structure is the organisation of the system according to how components interact. Physical structure tells us how components are packaged. Thus every system (device or component, etc.) is characterised by two distinct structural descriptions; *functional organisation* and *physical organisation*.

For each class of system there is a distinct basic level of description, e.g. digital circuits are built on three concepts: *modules*, *ports* and *terminals*. A module is a standard black box, a port is a place where information flows into or out of the module, and a terminal is a place where we can *probe* to examine the ingoing or outgoing information. Every port has at least two terminals. Two modules are connected through their terminals. Modules, or more generally components at any level may have substructure. That is the model has a *hierarchical* form. This makes it possible to probe for the fault without having to consider every single detailed component. On the contrary, it is desirable that the diagnosis search be constrained to go down the most likely branch. In digital circuits, the description ends at the *gate* level in the functional hierarchy and at the *chip* level in the physical hierarchy.

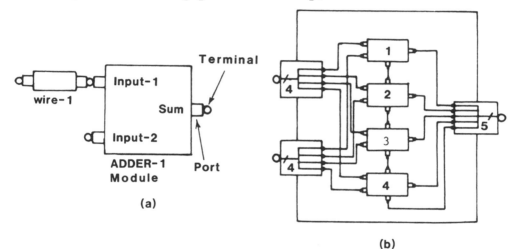

Figure 15-7(a): An adder module

          15-7(b): Next level structure of the adder showing that it is a ripple-carry adder.

```
definemodule adder NBitsWide
  (repeat NBitsWide i
    (part slice-i adder-slice)
    (run-wire (input-1 adder) (input-1 slice-i))
    (run-wire (input-2 adder) (input-2 slice-i))
    (run-wire (output slice-i) (sum adder))
  (repeat(-NBitsWide 1)i(run-wire (carry-out slice-i)
                                  (carry-in slice -[i+1])))
  (run-wire (carry-out slice-[NbitsWide-1])(sum adder)))
```

Figure 15-8: Structural LISP description of the adder module: Parts are described by a path name through the part hierarchy (e.g. input-1 adder).

The structural description of a component (module) is expressed as a *set of commands* for building it. For example, the adder of Figure 15-7 is described by the LISP program of Figure 15-8 (Davis (1984)). With *NBitsWide* bound to 4, the first expression tells that the following sequence of operations should be repeated four times:

(a) create an adder slice

(b) run a wire from input 1 of the adder to input 1 of the slice

(c) run a wire from input 2 of the adder to input 2 of the slice

(d) run a wire from the output of the slice to the output of the adder.

This program is executed by the system, resulting in the creation of data structures which model all the components and connections shown. Clearly, the data structures are connected in the LISP sense exactly as the objects are connected in Figure 15-7(b) (*isomorphicity* between diagram and program). Since the actual systems are very large use is made of lazy *instantiation*, i.e. when an instance of a module is created only the *shell* is actually built at that time (the *outer box and ports* in the adder) while the *substructure* (the slices) is not built until it is actually needed.

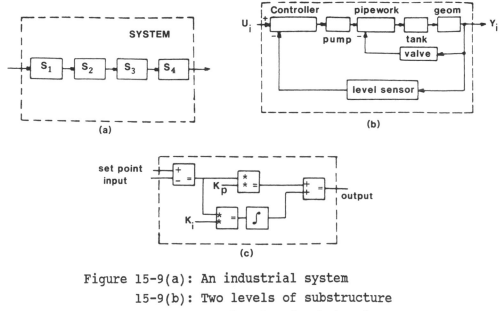

Figure 15-9(a): An industrial system

15-9(b): Two levels of substructure

15-9(c): Two levels of substructure.

Another example of the hierarchical description of system structure is shown in Figures 15-9 and 15-10 (Leary and Gawthrop (1987)). In Figure 15-9 the horizontal dotted lines indicate the connections between components. Connections are possible only at the same level and are not allowed between components at different levels.

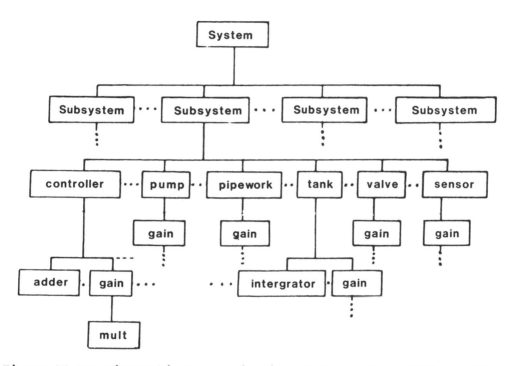

Figure 15-10: Hierarchical organisation of the system of Figure 15-9.

A LISP program, or a program in any object-oriented programming language (Goldberg and Robson (1983), Cox (1986), Keefe (1985), Rubenstein (1985), Sloman (1985), Shapiro and Takeuchi (1983), and Zaniolo (1984)) can be written to implement the building of this hierarchical model. A Prolog implementation (using OBJECTS (Keefe (1985)) is the following (Leary and Gawthrop (1987)), working from the top to the bottom of the hierarchy:

Top level (Figure 15-9(a))
```
system (Name, A, B) consists_of
  [
  subsystem (N1, A, B),
  subsystem (N2, B, C),
  subsystem (N3, C, D),
  subsystem (N4, D, E)
  ].
```
Second level (Figure 15-9(b))
```
subsystem (Name, A, B) consists_of
  [
  controller (Cname, A, B, V),
  pump (Pname, V, Fn1, Qin),
  pipework (PWname, Qin, Qout, Qnet),
  tank (Tname, Qnet, L),
  valve (Vname, L, Kv, Qout),
  sensor (Sname, L, Fn2, B)
  ].
```
Third level (Figures 15-9(b) and 15-9(c))
```
controller (Cname, A, B, V) consists_of
  [
  adder (_, A, B, E),
  gain (_, E, Kp, V)
  ].
pump (Pname, V, Fn1, Qin) consists_of
  [
  gain (_, V, Fn1, Qin)
  ].
pipework (PWname, Qin, Qout, Qnet) consists_of
  [
  adder (_, Qin, Qout, Qnet)
  ].
tank (Tname, Qnet, L) consists_of
  [
  integrator (_, Qnet, T, Vol),
  gain (_, Vol, InvArea, L)
  ].
valve (Vname, L, Kv, Qout) consists_of
  [
  gain (_, L, Kv, Qout)
  ].
sensor (Sname, L, Fn2, B) consists_of
  [
  gain (_, L, Fn2, B)
  ].
```

Fourth level

The description continues at the fourth, fifth and other levels in the same way until the hierarchy is complete.

For example the gain description is:

```
gain (Name, A, B, C) consists_of
  [
  mult (Name, A, B, C)
  ].
```

Using this hierarchical structure, one needs only to search the substructure of a suspect component and either detect the fault or ignore this component. The interconnections are implemented by stating that two or more ports (as the case may be) share the same signals.

System Behaviour: Here, the *behaviour* or *function* of a component is considered to be its black-box description, i.e. the relation(s) between the information entering and leaving the component. Among the methods of describing behaviour we mention input/output mapping rules, Petri-nets and unrestricted chunks of code. Simple mapping rules are used for devices with uncomplicated behaviour, Petri-nets are appropriate when the focus is on modelling parallel events, and unrestricted code is usually the final resort when more structured forms of expression are too limited. Any combination of these descriptions can also be used.

Constraints are relationships (or rules) that define the behaviour of the component at hand. Thus a complete description of a component is composed of its structural description (as explained before) and its behavioural description in terms of rules that interrelate the values at its terminals. In writing down the rules describing the behaviour of the various components of a system we must take into account whether a component is non-directional or not. Analogue devices are in most cases non-directional, but digital devices are not.

Let us for example consider the adder module of Figure 15-7(a). The rules that describe the relations of the logic values at its terminals are:

Rule 1: to get sum from (input_1 input_2) do (+input_1 input_2)

Rule 2: to get input_1 from (sum input_2) do (-sum input_2)

Rule 3: to get input_2 from (sum input_1) do (-sum input_1) .

Rule 1 is actually a rule that represents the real behaviour of the adder i.e. the flow of electricity and is called a *simulation rule*. Rule 2 and Rule 3 however do not model the real behaviour of this adder chip (which doesn't do subtraction), but they are rather inferences that we can make about the device. For this reason they are called *inference rules*. In fault diagnosis of digital and other systems this distinction, i.e. the use of a network containing simulation rules (modelling the causality) and a network containing the inference rules is very convenient. Note that each network works independently, employing the conventional *demon-like* way of propagating values through its set of slots. Also this machinery allows us to keep track of dependencies (i.e. how the value in a slot got there), and can be extended to model components with memory. One way to do this is to use a global clock and time stamp all values, enlarge the behaviour-rule vocabulary so that it can refer to previous values, and keep a history of values at each terminal (Davis and Shrobe (1983)).

Fault Detection by Interaction of Simulation and Inference: Having discussed the means for describing the *structure* (functional organisation and physical organisation) and *behaviour*, we now proceed to the main step of our work, i.e. to fault detection (trouble-shooting). We first note that fault detection is equivalent to the detection of inconsistent signals or constraint violation. This is best illustrated by an example.

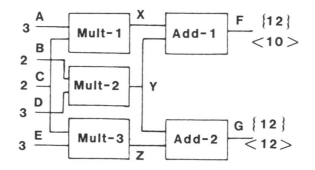

Figure 15-11: Fault location using constraint-violation detection
              (Expected { }, Actual < >).

Example 3

Consider the circuit of Figure 15-11. Given the input values shown in the figure the expected value at F is 12. If the value at F actually measured is 10, then we have a conflict between observed results and our model of correct behaviour. To find the possible sources of fault one must trace back through the dependency network. Here, at least one of three things: *Add-1, first input of Add-1* and *second input of Add-1* can be the source of conflict. The first possibility generates *Hypothesis-1: Adder-1 is broken.* Assuming that the second input to Adder-1 is good, the first input must be 4 i.e. different than the expected value 6. The only way to obtain this discrepancy is *Mult-1* to be broken (since its inputs are given externally and are correct). Hence: *Hypothesis-2: Adder-1 is good and Mult-1 is faulty.* Now, assuming that the first input of *Add-1* is good, then the second input must have the value 4 (different than the correct value 6) which implies that G should be 10. But the measured value at G is 12 as expected, and so *Mult-2* cannot be faulty. This kind of reasoning is known as *interaction of simulation and inference* (Davis (1984), and Jones and Burge (1987)). Simulation produces expectations about correct behaviour on the basis of known inputs and *component simulation (function) rules.* Inference generates conclusions regarding the actual behaviour (function) on the basis of measured outputs and *component inference rules.* The comparison of the two, in particular the difference between them provides the basis for fault diagnosis and has the following properties:

(a) It allows the system to deal with a wide range of faults, since it defines malfunction or failure functionally (as anything that does not match the expected behaviour).
(b) It allows a natural use of hierarchical descriptions, which is very useful when dealing with complex structures.
(c) It allows the system to yield symptom information about malfunction.

Fault Detection by Constraint Suspension: Although fault detection by constraint violation detection is good for simple systems, in a multicomponent system it may not be easy to locate the fault. This is facilitated by using the approach of suspending (disabling) constraints (i.e. rules of behaviour). The idea of constraint suspen-

sion is based on the observation that a faulty component can be
regarded as a new component with different and unspecified
properties. Then, constraint suspension implies the removal of all
the internal functional constraints imposed by the component at hand.
If after the suspension of the constraints imposed by a component the
observed signals are consistent with the remaining constraints, then
one concludes that a fault in that component justifies the observed
data. In general, more than one component may exist, whose
constraints when suspended, give a consistent set of signals. All
these components are equivalent candidates. But a different set of
observed data may rule out some of these candidates (simultaneous
candidates are not considered here). Constraint suspension provides a
mechanism for the careful treatment (management) of assumptions. The
traditional approach to diagnosis proceeds by assuming that the way
in which the component might be failing is known. It is displaying
one of the known malfunctions or malfunctions found in the set of
fault models (dictionary). Constraint suspension proceeds by simply
suspending all assumptions about how a component might behave. Then
the symptoms tell us what the component might be doing.

To illustrate the applicability of the constraint suspension
technique we consider again the circuit of Figure 15-11.

To check the global consistency of *Adder-1* we suspend all its
rules, assign the input values to input ports A through E, and assign
the observed values to output ports F and G. Then we try to see
whether there is some set of values on the ports of Adder-1
consistent with the inputs and observed outputs. If in fact the
circuit reaches a consistent state, then *Adder-1* can account for the
symptoms. Also looking at the resulting state we can see what values
the candidate (i.e. *Adder-1*) must have at its ports. Here we see that
the inputs of Adder-1 must be 6 and the output 10. This generation of
symptom information about malfunctioning, simultaneously with the
examination of global consistency is an additional advantage of the
technique. If there is no assignment of values to the component that
is consistent with all the symptoms, then there is no way for that
component to explain the observed malfunction. This, for example,
happens when we suspend the rules of *Mult-2* and insert the inputs and
observed outputs. Then always a contradiction results which shows
that *Mult-2* cannot be the faulty component.

Fault diagnosis by constraint suspension involves the following steps (Davis and Shrobe (1983), Hamscher and Davis (1984), and Davis (1984)).

15.4.2.2.1 Constraint suspension diagnosis procedure

Step 1: *Simulate and Collect Inconsistencies*

(a) Put component inputs into the constraint network inputs
(b) Compare predicted with observed outputs and collect inconsistencies.

Step 2: *Determine Potential Candidates Using Dependency Chain*

(a) For each inconsistency detected in Step 1: follow the dependency chain back from the predicted value to locate all components that contributed to that prediction.
(b) Find the components common to all discrepancies (i.e. the intersection of all sets determined in Step 2a). These are potentially able to explain all discrepancies.

Step 3: *Determine Candidate Consistency Using Constraint Suspension*

(a) For each component found in Step 2b:
   - Suspend (remove) the constraint describing its behaviour
   - Insert observed values at outputs of constraint network (inputs were put in at Step 1a)
   - If there is an assignment of values consistent with all symptoms (consistent state) then:

      (i)   the component is a globally consistent candidate
      (ii) its symptoms can be found at its ports
      (iii) the candidate and its symptoms are added to the candidate list, otherwise:
            the candidate is not globally consistent and is ruled out.
   - Retract the values at constraint network outputs
   - Reinsert (turn on) the constraint suspended at the beginning of Step 3a.
   Repeat step 3a for another component (of Step 2b).

Specialising this procedure to the circuit of Figure 15-11 we insert for example the values 3,2,2,3 and 3 at the primary inputs A through E and by simulation we predict the values at F and G. (Step 1a). Then we see that prediction and observation differ at F (Step 1b). At Step 2a, if we follow the dependency chain back from F=12, we find *Adder-1*, *Mult-1* and *Mult-2*. At step 2b we find only one discrepancy. At step 3a (third stage) we add for example *Adder-1* and its *symptoms values* 6,6 and 10. One candidate not globally consistent is *Mult-2* and is ignored.

A Prolog program that implements fault detection through consistency checking and constraint suspension is the following (Leary and Gawthrop (1987)):

```
detfault (List,F):-
  structure (List),
  ( ( constraint_check(List),F='none',!);
    (member(F,List),
     suspend_constraints(F,List,NewList),
     constraint_check(NewList),
    )
  ).
```

where detfault either detects no fault, in which case F='none' or it finds one. At first, instantiation of the component list and assignment of the measured values to the appropriate component ports is taking place, via structure (list). Then a check is made whether the component list is consistent using constraint_check (explained below). If it is consistent then there is no fault, but if it is not consistent then there is a fault. Possible explanations can be produced by searching the component list for likely candidates.

The routine constraint_check is defined as:

```
constraint_check ([ ]) :-!.
constraint_check (List) :-
  member (Node,List),
  (validate > — — > Node),
  remove_item (Node,List,NewList),!,
  constraint_check (NewList).
```

Given a list of components Y the goal constraint_check (Y) is satisfied only if all signal values passing through each component of Y satisfy all the constraints.

Structure is merely a quick way of instantiating the component list Y. Thus if there are no faults in our model Y, the response to:

```
?- structure(Y), constraint_check(Y).
```

is *true*, which means that the measured signals are consistent with the model. If there is a discrepancy, then the goal fails. This means that if there is a fault in the system model Y, it plays the role of an obstacle to the legal propagation of the constraints and thus *constraint_check* fails. To track down the most probable cause of the blockage, use is made of the fact that, if the constraints are suspended temporarily within the object causing the blockage, then the remaining constraints once again, propagate legally. Thus, to locate the fault we use the following goals:

```
?- member (F,List), suspend_constraints (F,List,NewList),
   constraint_check (NewList).
```

where *member (F,List)* is the standard Prolog function determining whether F belongs to a list (here the component list List) (Closksin and Mellish (1984)) and the algorithm *suspend_constraints* is defined as:

```
suspend_constraints (N,L1,L2):-remove_item (N,L1,L2).
```

To obtain an idea of the maximum number of steps needed to locate a single fault by this procedure, consider a hierarchical system consisting of A base level components, with K levels below the top level, where every component has exactly B components beneath it (obviously the bottom ones have none). Then $A=B^K$. The number of suspended constraints, at each node is B, and for each of them the remaining B-1 constraints are rechecked. Thus the total number of constraints checked throughout the search from top to base is equal to:

$$C = \sum_{i=1}^{K} Bx(B-1) = KxBx(B-1)$$

$$= \log_B (A)xBx(B-1)$$

i.e. for fixed B, C is proportional to the logarithm of A. When B=A (no hierarchy) the above equation gives C = Ax(A-1).
Two examples are:
(a) A=27, B=27, K=1: C=702
(b) A=27, B= 3, K=3: C= 18.

15.4.2.3 Governing equations technique

This technique was developed in the framework of locating faults in chemical process plants, but it is generally applicable to all cases where one can find an association of each quantitative constraint equation of the system at hand with a set of faults that are sufficient to cause violation of the constraint (Kramer (1986), and Kramer (1986(b))). If for example the constraint $F_1-F_2=0$ expresses the mass balance for flow into and out of a unit and the left-hand side of the constraint is found to be much less than zero, then we can infer that:

   (F_1-sensor-fails-low) V (F_2-sensor-fails-high)

In the opposite case, i.e. when the left-hand side of the expression is found to be much greater than zero, then we infer that

   (F_1-sensor-fails-high) V (F_2-sensor-fails-low) (system-leak)

where 'V' is the logical *OR* operator. By appropriately combining the inferences drawn from the complete set of process constraints one can arrive at the fault diagnosis.

15.4.2.3.1 Boolean logic inference

The first basic assumption of the governing equations technique is that all quantities entered in the constraint equations are either measured or nominally fixed. If they are not one can use *nodal aggregation* (Mah, Stanley and Downing (1976)) or *output set assignment* (Romagnoli and Stephanopoulos (1980)) to derive a set of constraints with all their variables measurable.

Assuming here that violation of each constraint can be represented by means of a fixed tolerance tol_i, then the conditions for positive constraint violation (S_i^+), negative violation (S_i^-) and constraint satisfaction (S_i^o) are:

$$S_i^+ <=> \{S_i > tol_i\}, \quad S_i^o <=> \{S_i < tol_i\}, \quad S_i^- <=> \{S_i < -tol_i\}$$

for $i=1,2,\ldots,N_c$ where N_c is the number of constraints.

The set of faults f that are sufficient to cause violation of the ith constraint is defined as:

$$(\forall \ f) \ [f \rightarrow S_i^+] \ \epsilon \ U_i^+, \ (\forall \ f) \ [f \rightarrow S_i^-] \ \epsilon \ U_i^-$$

The second basic assumption of the technique is that two or more faults having competing effects do not affect the same constraint simultaneously.

Now let S^* be the state of the system, i.e.

$$S_i^* = \begin{cases} S_i^+, & \text{if ith constraint is violated high} \\ S_i^o, & \text{if ith constraint is satisfied} \\ S_i^-, & \text{if ith constraint is violated low,} \end{cases}$$

and U_i^* be the fault set activated by the condition of the ith constraint, i.e.

$$U_i^* = \begin{cases} U_i^+ \ \text{if} \ S_i^* = S_i^+ \\ U_i^o \equiv \ \sim(U_i^+ \cup U_i^-) \ \text{if} \ S_i^* = S_i^o \\ U_i^- \ \text{if} \ S_i^* = S_i^- \end{cases}$$

Enumerating the elements of U_i^* as $U_i^* = \{u_{i1}, \ u_{i2}, \ \dots, \ u_{iI}\}$ and defining $\hat{U}_i^*$ as

$$\hat{U}_i^* <==> u_{i1} \lor u_{i2} \ \lor \dots \lor u_{iI}$$

then we see than the implication of the ith constraint is:

$$S_i^* ==> \hat{U}_i^* \ .$$

Thus the implication of all constraints together is

$$S^* ==> \hat{U}_1^* \land \hat{U}_2^* \land \dots \land \hat{U}_{N_c}^*$$

where "$\land$" is the logical "and" operator.

This equation is valid for both single and multiple faults under the condition imposed by the second basic assumption mentioned above, i.e. under the condition

$$\{(F^* \cap U_i^+) = \Phi\} \ \lor \ \{F^* \cap U_i^-) = \Phi\}; \ i=1,2,\dots,N_c$$

where F^* is the set of faults actually existing in the system, Φ is the empty set, and "$\bigcap$" denotes *set intersection*.

In the special case of single-fault hypotheses, which account for all violated constraints and lie in the intersection of the $U_i^*(i=1,2,\ldots,N_C)$ we find the formula:

$$S_1 = (U_1^* \bigcap U_2^* \bigcap \ldots \bigcap U_{N_C}^*)$$

where S_1 is the set of single fault hypotheses.

As an illustration consider a system with two constraints. If the possible faults on *high* and *low* violation of constraint 1 are $U_1^+=\{A,B\}$ and $U_1^-=\{C\}$ respectively, and those of constraint 2 are $U_2^+=\{B\}$ and $U_2^-=\{A,D\}$, then:

a) for *constraint 1 high* and *constraint 2 low*:
   $$S_1 = \{A,B\} \bigcap \{A,D\} = \{A\}$$
b) for *constraint 1 normal* and *constraint 2 low*:
   $$S_1 = \smallsmile\{A,B,C\}\bigcap\{A,D\} = \{D\}.$$

The above fault diagnosis results can be put in the form of expert system (IF ... THEN) rules as:

Rule 1: IF C1 is high AND C2 is low
       THEN the diagnosis is A
Rule 2: IF C1 is high AND C2 is high
       THEN the diagnosis is B
Rule 3: IF C1 is low AND C2 is normal
       THEN the diagnosis is C
Rule 4: IF C1 is normal AND C2 is low
       THEN the diagnosis is D
Rule 5: IF C1 is normal AND C2 is normal
       THEN the diagnosis is normal.

where C1 stands for *constraint-1* and C2 for *constraint-2*. The general form of the diagnostic rule is

   IF $C1^f$ AND $C2^f$ AND ... AND CN_C^f

   THEN the diagnosis is f

where the premises are the complete pattern of constraint violations and satisfactions which are the outcome of the fault at hand.

Using this rule form one can readily integrate the governing equations method to fault diagnosis with other pragmatic and heuristic approaches.

15.4.2.3.2 Non-Boolean logic inference

In the analysis of the previous section only two values (true and false) were allowed for a hypothesis, a fact which is not realistic in many practical situations where diagnosis cannot be rendered with full certainty. Also the classification of constraints as S^+, S^- and S^0 was precise (sharp), which, due to the infinite sensitivity of diagnosis to incremental changes of the system state around the threshold values, can lead to diagnostic instability. Thus, if for example there exists measurement noise near the threshold between *normal* and *high* states of constraint 1, in the illustrative example given at the end of the previous subsection, then the diagnosis would be fluctuating between A and D.

These and other difficulties can be overcome by using inference methods taking into account the existing uncertainty, i.e. non-Boolean logic methods.

The most common non-Boolean reasoning methods are the following:

(a) Bayesian inference
(b) Fuzzy logic
(c) Theory of confirmation
(d) Shafer-Dempster theory
(e) Theory of usuality.

A brief description of the above methods is given here for convenience.

Bayesian Inference

This is a probabilistic reasoning technique based on Bayes' formula:

$$P(f_i/S) = \frac{P(f_i)P(S/f_i)}{\sum_j P(f_j)P(S/f_j)} \quad , \quad i=1,2,\ldots,n$$

where $P(f_i)$ is the *a-priori* probability (i.e. in the absence of evidence) of the fault (hypothesis) f_i, $P(S/f_i)$ is the conditional probability of the symptom (event) S to occur if f_i holds, and $P(f_i/S)$ is the *a-posteriori* probability of the fault f_i given that symptom S has been observed.

Clearly, this formula provides a link between premises (faults) and conclusions (symptoms) defined by the weight with which the truth of premises supports the truth of conclusions. The degree of the conclusion changes proportionately according to the link and the certainty of the user's facts. Bayes' formula can only be used if $P(f_i)$ and $P(S/f_i)$, i=1,2,...,n are known (measured or estimated) in advance. Also commutativity of $P(f_i)$ and $P(S/f_i)$ must be admissible. Some expert system cores based on Bayesian inference are PROSPECTOR, M1 and EXPERT EDGE. Given a set of statistically independent symptoms $S = \{s_1, s_2, ..., s_n\}$, the probability of a fault f given S is found by the following form of Bayes formula (Charniak and McDermott (1985)):

$$P(f/S) = P(f)I(f/s_1)I(f/s_2) \; ... \; I(f/s_n)$$

where $I(f/s_i) = P(s_i/f)/(P(s_i)$. This follows from the independency of symptoms which implies the following relations:

$$P(s_i/s_j) = P(s_i), \; P(s_i,s_j) = P(s_i)P(s_j)$$

$$P(s_i/s_j,f) = P(s_i/f), \; P(s_i,s_j/f) = P(s_i/f)P(s_j/f),$$

and so on.

The above formula is useful because it suggests how to modify previous probability estimates of a fault each time we get a new symptom.

In chemical process and technological system applications the symptoms are not guaranteed to be independent, and also *a-priori* probabilities of faults and symptoms are not readily available. Hence the Bayesian inference approach may not be a convenient method for fault diagnosis.

Fuzzy Logic (Zadeh (1965), Baldwin (1986)):

A fuzzy set A is a set of ordered pairs:

$$A = \{(a_1, \alpha_1), (a_2, \alpha_2), ...\}$$

where $\alpha_i \in [0,1]$ is a number representing the grade of membership of a_i

in A. If $\alpha_i=0$ for some a_i then a_i does not belong to A, and if $\alpha_i=1$ then a_i fully belongs to A.

The fuzzy analogues of the *union* and *intersection* operations are as follows:

$$A \cup A' = \{(a_1,\max(\alpha_1,\alpha_1')), (a_2,\max(\alpha_2,\alpha_2')), \ldots\}$$

$$A \cup A' = \{(a_1,\min(\alpha_1,\alpha_1')), (a_2,\min(\alpha_2,\alpha_2')), \ldots\}$$

where the a_is include all elements of A and A'.

This means that in the fuzzy logic the degree of membership of an element "a" in the union of A and A' is the maximum of the degrees of membership of "a" in these sets, and the degree of membership in the intersection is the minimum of the two degrees. This fuzzy set framework is appropriate for introducing the uncertainty property in the governing equations technique discussed above.

Theory of Confirmation

In this theory the concepts of *measure of belief* (MB) and *measure of disbelief* (MD), graded on [0,1] are used to draw conclusions about the possibility that q occurs if p occurs:

$$\pi (q/p) = 1-MD(p,q)$$

$$\pi (\sim q/p) = 1-MB(p,q)$$

Observations and measurements of, or subjective opinion on, the occurrence of some events or symptoms guide the inference chain so that some conclusions are rejected and others are confirmed.

A model of inexact reasoning based on confirmation theory is that used in MYCIN and EMYCIN (Shortliffe (1976), Shortliffe and Buchanan (1975), and Hayes-Roth (1984)) which is based on *certainty factors* (CFs) for the quantification of the degree of confirmation of hypotheses. If MB[h;e] and MD[h;e] are the measures of *belief* and *disbelief* (respectively) of the hypothesis (or fault) h given the evidence (or symptom) e, then the certainty factor CF[h;e] of h given e is defined as:

$$CF[h;e] = MB[h;e]-MD[h;e]$$

Obviously CF takes any value between -1 (hypothesis completely not true) and +1 (hypothesis completely true).

The measure of belief MB has the following properties

$$MB[h_1 \wedge h_2;e] = min[MB[h_1;e], MB[h_2;e]]$$

$$MB[h_1 \vee h_2;e] = max[MB[h_1;e], MB[h_2;e]]$$

$$MB[h;e_1,e_2] = MB[h;e_1]+MB[h;e_2] [1-MB[h;e_1]]$$

Analogous properties are true for MD.

No complete application of EMYCIN in engineering diagnostics has been reported. However it can be shown that in the case of single faults the MYCIN/EMYCIN certainty factors approach is equivalent to the fuzzy set approach (see Example 4).

Shafer-Dempster Evidence Theory

In the theory of evidence (Shafer (1976), Dubois and Prade (1980), and Prade (1985)) as information we consider any logical proposition (fact, rule etc.) denoted by A,B and so on. A proposition that is always false is denoted as $\phi^* = A \wedge \sim A$ ($\sim$ means negation) and the always true proposition is denoted as $1^* = \sim(A \wedge \sim A)$. The logical implication "$\rightarrow$" is defined as:

$$(A \rightarrow B) = \sim A \quad V \quad B.$$

Let us consider a set P of propositions such that:

(a) if $A \epsilon P$ then $\sim A \epsilon P$
(b) if $A \epsilon P$, $B \epsilon P$ then $A \wedge B \epsilon P$.

Axiom: A map μ from P to [0,1] constitutes a measure of belief if:

(a) $\mu(\phi^*) = 0$,
(b) $\mu(1^*) = 1$,
(c) $A \rightarrow B ==> \mu(A) \leqslant \mu(B)$; A, $B \epsilon P$.

This axiom characterises a wide class of measures of belief (or uncertainty) and so many special cases have been proposed. Shafer has proposed the *credibility measure* (Cr) defined as:

$$Cr(A) = \sum_{B/B \rightarrow A} m(B) \quad \forall \quad A \epsilon P$$

where m is a function from P to [0,1] such that:

$$m(\phi *) = 0 \text{ and } \sum_{B \in P} m(B) = 1$$

It is remarked that the function m does not obey the above axiom and thus it is not a measure of belief, but it rather represents the distribution of a unit belief among the propositions of P. For this reason the function m is called the basic *probability mass distribution* (BPMD).

In the same way one can define the *plausibility measure* (Pl) as:

$$Pl(A) = 1 - Cr(\sim A)$$

$$= \sum_{B/B \to \sim A \neq 1*} m(B) \quad \forall \quad A \in P$$

Shafer has shown that Cr and Pl satisfy the above relations if and only if they are respectively *superadditive* and *subadditive* for any order n; for example if n=2:

$$Cr(A \lor B) \geqslant Cr(A) + Cr(B) - Cr(A \land B)$$
$$Pl(A \land B) \leqslant Pl(A) + Pl(B) - Pl(A \lor B)$$

Putting B = A in the last two inequalities yields

$$Cr(A) + Cr(\sim A) \leqslant 1, \quad Pl(A) + Pl(\sim A) \geqslant 1 \quad \forall \quad A \in P$$

which in the case of total ignorance implies that $Cr(A) = Cr(\sim A) = 0$ and $Pl(A) = Pl(\sim A) = 1$; i.e. two contradictory propositions can be plausible without being credible at all.

From the definitions of Cr(A) and Pl(A) it follows that:

$$Cr(A) \leqslant Pl(A) \quad \forall \quad A \in P$$

i.e. the plausibility of a proposition is always greater than or equal to its credibility.

The *uncertainty* of A denoted by u(A) is defined as:

$$u(A) = Pl(A) - Cr(A)$$

If $Pl(A) = Cr(A)$ for all $A \in P$, then the resulting unique measure of belief reduces to the probability measure, and the Bayesian inference results.

The set of propositions P is called the *frame of discernment*. In fault diagnostics P is the set of all faults, including the normal operation (no fault case, negation of all faults). The interval [Cr(A), Pl(A)] which is always a subset of [0,1] is called *evidential interval*. The basic probability mass distribution (or assignment) m(A) represents the portion of the unit of belief assigned to the proposition A, where A can be any subset of P. Belief assigned to P represents the residual belief that cannot be assigned to any subset of P on the basis of available evidence (thus introducing uncertainty into the diagnosis).

Dempster's Rule: Dempster's rule which generalises Bayes' rule, allows the combination of two basic probability mass distributions m_1 and m_2 of two knowledge sources on the same frame P into one basic assignment m_{12} such that:

$$m_{12}(A) = \frac{m(A)}{1-m(\phi*)}$$

where:

$$m(A) = \sum [m_1(A_1) \times m_2(A_2) | A_1 \cap A_2 = A] = [m_1 \cap m_2](A)$$

The rule is commutative and associative, and so in the case of multiple sources the order in which they are combined is indifferent.

In essence the rule states that the basic probability mass assigned to the intersection of A_1 and A_2 is the product of the basic probability masses of A_1 and A_2. The denominator $1-m(\phi*)$ is just a normalisation factor that keeps the total belief equal to unity and is due to the possible presence of propositions in m_1 and m_2 with empty intersection. Note that if m_1 and m_2 are probability measures referred to two sources that are mutually exclusive and exhaustive for the evidence, then Dempster's formula reduces to Bayes' formula.

The concepts of evidence theory are illustrated by the following reduces to Bayes' formula.

The concepts of evidence theory are illustrated by the following simple numerical examples:

(a) Given the frame P = [A,B,C] and the belief assignment
$m_1(A,B,P)$ = (0.3,0.5,0.2), then the evidential intervals
[Cr,Pl] of A,B and C are found to be:
A : [$m_1(A)$, 1-$m_1(\sim A)$] = [0.3,0.5], $\sim$A : [0.5,0.7]
B : [$m_1(B)$, 1-$m_1(\sim B)$] = [0.5,0.7], $\sim$B : [0.3,0.5]

C : $[m_1(C), 1-m_1(C)] = [0, 0.2], \sim C : [0.8, 1]$

(b) Given the frame $P=\{A,B,C\}$ and the belief assignment $m_2(A, \sim A, B, \sim B) = (0.1, 0.3, 0.4, 0.2)$, then the evidential intervals are:

A : [0.1,0.3], B : [0.4,0.7], C : [0,0.5]

Note that Pl(A) is actually given by

$1-m_2(\sim A)-m_2(\sim B)-m_2(\sim C)$ i.e. one minus the sum of the basic probability masses of $\sim A$ and all subsets of $\sim A$. Here $\sim A=P-\{A\}=\{B,C\}$ and its subsets are B and C. The plausibilities Pl(B) and Pl(C) can be calculated analogously.

(c) Given the above *two* knowledge sources m_1 and m_2 regarding the propositions A,B,C of the frame P, the combined belief assignment to $A,B,\sim A$ and $\sim B$ are found as follows.

First, from the given probability assignments $m_1(A,B,P)$ and $m_2(A,\sim A, B, \sim B)$ we construct the following Table 15-1 of combined probability assignments.

m_2 / m_1	A	$\sim A$: B∪C	B	$\sim B$: A∪C
	0.1	0.3	0.4	0.2
A 0.3	A(0.03)	Φ*(0.09)	Φ*(0.12)	A(0.06)
B 0.5	Φ*(0.05)	B(0.15)	B(0.20)	Φ*(0.10)
P 0.2	A(0.02)	B∪C(0.06)	B(0.08)	A∪C(0.04)

From the table we find that:

$m_{12}(\Phi*) = 0.05 + 0.09 + 0.12 + 0.10 = 0.36$

$m_{12}(A) = (0.03 + 0.02 + 0.06)/(1 - 0.36) = 0.17$

$m_{12}(B) = (0.15 + 0.20 + 0.08)/0.64 = 0.67$

$m_{12}(\sim A) = 0.093$, $m_{12}(\sim B) = 0.062$

The evidential intervals determined by the combined probability masses are:

A:[0.17, 0.23], B:[0.67, 0.83], C:[0, 0.16]

Theory of Usuality

The theory of *usuality* is a generalisation of Zadeh's theory of *possibility* and is based on the fact that the *usual value* of a variable in dispositional valuations is characterised via the implicit fuzzy quantifier *usually*. Dispositional valuations are propositions with implicit external fuzzy quantifiers and are employed for formalising the common-sense reasoning or as a tool for combining evidence.

The dispositional *modus ponens* rule is:

$$\frac{\text{(usually) (A is P)}\qquad\text{IF A is P, THEN (usually) (B is Q)}}{\text{(usually)}^2 \text{ (B is Q)}}$$

where B and Q are fuzzy predicates, and the quantifier (usually)^2 is less specific than the quantifier (usually).

The dispositional *entailment principle* is expressed as:

$$\frac{\begin{array}{ll}\text{(usually)} & \text{(A is P)}\\ \text{(usually)} & \text{(P} \subset \text{Q)}\end{array}}{\text{(usually)}^2 \text{ (A is Q)}}$$

Another important extension of Zadeh's basic work is Yager's quantifier approach to evidence aggregation (Yager). Suppose that F is a frame in which a variable x takes some unknown value, and R_i, i=1,2,...,n are opinions about the location of x. Here R_i is a fuzzy subset of values which, in the opinion of the ith expert, x will take within F. The weight of the opinion is a value denoted by W_i.

Yager's measures of aggregation of expert opinion on some subset $H \subset F$ are:

Possibility that x belongs to H
$$Poss \ [H|R_i] = \max_{x \in F} \ [H(x) \wedge R_i(x)]$$

Certainty that x belongs to H
$$Cert \ [H|R_i] = 1 - \max_{x \in F} \ [\sim H(x) \wedge R_i(x)]$$

Truth that x belongs to H
$$Truth \ [H|R_i] = \mathop{\mathbf{U}}_{x \in F} \left[\frac{R_i(x)}{H(x)} \right]$$

where "$\wedge$" stands for minimum.

The grade of membership $\alpha(r)$ in the fuzzy subset representing the linguistic quantifier $A(r)$ is given by:

$$\alpha(r) = \sum_{i=1}^{n} W_i \wedge Poss[H|R_i] / \sum_{i=1}^{n} W_i$$

This formula is also valid for the other two criteria. One has merely to replace *Poss* by *Cert* and *Truth*.

15.4.2.3.3. Fault detection under uncertainty

From among the various methods of fault detection under uncertainty we briefly describe the ones that are based on fuzzy logic and evidence theory.

The conventional way of classifiying the system constraints as *high* (S_i^+), *low* (S_i^-) and *normal* (S_i^0) is through the chi-squared distribution, assuming that the residual ϵ_i of the ith constraint is Gaussian with zero-mean and variance σ_i^2 (which is measured experimentally or estimated using standard statistical methods). The chi-squared distribution gives the probability that $x^2 = \epsilon_i^2 / \sigma^2$ exceeds x_C^2. The belief essigned to S_i^0 is one if $x^2 < x_C^2$ and zero otherwise, for a desired level of significance, e.g. $P(x_C^2) = 0.1$ (90% confidence) (see Figure 15-12). This technique is appropriate for choosing the tolerances in the Boolean reasoning approach. But as it was discussed already such sharp description of beliefs leads to diagnostic instability.

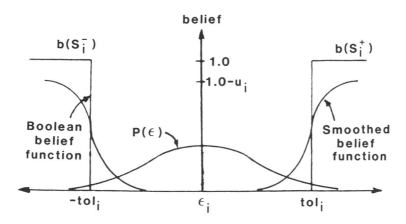

Figure 15-12: Belief functions (Boolean and sigmoidal).

The smooth belief function worked-out by Kramer is the sigmoidal one and has round corners as shown in Figure 15-12. Mathematically this function is defined as (Kramer (1986)):

$$b_i = (1-u_i) \; \xi_i^n/(1+\xi_i^n), \quad \xi_i = \epsilon_i^2/(\sigma_i^2 x_c^2)$$

where $b_i = b_i(\epsilon_i)$ is the belief function of the ith constraint, and n represents the degree of smoothing the step belief function with extreme values $n=\infty$ (no smoothing) and $n=0$ (full smoothing, b_i=constant). The factor $1-u_i$ allows the introduction of uncertainty in b_i when the reliability of the information resulting from the constraint is less than one.

   Using this belief function one finds the following evidential intervals which have obvious meaning:

For $\epsilon_i > 0$; $U_i^+:[b_i,b_i+u_i]$, $U_i^-:[0,u_i]$, $U_i^o:[1-b_i-u_i,1-b_i]$
For $\epsilon_i < 0$; $U_i^+:[0,u_i]$, $U_i^-:[b_i,b_i+u_i]$, $U_i^o:[1-b_i-u_i,1-b_i]$

In compatibility with these evidential intervals the probability masses assigned to the fault sets U_i^+, U_i^- and U_i^o are:

$$m \; (U_i^+, \; U_i^-, \; U_i^o, \; P) = \begin{cases} (b_i, \; 0, \; 1-b_i-u_i, \; u_i), & \epsilon_i > 0 \\ (0, \; b_i, \; 1-b_i-u_i, \; u_i), & \epsilon_i < 0 \end{cases}$$

The uncertainty is introduced by assigning the probability mass u_i directly to P, which means that u_i refers to all hypotheses and cannot be assigned to any particular subset of P.

   Now, given several probability assignments to U_i^+, U_i^- and U_i^o, coming from different sources of information, one can determine an overall assignment using Dempster's rule and then find the corresponding evidential intervals.

   To facilitate the integration of the method into rule-based expert system cores, one needs to convert the procedure into rule-based format.

Two ways of doing this are via *fuzzy logic* and *evidence theory*.

(a) **Rule-based diagnosis using fuzzy logic:** Let S_i^f be the condition of the ith constraint corresponding to the fault set U_i^f (e.g. $S_i^f = S_i^+$ if $f \epsilon U_i^+$). Obviously S_i^f are the symptoms of the fault $f \epsilon U_i^f$ expressed as violations or satisfaction of the constraint. The certainty factor $CF(U_i^f)$ for U_i^f can be defined as:

$$CF(U_i^f) = m(U_i^f) + u_i$$

i.e. as the sum of the uncertainty u_i in the constraint, and the basic probability mass assigned to U_i^f.

A reasonable choice for the membership grade $\alpha_i(f)$ of the fault f in U_i^f is:

$$\alpha_i(f) = CF(U_i^f)$$

The diagnosis under fuzzy set theory is the fuzzy intersection of the fault sets U_i^f, $i=1,2,\ldots,N_c$. Thus the single fault diagnosis S_1 does not simply contain one element, but a set of possible faults together with the corresponding membership grades.

Now, given the diagnostic rule,
"**IF** C_1^f and C_2^f and ... and C_{NC}^f
THEN f",
the belief in fault f is equal to
$$CF(f) = \min\{CF(C_1^f), CF(C_2^f),\ldots\}$$

(b) **Rule-based diagnosis using evidence theory:** We first introduce a new function q(f) of the fault (hypothesis) f defined as

$$q(f) = \prod_i CF(U_i^f)$$

This function is called *supportability* (support + plausibility) and satisfies the relation

$$Cr(f) \leqslant q(f) \leqslant Pl(f)$$

Thus, given a diagnostic rule of the above type the supportability of f is equal to the product of the certainty factors of the rule antecedents.

Example 4: Consider a system with two constraints S_1, S_2 and fault sets:

$$U_1^+ = \{A,B\}, \quad U_1^- = \{C\}, \quad U_2^+ = \{B\}, \quad U_2^- = \{A,D\}$$

The frame of discernment is $P=\{A,B,C,D,F\}$ where F corresponds to fault-free operation.

If $b_1=0.8$ and $b_2=0.7$ with $\epsilon_1>0$ and $\epsilon_2>0$, respectively, and $u_1=0$, $u_2=0.2$, then the assigned probability masses are:

$$m_1(U_1^+, {\sim}U_1^-, \quad (U_1^+ \lor U_1^-), \ P) = (0.8,0,0.2,0)$$
$$m_2(U_2^+, {\sim}U_2^-, \quad (U_2^+ \lor U_2^-), \ P) = (0,0.7,0.1,0.2).$$

Then, Dempster's rule gives,

$$m(A,B,C,D, \ A \lor B, \ D \lor F, \ F) = (0.61,0,0,0.15,0.17,0.04,0.02)$$

and the corresponding evidential intervals are:

A:[0.61,0.78], B:[0,0.17], C:[0,0], D:[0.15,0.19], F:[0.02,0.06]

One can see that A is the most likely fault.

The rule-based formulation of the fault sets is:
IF C1-high **and** C2-low **THEN** A
IF C1-high **and** C2-high **THEN** B
IF C1-low **and** C2-normal **THEN** C
IF C1-normal **and** C2-low **THEN** D
IF C1-normal **and** C2-normal **THEN** F

The certainty factors of the rule antecedents are:
CF(C1-high)=0.8, CF(C1-normal)=0.2, CF(C1-low)=0
Cf(C2-high)=0.2, CF(C2-normal)=0.3, CF(C2-low)=0.9

The supportability of each fault is calculated by the products of the CF's of the antecedents in each rule, i.e.
$q(A)=0.72$, $q(B)=0.16$, $q(C)=0$, $q(D)=0.18$, $q(F)=0.06$
One can check that the q's indeed lie in the evidential intervals calculated above.

Now, writing the fuzzy set form of the diagnosis we obtain

$$U_1=\{(A,0.8), \ (B,0.8), \ (C,0), \ (D,0.2), \ (F,0.2)\}$$
$$U_2=\{(A,0.9), \ (B,0.2), \ (C,0.3), \ (D,0.9), \ (F,0.3)\}$$
$$S_1=U_1 \cap U_2=\{(A,0.8), \ (B,0.2), \ (C,0), \ (D,0.2), \ (F,0.2)\}$$

where for example $\alpha(A)=\min\{\alpha_1(A), \ \alpha_2(A)\}=\min\{0.8,0.9\}=0.8$ etc.

The same result might be obtained by computing the minimum of the certainty factors in the rule antecedents (if the rule format is used). From this example one can see that in the context of rule formulation, fuzzy set theory and confirmation theory lead to identical fault diagnosis results.

15.5 DESIGN EXAMPLES OF FAULT DIAGNOSIS EXPERT SYSTEMS

To see how *Fault Diagnosis Expert Systems (FDES)* are built and implemented we describe here three particular FDES's from the open literature. Each of them is based on a different knowledge processing approach. Many other FDES's are available and can be found in the technical literature and/or the market.

15.5.1 FDES based on inference engine GENIE

At Vanderbilt University a general purpose FDES was designed (called *Fieldserve*) which is suitable for complex systems involving electronic and mechanical components (Hofmann et al. (1986), Caviedes and Bourne (1986), and Caviedes et al. (1988)). Most knowledge was encoded in rules and frames using the general purpose inference engine GENIE (Sandell et al. (1984)). Also the Franzlisp dialect of Lisp was used for implementing (a) the agenda that specifies the flow of control in any particular system, (b) a set of auxiliary functions, and (c) demons that are used mainly for data acquisition. All data are stored in frames and contain the knowledge and facts about the system under repair. Frames provide prototype structures that can be interactively pruned for building a model of the specific domain. Frame-based agendas list tasks to carry out pruning of trees of alternatives, input of data, application of rules, and so on. The rule interpreter allows backward-chaining, forward chaining or event-driven rule application to be used as a strategy. Data types include *active knowledge* (encoded into rules mainly used for reasoning and data acquisition) and *passive knowledge* (e.g. menu input, parameter specifications and facts concerning the system under repair).

```
(symptoms
    (faulty-unit
        (AC-power
            (wiring)
            (breaker)
            (fuse)
        )
        (DC-voltages
            (some-incorrect (if-needed (d-measure-voltages)
                    (regulator))
            (some-zero
                    (fuses))
                    :
        )
        (digital-logic
                    :
        )
        (analog-section
        )
                    :
    )
)
```

Figure 15-13: Typical skeletal frame for one symptom.

A typical skeletal frame for one symptom *fault-unit* and two subslots containing potential causes of this symptom (*AC-power*, *DC-power*) is shown in Figure 15-13. The demon *d-measure-voltages* is a procedure that

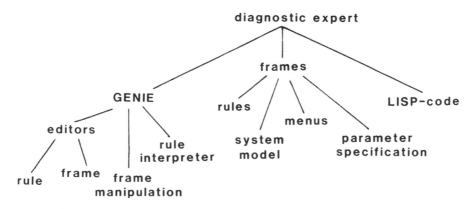

Figure 15-14(a): General organisation of FDES

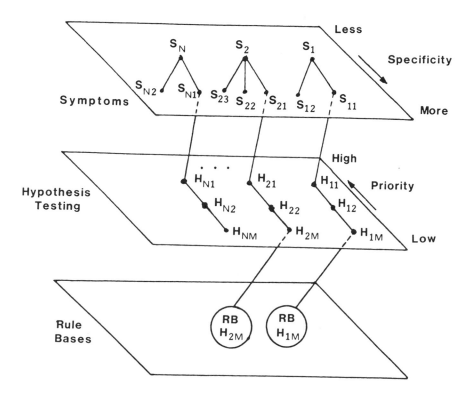

Figure 15-14(b): Layered architecture of FDES
(H: Hypothesis, S: Symptom, RB: Rule base)

waits for the path (symptoms faulty-unit DC-voltages some-incorrect)
to be fetched before it is activated.

```
(system model
  (symptoms
    (main symptom
      (specific type
        (hypotheses
          (H1 (RB1))
          :
        )
      )
    )
  )
  (H1 data
    ( )
    ( )
  (H2 data
    ( )
  )
  (common data
    ( )
  )
)
```

Figure 15-14(c): Outline of system-model frame structure

The general organisation of the FDES is shown in Figure 15-14(a). The system has a hierarchical (layered) architecture as shown in Figure 15-14(b). The top level contains a set of symptom classifications from which the actual symptoms can be selected using menu input. On the basis of the symptoms a list of possible hypotheses (faults) contained at the second level is tested. For each testable hypothesis there is a rule-base. This hierarchical form of the system allows a straightforward description of the inherent hierarchical structure of most real machines and industrial processes. An outline of the system-model frame structure for the symptom and hypothesis abstractions is shown in Figure 15-14(c).

A monotonic control strategy (from the top level) was employed, which becomes more powerful if the rule-bases are allowed to modify the hypothesis list. If the rule bases are recursive then the subproblems can be solved with the same technique as the top level.

A set of rules of *Fieldserve* is the following:

Operating Switch Rule 1

IF the operating switch is replaced and the user responds "true" to the question "Did this fix the problem?"

THEN the switch is fixed, print "Thanks".

Operating Switch Rule 2

IF the operating switch resistance does not test ok or the user responds "true" to the question "Is the operating switch defective?"
THEN replace the operating switch, print "Replace the operating switch and reapply power!"

```
OP_switchRB_rule1
    ifall
        (device_model op_switch replace operating_switch") = true
        (device_model op_switch "Did this fix the problem?) = true
    then
        conclude: (device_model op_switch fixed)
            with value: true
        print "Thanks!"

OP_switchRB_rule2
    ifall
        or
    (device_model op_switch resistance_test_ok?) = false
    (device_model op_switch "Is the operating switch defective?")=true
    then
        conclude: (device_model op_switch replace operating_switch)
            with value: true
        print "Replace the operating switch and reapply power!"

OP_switchRB_rule 3
    ifall
        or
        (device_model op_switch "Is the "ON" resistance OK?")=false
        (device_model op_switch "Is the "OFF" resitstance OK?")false
    then
        conclude: (device model op_switch resistance_test_ok?)
            with value: false
```

Figure 15-15: Fieldserve rules in Genie form.

Operating Switch Rule 3

IF the "ON" resistance is not OK or the "OFF" resistance is not OK
THEN the operating switch resistance does not test ok.
These rules implemented in GENIE take the form shown in Figure 15-15.
Hi. I'd like to help you repair your equipment
CAUTION

Before beginning to service unit, record all critical data stored
in volatile memory to prevent inadvertent loss or modification.
CAUTION
Always remove power from the unit before you replace any board or
connectors! When you are ready for further text display, hit the
"return key: Model symptoms
        1) No display on power up
        2) Incorrect display
        3) Command module will not function properly
        4) No lamp test (8's) at turn on
        5) Does not respond to console
        6) No product delivered
        7) Product delivered but not totalised
        8) Battery back-up inoperative
        9) Improper data displayed
        10) Console fails to select proper unit
        11) Low speed delivery only
Please enter choice(s): [6]
        Let's get to work!
Enter "why?" to see an explanation screen!
Did you hear relay pulling in?
If not, open the C/D module and check it!
Is the relay pulled in? Is the above true Y/N? [n]
Do you measure 120 VAC across the relay?
        Is the above true Y/N? [n]
We first must correct the subsymptom:
No 120 VAC at the power relay
Check the connections in the 120 VAC line to the power relay!
Did you have to secure J4, J16, or J17?
        Is the above true Y/N? [y]
Check the subsymptom: No 120 VAC at the power relay
Did we fix this symptom? Is the above true Y/N? [n]
Replace the R/I-board
Did this fix the problem?
        Is the above true Y/N [why?]
SYMPTOM: No product delivered
SUBSYMPTOM: No 120 VAC at the power relay
CURRENT TEST GOAL: Establish if the defect is a defective R/I-board.
PERFORMED TESTS
NO DEFECT WAS FOUND WHILE TESTING IF THE DEFECT
        is that no unit price had been entered.
        is that the power relay is defective.
A DEFECT WAS FOUND WHILE TESTING IS THE DEFECT
        is a defective R/I-board.
        is that connectors J4, J16, or J17 are not plugged in correctly.
Did this fix the problem? Is the above true Y/N? [y]
Thanks!

        Figure 15-16: Example of interaction session with Fieldserve.

The system has been ported to an IBM PC running IQLisp and work is in progress for implementing it on 68000-based machines. It needs a highly effective user interface and some keyboard operations. Presently a meta-level architecture is being developed to eliminate most Lisp from the implementation. In this architecture the control strategies are declaratively implemented to allow a more structured representation of the deep and shallow knowledge levels. With this representation, an explanation system based on a dynamically-generated proof tree naturally becomes a strategy-based explanation system which improves the *friendliness* of Fieldserve (Caviedes et al. (1988)). The transcript of an example session using Fieldserve is shown in Figure 15-16 (Hofmann et al. (1986)) where the user's response is given in brackets.

15.5.2 The Lockheed expert system: LES

This system (Perkins and Laffey (1984), and Laffey et al. (1986)) is presently used for fault monitoring a *baseband distribution system (BDS)* that accepts 40 baseband input signals and connects them, under computer control, to any one of up to 304 baseband signal output ports without signal degradation under normal conditions. The BDS contains 16 device cabinets, a terminal and a line printer. The FDES built using LES performs repair maintenance on the BDS by quickly isolating the faulty printed circuit board or other sections of the chassis-mounted piece that is the source of the fault. The BDS contains 3000 printed circuit boards, and 1000 cables or other devices. LES employs an attribute database (or frame structure) to represent the structure and behaviour of the BDS (electrical connections, properties etc.). The approach adopted is a combination of the deep and shallow approaches, where a set of categories and a set of attributes for each object is selected. Typical categories (modules) are printed circuit cards, cables, frequency synthesisers and spectrum analysers. Typical attributes of an object in these categories are: name, location, origin of a signal, destination of a signal, whether a signal passes through the object, the object's likelihood of being faulty, etc.

Category: CARD

Slot	Value
NAME	"16x1 switch"
TYPE_OF	"16x1 card"
LOCATION	"cabinet 302 A1 A4 A20"
FAULTY_LIKELIHOOD	"0.1"
SIGNAL_PASSING	"TRUE"
ELECTRICALLY_CONNECTED_INPUT	"CARD(NAME = 1x16 buffer)"
ELECTRICALLY_CONNECTED_OUTPUT	"CARD(NAME = 8x1 second-level)"
ELECTRICALLY_SUPPORTED_BY	"CARD(NAME = 16x16 matrix)"
ELECTRICALLY_SUPPORTS	"
HAS_FAULT_LIGHT	"TRUE"
TESTABLE_INPUT_POINT	"FALSE"
TESTABLE_OUTPUT_POINT	"TRUE"
GIVES_ALARM	"FALSE"
FAULTY	"
INPUT_SIGNAL_PRESENCE	"
OUTPUT_SIGNAL_PRESENCE	"

Figure 15-17(a): LES frame representation of a 16x1 switch card

Slot	Value
NAME	"16x16 switch section faults"
SUBJECT_CATEGORY	"diagnose: 16x16 switch section"
PRIORITY	"0"
OUTPUT_RESULT	"true"
DEFINITELY_KNOWN	"80"
DEFINITELY_UNKNOWN	"20"
HYPOTHESIS	"input regulator card faulty OR cable J236 is loose OR 16x1 switch card is faulty OR ...

Figure 15-17(b): Example of BDS goal frame

An example of a typical object represented in LES frame structure is shown in Figure 15-17(a), where some slots do not have values and are filled by rule instantiation or user replies during the fault finding session. All objects and goals in LES are represented uniformly in frame structures. An example of a BDS goal frame is shown in Figure 15-17(b).

The knowledge in LES is captured in *IF-THEN* and *WHEN-THEN* rules derived from BDS maintenance manuals, as well as from the experience of experts. Presently LES contains 110 rules and 1200 facts. These rules are represented in the *case-grammar* format (Fillmore (1968)), extended so as to handle *Actors* that are attributes of objects. The case-grammar format for rules is consistent with the format of the attribute database. This case-grammar structure offers greater versatility in describing the relationships between objects under analysis, and in using natural language for the explanation facilities. Variables can be represented very easily, e.g.

    CARD (NAME=?X)

where ?X is the variable that can be bound to any object (frame) or type Card. LES regards the constraints existing in rules as *descriptors* and uses them to do a constrained looking in its frame structured attribute database. For example the rule:

WHEN *<some condition(s) become true>*

THEN *mark all components in 16x16 switch section "not faulty"*

has the following Case-grammar representation:

WHEN: *<some condition(s) become true>*

THEN: TYPE_ENTRY: "STATE"

      ACTOR: "FAULTY [THING (IS_PART_OF
=SECTION (NAME =16x16
switch))]"

      ACTION_VERB: "="

      OBJECT: "FALSE"

LES actually scans all objects in its attribute database (since THING was specified) and marks all those that satisfy the given constraints as *not faulty*. It is mainly a goal-driven, backward chaining system like EMYCIN and uses standard AND/OR goal trees to represent goals. In addition to goal-driven reasoning, forwardchaining reasoning is allowed via WHEN-THEN rules, which are a slight variation of procedural attachment and event driven processing. These rules act as *demons* which are activated when certain events occur (i.e. when slots in the database take specified values). Attaching the WHEN rules to slots in the frames increases the efficiency especially when dealing with large rule bases, since cycling through all rules is not required. LES can automatically switch goals by increasing the priority of goals that look more promising (metalevel reasoning).

This metalevel reasoning is made explicit by the WHEN rules. A WHEN rule example (in English-like representation) is the following:

WHEN the output_signal_presence of 16x1 switch card=true,

> **THEN** mark all components in 16x16 section "not faulty", AND
> inform user "problem is not in 16x16 switch section"=0, AND
> make priority of diagnosing 8x1 second-level switch
> section=50.

In this rule, when a signal is detected at the output of the 16x1
card, the four actions of the THEN part are fired. The first two
concern the procedural knowledge, whereas the last two actions deal
with metalevel reasoning. In standard production rule-based expert
systems the metalevel reasoning is controlled using obscure ways such
as flags. The WHEN rules are a natural way to factor out the control
information from goal-driven rules, and allow a significant increase
of the number and complexity of rules that can be handled.

*[LES calculates a single test point for the spectrum analyser from well over
10,000 possible test points]*
USE SPECTRUM ANALYSER
  You now need to connect the spectrum analyser to Input Connector/
Regulator Card at cabinet 302 A1 A5 A18 (jack 12) and compare the
signal detected with that generated by Frequency Synthesiser A at
cabinet 300 A1 A6.
  First, you must connect the spectrum analyser to the input of the
Input Connector/Regulator Card.
***WHY

*[User asks why LES is suggesting this test, and the system explain its line of
reasoning]*
IF we can
   (1) show that INPUT_SIGNAL_PRESENCE of the Input
   Connector/Regulator Card is TRUE; and
   (2) show that OUTPUT_SIGNAL_PRESENCE of the 16x1
   Switch Card is DISTORTED;
THEN it follows from Rule 31 that
   FAULTY of the 1x16 Buffer Control Card is TRUE.
   Does the signal detected by the spectrum analyser agree with the
generated signal? (Y or N)
*** Y

*[At this point a WHEN rule fires and the user is informed that the problem is
not in the Secondary Multicoupler cabinet. Also, the priority of one goal is
lowered to 0, while another is raised to 50]*
  The faulty component does not lie in the Secondary Multicoupler
cabinet nor with the power splitter.
[LES suggests another test point]
USE SPECTRUM ANALYSER
  You now need to connect the spectrum analyser to 16x1 Switch Card
at cabinet 302 A1 A5 A33 and compare the signal detected with that

generated by Frequency Synthesiser A at cabinet 300 A1 A6.
First, you must connect the spectrum analyser to the output of the 16x1 Switch Card.
Is the signal detected by the spectrum analyser: Distorted, Missing completely, or Present and good (D,M, or P)?
*** INFO

[INFO takes the user into the database retrieval mode, where the user can type questions in English, which LES will try to answer]
What is your question? (enter HELP for options)
*** WHERE IS SW_16X1_SWITCH CARD?

The 16x1 Switch Card is located at cabinet 302 A1 A5 A33.
What is your question? (enter HELP for options)
*** WHAT IS ITS INPUT?

The 1x16 Buffer Control Card is the input to the 16x1 Switch Card.
What is your question? (enter HELP for options)
*** WHY IS FAULTY OF THE PRIMARY_MULTICOUPLER CARD FALSE?

[LES can answer questions about why a particular component is not faulty, since it keeps track of everything that has happened (and why it happened) in its Chronological-Event database]
FAULTY of the Primary Multicoupler Card is FALSE for the following reasons:
  (1) The OUTPUT_SIGNAL_PRESENCE of the Network
  Analyser (channel A) is TRUE, and
  (2) The PATH_POSITION of the Network Analyser
  (channel A) is greater than the PATH_POSITION of the
  Primary Multicoupler Card.
What is your question? (enter HELP for options)
*** QUIT

Is the signal detected by the spectrum analyser: Distorted, Missing completely, or Present and good? (D,M, or P)
*** P

[After several more tests are run, LES reaches a conclusion and informs the user]
I have reached the conclusion that the 8x1 Second-level Switch Card at cabinet 302 A1 A6 A2 is faulty!

Figure 15-18: Log of an interaction session with LES.

The log of an interaction session with LES is shown in Figure 15-18 where user responses are preceded by asterisks and comments are enclosed in brackets (Laffey et al. (1986)).

15.5.3 Attribute grammar-based fault diagnosis tool

At the National Technical University of Athens the attribute grammar approach has been adopted for developing a knowledge representation fault diagnosis tool (Papakonstantinou and Kontos (1986), Moraitis et al. (1987), Moraitis (1986), Papakonstantinou et al. (1988), and Panayiotopoulos et al. (1987)). Employing the attribute grammar model for knowledge representation one can combine both *declarative* (factual) knowledge and *inferential* (procedural) knowledge in a single tool. Since presently many implementations of compilers and interpreters for attribute grammars' processing exist, this approach promises a lot for the knowledge engineers and the fault diagnosis experts. A practical implementation of an attribute grammar interpreter for cases where knowledge can be expressed in the form of logic rules is described in (Papakonstantinou et al. (in press)), while its extensions required to incorporate uncertain and imprecise reasoning are discussed in (Moraitis et al. (1987), and Moraitis (1986)).

In this approach the knowledge-base is written in the form of logic rules in a PROLOG-like way, i.e.

$$R_0(t_{o1}, t_{o2}, \ldots, t_{ok_o}) \text{ is true if}$$

$$R_1(t_{11}, t_{12}, \ldots, t_{1k_1}) \text{ is true and}$$

$$\begin{array}{cccccc} . & . & . & . & . \\ . & . & . & . & . \end{array}$$

$$R_{m_e}(t_{me_1}, t_{me_2}, \ldots, t_{me_{km}}) \text{ is true,}$$

where t_{ij} ($0 \le i \le m_e$, $i \le j \le k_i$, $1 \le e \le n$) is a constant or a variable, and e is the rule number (assuming that the rules are numbered from 1 to n). If $m_e = 0$, then the rule reduces to a fact.

A syntax rule corresponding to the above logic rule for an *homologous attribute grammar* (i.e. a grammar which when processed by its interpreter will give the same results as those coming from the successful application of the logic rules) has the form:

$$<R_0>::=<R_1><R_2>\ldots<R_{me}>|{}^-|$$

where the combination of the last two characters means the end of the syntax rule.

The parser of the homologous attribute grammar has the following features:

(a) No terminal symbols are used (i.e. it is degenerate).
(b) An extended stack is used for saving the attribute value as well.
(c) Calls to the attribute evaluator are included.
(d) A meta-variable, named FLAG, is used to show variable matching (when a value mismatch occurs, the FLAG takes the value false).

A false value of FLAG results in a backtracking of the parsing process. The semantic rules that perform the unification of variables can be written in a straightforward way. The interpreter includes a facility of providing *why* and *how* explanations.

The standard Prolog interpreters as well as the basic attribute grammar interpreter presented in (Papakonstantinou et al. (in press)) do not meet the requirements of a complete theorem-prover, since they are based only on a suitable subset of first-order logic. Of course the attribute grammar interpreter is better than the Prolog interpreter since it combines in one tool both the declarative properties of Prolog and the procedural characteristics of production systems.

By including the *Model Elimination* (ME) operation in them one obtains complete inference engines for first-order calculus. The ME *reduction* operation is based on the following rule:
"If the current goal matches the complement of one of its ancestor goals, then apply the matching substitution and treat the current goal as if it were solved" (Loveland (1969), and Stickel (1984)).

The resulting complete attribute-grammar theorem-prover has been employed as a fault diagnosis tool for digital circuits (Panayiotopoulos et al. (1987)). Its principal properties are:

(a) It accepts and treats non Horn clauses
(b) It possesses full theorem prover capabilities
(c) It is simple and portable
(d) It is extensible (one can add user defined calls written in the host language).

Reasoning under uncertainty can be easily implemented using this attribute grammar interpreter/prover by incorporating any method of treating inexact knowledge (Bayesian, theory of evidence, etc.).

Fault diagnosis examples worked out by the full theorem prover and imprecise/uncertain reasoning using possibility/necessity measures and fuzzy sets can be found in (Moraitis et al. (1987), Moraitis (1986), and Panayiotopoulos et al. (1987)).

15.6 CONCLUSIONS

In this chapter an attempt has been made to provide a comprehensive review of the knowledge-based methodology for fault diagnosis in technological systems. The review is unavoidably non-exhaustive, but the material included has been selected so as to be representative and to cover all the basic concepts, issues and tools.

Research on computer-based automated diagnosis is receiving increasing attention and the currently available numeric and non-numeric (symbolic) tools are already sufficient for developing practical systems for on-line and off-line automated diagnosis and supervision of electronic, mechanical, chemical, aerospace and other devices and processes. A commercially available knowledge-based diagnosis/supervision system is PICON (*Process Intelligent Controller*), developed by LMI (LISP Machine Inc.) in the USA, which operates on a LISP machine interfaced with a conventional distributed control system. The PICON expert system can handle up to 20,000 measurement points and alarms, uses both *deep* and *shallow* reasoning, it can work in real and simulated time, and includes a natural language menu as well as a simulation facility. PICON together with other existing fault monitoring/ diagnosis expert systems show that the knowledge-based technology has now reached the level of full-scale efficient and productive utilisation in industrial and other complex modern life systems.

References

Abramowitz, M. and Stegun, I.A. (eds), *Handbook of Mathematical Functions*, National Bureau of Standards, 1965.

Akaike, H., On the Use of a Linear Model for the Identification of Feedback Systems, *Annals Inst. Statist. Math.*, Vol. 20, 425-439, 1968.

Akaike, H., A New Look at the Statistical Model Identification, *IEEE Trans. Automatic Control*, Vol. AC-19, 716-723, 1974.

Akaike, H., Time Series Analysis and Control through Parametric Modelling, in Findley, D.F. (ed.), *Applied Time Series Analysis*, Academic Press, 1978.

Akpan, I., *An Approach for Including Reliability in Control System Design*, Msc thesis, Department of Engineering Mechanics, Old Dominion University, Virginia, Aug. 1982.

Aldeen, A. and Blitz, J., Eddy Current Investigations of Oblique Longitudinal Cracks in Metal Tubes Using a Mercury Model, *NDT International*, 211-216, November 1979.

Aldeen, A. and Blitz, J., Assessment of Oblique Cracks in Metal Tubes with Eddy Currents, *World Conference on Non-destructive Testing, Melbourne, Australia*, Session 4B-1, Nov. 1979.

Anderson, B.D.O. and Moore, J.B., *Optimal Filtering*, Prentice Hall, Englewood Cliffs, New Jersey, 1979.

Anderson, B.D.O., Moore, J.B. and Hawkes, R.M., Model Approximations via Prediction Error Identification, *Automatica*, Vol. 14, 615-622, 1978.

Anderson, T.W., *An Introduction to Multivariate Statistical Analysis*, John Wiley, 1985.

Andow, P.K. and Lees, F.P., Process Computer Alarm Analysis: Outline of a method based on list processing, *Trans. Inst. Chemical Engineers (UK)*, Vol. 53, 195, 1975.*

Andow, P.K., Fault Diagnosis using Intelligent Knowledge-Based Systems, *Chem. Eng. Res. Des.*, Vol. 63, 368, 1985.*

Armstrong, E.A., *ORACLS - A Design System for Linear Multivariable Control*, Marcel Dekker, Inc., New York, pp. 63-66, 83-91, and 99-104, 1980.

Aslin, P.P and Patton, R.J., *The Development of a Full Force Digital Simulation of a Remotely Piloted Vehicle for Application to Flight Control System Design and Assessment*, Research Report No. YEE 1, University of York, Department of Electronics, UK, 1983.

Aström, K.J. and Eykhoff, P., System Identification - a Survey, *Automatica*, Vol. 7, 123-162, 1971.

Athans, M. and Chang, C.B., *Adaptive Estimation and Parameter Estimation using Multiple Model Estimation Algorithms*, MIT Lincoln Lab., Lexington, Mass., Report no. TN-1975-59, 24 Nov. 1975.

Basille, G. and Marro, G., *On the Observability of Linear Time Invariant Systems with Unknown Inputs*, Journal of Optim. Theory and Application, Vol.3, 411-415, 1969.

References marked * are knowledge-based, expert systems or A.I. - specific publications referred to in Chapter 15.

Baldwin, J.F., Automated Fuzzy and Probabilistic Inference, *FSS*, Vol. 18, 219-235, 1986.*

Basden, A. and Kelly, B.A., DART: An Expert System For Computer Fault Analysis, *Proc. 7th IJCAI*, 843-845, 1981.*

Baskiotis, C., Raymond, J. and Rault, A., Parameter Identification and Discriminant Analysis for Jet Engine Mechanical State Diagnosis, *Proc. 15th IEEE CDC Conference, Fort Lauderdale, Fla*, Dec. 1979.

Basseville, M. and Benveniste, A., (eds), Detection of Abrupt Changes in Signals and Dynamical Systems, *Lecture Notes in Control and Information Sciences*, Vol. 77, Springer-Verlag, 1985.

Bastl, W. and Wach, D., Experiences with Noise Surveillance Systems in German LWRs, *Progr. Nucl. Energy*, Vol. 9, 505-516, 1982.

Beard, R.V., *Failure Accommodation in Linear Systems through Self-Reorganization*, Man Vehicle Lab., MIT, Cambridge, Mass., Report no. MVT-71-1, Feb. 1971.

Beazley, W.G., Prevention of Chemical Leaks using Expert Systems, *Proc. API Spring Meeting, San Diego, Calif.*, 1986.*

Belkoura, M., *Detection of Instrument Failures using a Kalman Filter*, Ph.D. Thesis (in German), Univ. Duisburg, FRG, 1983.

Bengtsson, G., Minimal System Inverses for Linear Multivariable Systems, *Journal of Math. Analysis and Appl.*, Vol. 46, 261-274, 1974.

Benzecri, J.P., Analyse Discriminante et Analyse Factorielle, *Chiers Ana. Donnees*, Vol. 2, 369, 1977.

Bhattacharyya, S.P., Observer Design for Linear Systems with Unknown Inputs, *IEEE Trans. Automatic Control*, Vol. AC-23, 483-484, 1978.

Billmann, L., *Methoden zur Lecküberwachung und Regelung von Gasfernleitungen*, Fortschrittberichte VDI, Reihe 8, Nr. 85, 1985.

Blitz, J. and Rowse, A.A., The Investigation of Defects in Non-ferromagnetic Metal Tubes using Eddy Currents, *Applied Materials Research*, Vol. 3, 82-87, 1964.

Bonivento, C. and Tonielli, A., A Detection Estimation Multifilter Approach with Nuclear Application, *Proc. 9th IFAC World Congress*, Vol.3, *Budapest*, 2nd-7th July 1984.

Breinl, W. and Müller, P.C., State-observer with Low Sensitivity and its Application to Maglev Vehicle Suspension Control, *Regelungstechnik*, Vol. 30, 403-411, 1982.

Brillouin, L., *Science and Information Theory*, Academic Press, 1962.

Brockhaus, R., Analytical Redundancy through Nonlinear Observers, *7th IFAC Symposium on Identification and System Parameter Estimation, York, UK*, July 1986, Pergamon Press, Oxford.

Brogan, W.L., *Modern Control Theory*, Quantum, New York, 1974.

Brown, H., Corley, R.C., Elgin, J.A. and Spang, H.A., *Sensor Failure Detection and Isolation in Multiengine Aircraft*, General Electric Co., Aircraft Engine Business Group, Cincinnati, OH, NASA Contractor Report no. CR-174846, 1985.

Burbea, J. and Rao, C.R., On the Convexity of Some Divergence Measures based on Entropy Functions, *IEEE Trans. Information Theory*, Vol. IT-28, 489-495, 1982.

Burrows, S.P. and Patton, R.J., Robust Eigenstructure Assignment using the CTRL-C Design Package, *Proc. 6th International Conference on Systems Engineering, Coventry, England*, 13-15th Sept. 1988.

Caglayan, A.K. and Lancraft, R.E., A Separated Bias Identification and State Estimation Algorithm for Nonlinear Systems, *Automatica*, Vol. 19, 561-570, 1983.

Carignan, C.R. and Vander Velde, W.E., *Number and Placement of Control System Components Considering Possible Control System Failure*, Space Systems Laboratory Report SSL 5-82, Dept. of Aeronautics and Astronautics, MIT, USA, Mar. 1982.

Castanon, D.A. and Teneketzis, D., Distributed Estimation Algorithms for Nonlinear Systems, *IEEE Trans. Automatic Control*, Vol. AC-30, 418-425, 1985.

Caviedes, J. and Bourne, J., Knowledge Engineering in Repair Domains: A Characterisation of the Task, *Proc. 6th Intl. Workshop on Expert Systems and their Applications, Avignon, France*, 479-490, Apr. 1986.*

Caviedes, J., Bourne, J., Brodersen, A., Osborne, P., Ross, A., Schaffer, J.D. and Bengtson, G., A Meta-Knowledge Architecture for Planning and Explanation in Repair Domains, in Tzafestas, S.G. (ed.) *Knowledge Based Systems Diagnosis, Supervision and Control*, Plenum, New York, 1988.*

Charniak, E. and McDermott, D., *Introduction to Artificial Intelligence*, Addison-Wesley, 1985.*

Chen, W.K., *Instrument Failure Detection for a Linear Uncertain System*, Ph.D. Thesis, Univ. of Washington, Seattle, USA, 1981.

Chester, D., Lamb, D. and Dhurjati, P., Rule-based Computer Alarm Analysis in Chemical Process Plants, *Proc. Micro-Delcon, Newark*, Vol. 22, 1984.

Chien, T.T. and Adams M.B., A Sequential Failure Detection Technique and Its Application, *IEEE Trans. Automatic Control*, Vol. AC-21, 750-757, 1976.

Chow, E.Y. and Willsky, A.S., Analytical Redundancy and the Design of Robust Failure Detection Systems, *IEEE Trans. on Automatic Control*, Vol. AC-29, 603-614, 1984.

Clancey, W.J. and Letsinger, R., NEOMYCIN: Reconfiguring a Rule-based Expert System for Application to Teaching, *Proc. IJCAI*, 829-836, 1981.*

Clark, R.N., Instrument Fault Detection, *IEEE Trans. on Aerospace and Electronic Systems*, Vol. AES-14, 456-465, 1978.

Clark, R.N., A Simplified Instrument Failure Detection Scheme, *IEEE Trans. on Aerospace and Electronic Systems*, Vol. AES-14 558-563, 1978.

Clark, R.N., The Dedicated Observer Approach to Instrument Failure Detection, *Proc. 15th IEEE-CDC Conference, Fort Lauderdale, Fla*, 237-241, Dec. 1979.

Clark, R.N., A Missing Term in the Linear System Equations, *IEEE Control Systems Magazine*, Vol. 5, 1985.

Clark, R.N. and Campbell, B., Instrument Fault Detection in a Pressurised Water Reactor Pressuriser, *Nucl. Tech.*, Vol. 56, 23-32, 1982.

Clark, R.N., Fosth, D.C. and Walton, W.M., Detecting Instrument Malfunctions in Control Systems, *IEEE Trans. Aerospace and Electronic Systems*, Vol. AES-11, 465-473, 1975.

Clark, R.N., Masreliez, C.J. and Burrows, J.W., A Functionally Redundant Altimeter, *IEEE Trans. Aerospace and Electronic Systems*, Vol. AES-12, 459-463, 1976.

Clark, R.N. and Setzer, W., Sensor Fault Detection in a System with Random Disturbances, *IEEE Trans. Aerospace Electronic Systems*, Vol. AES-16, 468-473, 1980.

Closkin, W.F. and Mellish, C.S., *Programming in Prolog*, Springer-Verlag, 1984.*

Cox, B.J., *Object Oriented Programming: an Evolutionary Approach*, Addison Wesley, 1986.*

Cruz, J.B. and Perkins, H.R., Conditions for signal and Parameter Invariance in Dynamical systems, *IEEE Trans. Automatic Control*, Vol. AC-11, 614-615, 1966.

Cunningham, T.B. and Poyneer, R.D., Sensor Failure Detection using Analytical Redundancy, *Proc. Joint Automatic Control Conference*, Vol. 1, *San Francisco, Ca*, 278-287, 1977.

Dan, G.J., The EPRI NDE Program, *9th World Conference on Nondestructive Testing, Melbourne, Australia*, Paper 2b-2, Nov. 1979.

Davis, J.F., Punch III, W.F., Shum, S.K. and Chandrasekaran, K., Application of Knowledge-Based Systems for the Diagnosis of Operating Problems, *Proc. AI ChE Natl. Meeting, Chicago, Ill.*, (Paper 70e) 1985.*

Davis, R., Diagnostic Reasoning Based on Structure and Behaviour, *Artificial Intelligence*, Vol. 24, 347-410, 1984.*

Davis, R. and Shrobe, H., Representing Structure and Behaviour of Digital Hardware, *IEEE Computer*, 75-82, Oct. 1983.*

Davis, T.J., A Multifrequency Eddy Current System for the Inspection of Steam Generator Tubing, Nondestructive Evaluation in the Nuclear Industry, *Proc. International Conference, Salt Lake City, Utah*, 13-15 February 1978, R. Natesh (ed.), American Society for Metals, Metals Parks, Ohio, 421-436, 1978.

Deckert, J.C., *Definition of the F-8 DFBW Aircraft Control Sensor Analytic Redundancy Management Algorithm*, C.S. Draper Laboratory Report R-1178, Cambridge, Mass, Aug. 1978.

Deckert, J.C., Desai, M.N., Deyst, J.J. and Willsky, A.S., F-8 DFBW Sensor Failure Identification Using Analytic Redundancy, *IEEE Trans. Automatic Control*, Vol. AC-22, 795-803, 1977.

De Graaf, E.A.B. and De Rijk, P., Evaluation and Comparison of Non-Destructive Service Inspection Methods, *Ninth World Conference on Non-Destructive Testing, Melbourne, Australia*, Session 4A, Nov. 1979.

De Hoff, R.L. and Hall, W.E., *Advanced Fault Detection and Isolation Methods for Aircraft Turbine Engines*, System Control Inc., Technical Report ONR-CR-215-245-1 (AD-A058891), 1978.

De Laat, J.C. and Merrill, W.C., A Real-Time Implementation of an Advanced Sensor Failure Detection, Isolation and Accommodation Algorithm', *Proc. AIAA Guidance, Control and Navigation Conference*, Paper No. 84-0569, 1984.

Desai, M. and Ray, A., A Failure Detection and Isolation Methodology, Theory and Application, *Proc. American Control Conference*, 262-270, 1984.

De Silva, C.W., Real-Time Failure Detection of Aircraft Engine Output Sensor, *Arabian Journal for Science and Engineering*, Vol. 7, 45-53, 1982.

Deyst, J.J. and Deckert, J.C., Maximum Likelihood Failure Detection Techniques Applied to the Shuttle RCS Jets, *J. Spacecraft and Rockets*, Vol. 13, 65-74, 1976.

Dodd, C.V., The Use of Computer Modelling for Eddy Current Testing, *Research Techniques in Nondestructive Testing*, Vol. 3, R.S. Sharpe (ed.), Academic Press, New York, 1977.

Dubois, D. and Prade, H., *Fuzzy Sets and Systems: Theory and Applications*, Academic Press, New York, 1980.*

Duda, R., Hart, P., Barrett, P., *Development of the PROSPECTOR System for Mineral Exploration*, SRI Report Projects 5822 and 6415, Stanford Research Institute, Palo Alto, Calif., 1978.*

Ellis, S.H., Self-Correcting Control for Turbofan Engine, *Proc. 3rd. International Symposium on Air Breathing Engines, Munich, FRG*, 171-186, 1986.

Elmadbouly, E., Keller, L, Frank, P.M., Application of Observers to Instrument Failure Detection, *Proc. IFAC Symposium on Theory and Applications of Digital Control, New Dehli*, pp. 7-13, 1982.

Elmadbouly, E., Frank, P.M., Robust Instrument Failure Detection via Luenberger Observers in Nuclear Power Plants, *CIGRE Symposium 39-83 on Control Applications for Power System Sensitivity, Florence*, 1-6, 1983.

Emami-Naeini, A.et al., *Robust Detection, Isolation, and Accommodation for Sensor Failures*, NASA Contractor Report no. CR-174825, July 1986.

Engell, S., Konik, D., Zustandsermittlung bei unbekanntem Eingangs-signal, *Automatisierungstechnik 34*, 38-42 and 247-251, 1986.

Fairman, F.W., Mahil, S.S., and Luk, L., Disturbance Decoupled Observer Design via Singular Value Decomposition, *IEEE Trans. Automatic Control*, Vol. AC-29, 84-86, 1984.

Fang, C.Z. and Ge, W., Failure Isolation in Linear Systems, *Proc. 12th IMACS World Congress on Scientific Computation, Paris*, 1988.

Fessas, P., An Analytic Determination of the (A-B)-invariant and Controllability Subspace, *Int. J. Control*, Vol. 30, 491-512, 1979.

Filbert, D. and Metzger, K., Quality Test of Systems by Parameter Estimation, *Proc. 9th IMEKO-Congress, Berlin*, 1982.

Fillmore, C.J., *The Case for Case*, in Universals in Linguistic Theory Bach and Harms (eds), Holt, Rinehart & Winston, Chicago, 1968.*

Finin, T., McAdams, J. and Kleinosky, P., FOREST: An Expert System for Automatic Test Equipment, *Proc. 1st. AIA Conf. (IEEE Computer Soc.)*, Dec. 1984.*

Forbus, K., Qualitative Physics, *Sigart Newsletter*, No. 93, July 1985.*

Forster, F., Theoretische und Experimentelle Grundragin der Zerstörungsfreien Werkstoffprüfung mit Wirbelstromverfahren, *Z. Metallkunde*, Vol. 43, 1952.

Forster, F., Nondestructive Inspection of Tubings for Discontinuities using Electromagnet Test Equipment, *Materials Evaluation*, Vol. 28, 19A-23A, 26A-28Al, 1970.

Forster, F., Sensitive Eddy Current Testing of Tubes for Defects on the Inner and Outer Surfaces, *Non Destructive Testing*, Vol. 10, 28-35, 1974.

Foster, G.W., *Examples of Low Altitude Atmospheric Turbulence Measurements*, RAE Technical Report 83026, 1983, Farnborough, U.K.

Frank, P.M., Robustness and Sensitivity: A comparison of the two methods, *Proc. ACI 83 IASTED Symposium on Applied control and Identification, Copenhagen*, Vol. 1, 13-31, June 28 - July 1, 1983.

Frank, P.M., Advanced Fault Detection and Isolation Schemes Using Nonlinear and Robust Observers, *10th IFAC World Congress, München*, 1987.

Frank, P.M., Fault Diagnosis in Dynamic Systems Via State Estimation - A survey, First European Workshop on Fault Diagnostics, etc., Rhodes, Greece, August 31-September 3, 1986: in *System Fault Diagnostics, Reliability and Related Knowledge-Based Approaches*, Vol. 1, Tzafestas, Singh and Schmidt (eds), Reidel Press, 1987.

Frank, P.M. and Janssen, K., Entdeckung von Komponentenfehlern in Dynamischen Systemen durch Hierarchische Zustandsschätzung, *Automatisierungstechnik*, Vol. 44, 23-31, 1986.

Frank, P.M. and Keller, L., Sensitivity Discriminating Observer Design for Instrument Failure Detection, *IEEE Trans. Aerospace and Electronic Systems*, Vol. AES-16, 460-467, 1980.

Frank, P.M. and Keller, J., Entdeckung von Instrumentenfehleranzeigen mittels Zustandsschätzung in technischen Regelungssystemen, *VDI-Fortschrittsberichte*, Reihe 8, Nr. 80, 1984.

Frank, P.M., Fault Diagnosis on the Basis of Dynamic Process Models-Survey Paper, *12th IMACS World Congress on Mathematical Modelling and Scientific Computation, Paris, France*, July 18 - 22, 1988.

Friedland, B., Maximum-Likelihood Estimation of a Process with Random Transitions (Failures), *IEEE Trans. Automatic Control*, Vol. AC-24, 932-937, 1979.

Friedland, B., Multidimensional Maximum-Likelihood Failure Detection and Estimation, *IEEE Trans. Automatic Control*, Vol. AC-26, 567-570, 1981.

Friedland, B., Maximum Likelihood Failure Detection of Aircraft Flight Control Sensors, *J. Guidance, Control and Dymanics*, Vol. 5, 498-503, 1982.

Friedland, B. and Grabousky, S., Estimating Sudden Changes of Biases in Linear Dynamic Systems, *IEEE Trans. Automatic Control*, Vol. AC-27, 237-240, 1982.

Fukunaga, K., *Introduction to Statistical Pattern Recognition*, Academic Press, New York, 1979.

Gabor, D., Theory of Communication, *J. Inst. Elec. Engrs.*, (UK), Vol. 93, 429-497, 1946.

Gai, E., Adams, M.B. and Walker, B.K., Determination of Failure

Thresholds in Hybrid Navigation, *IEEE Trans. Aerospace & Elect. Sys.*, AES-12, 744-755, 1976.

Gai, E., Adams, M.B., Walker, B.K., Smestad, T., Correction to Determination of Failure Thresholds in Hybrid Navigation, *IEEE Trans. Aerospace & Electronic Systems*, AES-14, 696-697, 1978.

Gai, E.G. and Curry, R.E., A Model of the Human Observer in Failure Detection Tasks, *IEEE Trans. Systems, Man, and Cybernetics*, Vol. AMC-6, 85-94, 1976.

Gantmacher, F.R., The Theory of Matrices, *Chealsea Publishing Company*, USA, 1959.

Gazdik, I., RELSHELL: An Expert System Shell for Fault Diagnosis, in *System Fault Diagnostics, Related Knowledge-Based Approach*, S. Tzafestas, M. Singh, G. Schmidt (eds), Vol. 2, pp 496-518, Reidel, Dordrecht, 1987.*

Ge, W. and Fang, C.Z., Detection of Faulty Components via Robust Observation, *Int. J. Control*, Vol. 47, 581-599, 1988.

Geiger, G., Fault Identification of a Motor-Pump System using Parameter Estimation and Pattern Classification, *Proc. 9th IFAC World Congress, Budapest*, July 1984, Pergamon Press, Oxford.

Geiger, G., *Technische Fehlerdiagnose mittels Parameterschätzung und Fehlerklassifikation am Beispiel einer Elektrisch Angetriebenen Kreiselpumpe*, Dissertation Technische Hochschule Darmstadt, VDI-Fortschrittbericht, Reihe 6, VDI-Verlag, 1985.

Genesereth, M.R., Diagnosis using Hierarchical Design Methods, *Proc. AAAI*, 178-183, Aug., 1982.*

Genesereth, M.R., Greiner, R. and Smith, D.E., *A Meta-Level Representation System*, HPP-83-28, Stanford University Heuristic Programming Project, Stanford, 1983.*

Genesereth, M.R., The Use of Design Descriptions in Automated Diagnosis, *Artificial Intelligence*, Vol. 24, 411-436, 1984.*

Georgeff, M.P. and Firschein, O., Expert Systems for Space Automation, *IEEE Control System Society Magazine*, Vol. 5, 3-8, 1985.*

Gertler, J. and Singer, D., Augmented Models for Statistical Fault Isolation in Complex Dynamic Systems, *Proc. American Control Conference, Boston, Ma.*, 1985.

Goedecke, W., Fault Detection in a Turbular Heat Exchanger based on Modelling and Parameter Estimation, *Proc. 7th IFAC-Symposium on Identification and System Parameter Estimation, York*, 3-7 July 1985.

Goedecke, W., *Fehlererkennung an einem Thermischen Prozeß mit Methoden der Parameterschätzung*, Internal research report, Institut für Regelungstechnik, Technische Hochschule Darmstadt.

Goldberg, A. and Robson, D., *Smalltalk-80: The Language and its Implementation*, Addison-Wesley, 1983.*

Golub, G., Numerical Methods for Solving Linear Least-Squares Problems, *Numer. Math., Vol. 7, 206-216*, 1965.

Golub, G.H. and Van Loan, C.F., *Matrix Computations*, North Oxford Academic, 1983.

Gray, R.M., Bugo, A. and Gray Jr, A.H., Distortion Measures for Speech Processing, *IEEE Trans. Acoustics, Speech & Signal Proc.*, Vol. ASSP-28, 367-376, 1980.

Grover, F.W., *Inductance Calculations - Working Formulas and Tables*, New

York: Dover Press, 1946. (Reprinted by the Instrument Society of America, ISBN: 0-87664-557-0)

Gustafson, D.E., Willsky, A.S., Wang, J.Y., Lancaster, M.C. and Triebwasser, J.H., ECG/VCG Rhythm Diagnosis Using Statistical Signal Analysis-I. Identification of Persistent Rhythms, *IEEE Trans. Biomedical Engineering*, Vol. BME-25, 344-353, 1978(a).

Gustafson, D.E., Willsky, A.S., Wang, J.Y., Lancaster, M.C. and Triebwasser, J.H., ECG/VCG Rhythm Diagnosis Using Statistical Signal Analysis - II. Identification of Transient Rhythms, *ibid.*, Vol. BME-25, 353-361, 1978(b).

Hägglund, T., *New Estimation Techniques for Adaptive Control*, Doctoral Dissertation, Dept. of Automatic Control, Lund University of Technology, Lund, Sweden, 1983.

Hamscher, W. and Davis, W., Diagnosing Circuits with State: An Inherently Underconstrained Problem, *Proc. AAAI Conf.*, 142, 1984.*

Hannan, E.J. and Quinn, B.G., *The Determination of the Order of an Autoregression*, J. R. Statist. Soc. B, Vol. 41, 190-195, 1979.

Harvey, C.A. and Stein, G., Quadratic Weights for Asymptotic Regulator Properties, *IEEE Trans. Automatic Control*, Vol. AC-23, 378-387, June 1978.

Hayes-Roth, F., The Knowledge-Based Expert System: A Tutorial, *IEEE Computer*, 11-28, Sept. 1984.*

Hayes-Roth, F., Rule-based systems, *Communications ACM*, Vol. 28, 921-932, 1984.*

Hertel, J.E. and Clark, R.N., Instrument Failure Detection in Partially Observable Systems, *IEEE Trans. Aerospace and Electronic Systems*, Vol. AES-18, 310-317, 1982.

Himmelblau, D.M., *Fault Detection and Diagnosis in Chemical and Petrochemical Processes*, Elsevier Press, Amsterdam, 1978.

Himmelblau, D.M., Fault Detection and Diagnosis - Today and Tommorrow, *Proc. IFAC Workshop on Fault Detection and Safety in Chemical Plants, Kyoto, Japan*, 95-105, 28th Sept. - 1st Oct. 1986.

Hochschild, R., Electromagnetic Methods of Testing Metals, *Progress in Non-Destructive Testing*, Vol. 1, E. G. Standford and J. H. Fearan (eds), pp 59-109, 1961.

Hofmann, M., Caviedes, J., Bourne, J., Beale, G. and Brodersen, A., Building Expert Systems for Repair Domains, *Expert Systems: The International Journal of Knowledge Engineering*, Vol. 3, 4-12, 1986.*

Hohmann, H., *Automatic Monitoring and Failure Diagnosis for Machine Tools*, (in German), Dissertation, Technische Hochschule Darmstadt, 1977.

Horak, D.T. and Allison, B.H., Experimental Implementation and Evaluation of the RMI Failure Detection Algorithm, *Proc. 1987 American Control Conference, Minneapolis, Minn.*, 1987.

Horner, G.C., Optimum Actuator Placement, Gain, and Number for a Two-Dimensional Grillage, *Proc. AIAA/ASME/ASCE/AHS 24th Structural Dynamics and Materials Conference*, AIAA Paper no. 83-0854, May 2-4, 1983.

Howard, R.A., *Dynamic Probabilistic Systems: Vol. 1, Markov Models*, Wiley & Sons, 1971.

Hrach, F.J., Arpasi, D.J. and Bruton, W.M., *Design and Evaluation of a Sensor Fail-Operational Control System for a Digitally Controlled Turbofan*

Engine, NASA Report No TM-X 3260, 1975.

Hughes, D.E., Induction Balance and Experimental Researches Therewith, *Philosophical Magazine, Series 5*, Vol. 8, 50, 1879.

Isermann, R., Einfache Mathematische Modelle für das Dynamische Verhalten beheizter Rohre, *Wärme*, Vol. 75, 89-93, 1969.

Isermann, R., Methoden zur Fehlererkennung für die Überwachung technischer Prozesse, *Regelungstechnische Praxis*, Vol 22, 322-325 and 363-368, 1980.

Isermann, R., Fehlerdiagnose mit Prozeßmodellen, *Technisches Messen 51*, 345-355, 1984a.

Isermann, R., Process Fault Detection Based on Modelling and Estimation Methods: A Survey, *Automatica*, Vol. 20, 387-404, 1984b.

Ishii, N., Sugimoto, H., Iwata, A. and Suzumura, N., Computer Classification of EEG Time Series by Kullback Information Measure, *Int. J. Systems Science*, Vol. 11, 677-687, 1980.

Jacquot, J.P., Poujol, A., Beaubatie, J. and Ciaramitaro, W., Operating Results Obtained in a Nuclear Power Plant with a Sensor Surveillance Prototype, paper presented at the *Fifth Power Plant Dynamics, Control and Testing Symposium, Knoxville*, 1983.

Janssen, K. and Frank, P.M., Component Failure Detection via State Estimation, *Proc. 9th IFAC World Congress, Budapest*, Vol. 1, 2nd - 7th July 1984.

Jones, A.H. and Burge, S.E., An Expert System Design using Cause-Effect Representations and Simulation for Fault Detection, in *System Fault Diagnostics, Reliability and Related Knowledge-Based Approaches*, S. Tzafestas, M. Singh, G. Schmidt (eds), Vol. 2, pp. 485-495, Reidel, 1987.*

Jones, H.L., *Failure Detection in Linear Systems*, Ph.D. dissertation, Dept. Aero. and Astro., M.I.T., Cambridge, Mass., Sept. 1973.

Jones, J.G., Modelling of Gusts and Windshear for Aircraft Assessment and Certification, *Proc. Indian Acad. Sci. (Engng. Sci.)*, Vol.3, Part 1, 1-30, 1980.

Jones, J.G., Processing of Fractal Functions by Linear and Nonlinear Operators, *RAE Technical Report, 85079*, 1985.

Jones, J.G., Response of Linear Systems to Gaussian Process Inputs in Terms of Probability Functionals, *RAE Technical Report 86028*, 1986a, Farnborough, UK.

Jones, J.G., A Unified Procedure for Meeting Power-Spectral-Density and Statistical-Discrete-Gust Requirements for Flight in Turbulence, *AIAA/ASME/ASCE/ AHS 27th Structures and Materials Conference, Part 1, San Antonio, Texas*, May 1986b.

Jones, J.G. and Haynes, A., A Peakspotter Programme applied to the Analysis of Increments in Turbulence Velocity, *RAE Technical Report 84071*, 1984, Farnborough, UK.

Kahn, A.H., Spal, R. and Feldman, A., Eddy Current Losses due to a Surface Crack in Conducting Material, *Journal of Applied Physics*, Vol. 48, 4454-4459, 1977.

Kanemoto, S., Ando, Y., Yamamoto, F., Kitamoto, K. and Nunome, K., Identification of Pressure Control System Dynamics in BWR Plant by Multi-Variate Autoregressive Modeling Technique, *J. Nucl. Sci. Tech.*, Vol. 19, 58-68, 1982.

Kanemoto, S., Ando, Y., Yamamoto, F., Idesawa, M. and Itoh, K., Identification of BWR Feedwater Control System using Autoregressive Integrated Model, *J. Nucl. Sci. Tech.*, Vol. 20, 105-116, 1983.

Kanemoto, S., Tsunoyama, S., Ando, Y., Yamamoto, F. and Sandoz, S.A., Noise Source and Reactor Stability Estimation in a Boiling Water Reactor using a Multivariate Autoregressive Model, *Nuclear Technology*, Vol. 67, 23-37, 1984.

Karcanias, N. and Kouvaritakis, B., The use of Frequency Transmission Concepts in Linear Multivariable System Analysis, *Int. J. Control*, Vol. 28, 195-240, 1978.

Karcanias, N. and Kouvaritakis, B., The Output Zeroing Problem and its Relationship to the Invariant Zero Structure, *Int. J. Control*, Vol. 30, 395-415, 1979.

Keefe, M., OBJECTS: Object Oriented Programming in Prolog, *POPLOG library*, University of Sussex, UK, 1985.*

Kerlin, T.W., Dynamic Analysis and Control of Pressurised Water Reactors, *Advances in Control and Dynamic Systems*, Vol. 14, 1978.

Kerr, T.H., A Two Ellipsoid Oberlap Test for Real Time Failure Detection and Isolation by Confidence Regions, *Proc. 13th IEEE Conference, Phoenix, Ariz.*, 1974.

Kerr, T.H., Statistical Analysis of Two-Ellipsoid Overlap Test for Real-Time Failure Detection, *IEEE Transactions on Automatic Control*, Vol. AC-25, Nr. 4, 762-772, 1980.

Kerr, T.H., Decentralized Filtering and Redundancy Management for Multisensor Navigation, *IEEE Transactions on Aerospace and Electronic Systems*, Vol. AES-23, 83-119, 1987.

Kincaid, T.G. and Chari, M.V.K., The Application of the Finite Element Method Analysis to Eddy Current Nondestructive Evaluation, *IEEE Transactions on Magnets*, Vol. MAG-15, 1956-1960, 1979.

Kitagawa, G. and Akaike, H., A Procedure for the Modelling of Non-stationary Time Series, *Annals Inst. Statis. Math.*, Vol. 30B, 351-363, 1978.

Kitamura, M., Detection of Sensor Failures in Nuclear Plant using Analytic Redundancy, *Trans. Am. Nucl. Soc.*, Vol. 34, 581-583, 1980.

Kitamura, M., Matsubara, K. and Oguma, R., Identifiability of Dynamics of a Boiling Water Reactor using Autoregressive Modelling, *Nucl. Sci. Engng.*, vol. 70, 106-110, 1979.

Kitamura, M., Washio, T., Kotajima, K. and Sugiyama, K., Small-Sample Modelling Method for Nonstationary Reactor Noise Analysis, *Annals Nucl. Energy*, Vol. 12, 399-407, 1985.

Klein, G. and Moore, B.C., Eigenvalue-Generalized Eigenvector Assignment with State Feedback, *IEEE Trans. Automatic Control*, Vol. AC-22, 140-141, 1977.

Kobayashi, N. and Nakamizo, T., An Observer for Linear Systems with Unknown Inputs, *Int. J. Control*, Vol. 35, 605-619, 1982.

Kokawa, M., Miyazaki, S. and Shingai, S., Fault Location using Digraph and Inverse Direction Search with Application, *Automatica*, Vol. 19, 729-738, 1983.*

Konik, D., Zur Analyse und Synthese Zentral und Dezentral Geregelter Linearer Mehrgrössensysteme, *VDI-Fortschrittberichte, Reihe 8, Nr. 123*,

VDI-Verlag, 1986.

Kouvaritakis, B. and MacFarlane, A.G.J., Geometric Approach to the Analysis of System Zeros: I and II, *Int. J. Control*, Vol. 23, 145-177, 1976.

Kramer, M.A., Integration of Heuristic and Model-Based Inference in Chemical Process Fault Diagnosis, *Proc. IFAC Workshop on Fault Detection and Safety in Chemical Plants, Kyoto*, Sept. 1986.*

Kramer, M.A., Malfunction Diagnosis using Quantitative Models and Non-Boolean Reasoning in Expert Systems, *AIChE Journal*, 1986.*

Kramer, M.A. and Palowitch Jr., B.L., A Rule-Based Approach to Fault Diagnosis using the Signed Directed Graph, *AIChE Journal*, 1988.*

Kucera, V., The Discrete Riccati Equation of Optimal Control, *Kybernetika*, Vol. 8, 430-447, 1972.

Kullback, S., *Information Theory and Statistics*, New York, John Wiley & Sons, 1959.

Kumamaru, K., Statistical Failure Diagnosis for Dynamical Systems, *Systems and Control*, Vol. 28, 77-86, 1984.

Kumamaru, K., Sagara, S., Fujise, S. and Fujii, T., A Hierarchical Diagnosis of System Failures using Parameter Monitoring Scheme, *18th JAACE Symp. on Stochastic Stystems Theory and Its Applications, Tokyo, Japan*, 175-179, 1986.

Kumamaru, K., Sagara, S., Nishimura, Y., Ono, T. and Kumamaru, T., *A Hierarchical Diagnosis for Failure Detection in Dynamical Systems Based on Linear Modelling with Oriented Additive Jumps (in Japanese)*, Technology Report of the Kyushu University, Vol. 56, pp 629-636, 1983.

Kumamaru, K., Sagara, S., Yanagida, H. and Söderström, T., Fault Detection of Dynamical Systems Based on a Recognition Approach to Model Discrimination, *Proc. 7th IFAC Symposium on Identification and System Parameter Estimation, York*, H.A. Barker, P. Young (eds), Pergamon Press, UK, 1986.

Kumamaru, K. and Söderström, T., Fault Detection and Model Validation using the Index of Kullback Discrimination Information (in Japanese), *Trans. Society of Instrument and Control Engineering*, Vol. 22, 52-59, 1986.

Kumamaru, K., Söderström, T., Sagara, S. and Yanagida, H., *Fault Detection of Dynamical Systems Based on a Model Discriminating approach*, Report UPTEC 84123R, Institute of Technology, Uppsala University, Uppsala, Sweden, 1984.

Kumamoto, H., Ikenchi, K., Inoue, K. and Henley, E.J., Application of Expert System Techniques to Fault Diagnosis, *Chem. Eng. J.*, Vol. 29, 1, 1984.

Kwakernak, H. and Sivan, R., *Linear Optimum Control Systems*, John Wiley, New York, 1972.

Laffey, T.J., Perkins, W.A. and Nguyen, T.A., Reasoning about Fault Diagnosis with LES', *IEEE Expert*, 13-20, 1986.*

Lahey, R.T. and Moody, E.J., The Thermal-Hydraulics of a Boiling Water Nuclear Reactor, *Trans. Am. Nucl. Soc.*, 1977.

Lainiotis, D.G., Optimal Adaptive Estimation: Structure and Parameter Adaption, *IEEE Trans. Automatic Control*, Vol. AC-16, 160-170, 1971.

Lainiotis, D.G., Partitioning: A Unifying Framework for Adaptive

Systems I: Estimation, *Proc. IEEE,* Vol. 64, 1126-1143, 1976.

Larsson, H., Multifrequency Eddy Current Testing of Case Hardened Cylinders - a Theoretical and Experimental Analysis, *Proc. 1st European Conference on Non-Destructive Testing, Mainz, Germany,* pp 24-26 Apr. 1978.

Leary, J.J. and Gawthrop, P.J., Process Fault Detection using Constraint Suspension, *Proc. IEE,* Vol. 134, Pt. D, 264-271, 1987.*

Leden, B., Multivariable Dead-Beat Control, *Automatica,* Vol. 13, 185-188, 1977.

Liporace, L.A., Variance of Bayes Estimates, *IEEE Trans. Information Theory,* Vol. IT-17, 665-669, 1971.

Ljung, L., On the Consistency of the Prediction Error Identification Method, in R.K. Mehra and D.G. Lainiotis (eds), *System Identification Advances and Case Studies,* Academic Press, New York, 1976.

Ljung, L. and Caines, P.E., Asymptotic Normality of Prediction Error Estimation for Approximate Models, *Stochastics,* Vol. 3, 29-46, 1979.

Lou, X.C., *A System Failure Detection Method - Failure Projection Method M.S. - Thesis,* Institute of Technology, Dept. of Electrical Engineering, Cambridge, Ma., USA, 1982.

Lou, X.C., Willsky, A.S. and Verghese, G.C., Optimally Robust Redundancy Relations for Failure Detection in Uncertain Systems, *Automatica,* Vol. 22, 333-344, 1986.

Loveland, D., A Simplified Format for the Model Elimination Procedure, *J. ACM,* Vol. 16, 349-363, 1969.*

Luenberger, D.G., An Introduction to Observers, *IEEE Trans.,* Vol. AC-16, 596-602, Dec. 1971.

Luppold, R.H., *Reliability and Availability Models for Fault-Tolerant Systems,* S. M. Thesis, Department of Aero. & Astro., MIT, Cambridge, Ma., Aug. 1983.

Luppold, R.H., Gai, E. and Walker, B.K., Effects of Redundancy Management on Reliability Modelling, *Proc. American Control Conf., San Diego, Calif.,* June 1984.

Magill, D.T., Optimal Adaptive Estimation of Sampled Stochastic Processes, *IEEE Trans. Automatic Control,* Vol. AC-10, 434-439, 1965.

Mah, R.S.H., Stanley, G.M. and Downing, D.M., Reconcilation and Certification of Process Flow and Inventory Data, *Ind. Eng. Chem. Proc. Des. Dev.,* Vol. 15, 175, 1976.*

Marr, D., Poggio, T. and Ullmann, S., Bandpass Channels, Zero-Crossings, and Early Visual Information Proc., *Journal of the Optical Society of America,* Vol. 69, 914-916, 1979.

Massoumnia, M.A., *A Geometric Approach to Failure Detection and Identification in Linear Systems;* Ph.D. Thesis, MIT, Cambridge, Ma., 1986a.

Massoumnia, M.A., A Geometric Approach to the Synthesis of Failure Detection Filters, *IEEE Trans. Automatic control,* Vol. AC-31, 839-846, 1986b.

Massoumnia, M.A. and Vander Velde, W.E., Generating Parity Relations for Detecting and Identifying Control System Component Failures, *AIAA J. of Guidance Control and Dynamics,* Vol. 11, 60-65, 1988.

Matthews, V.J. and Tugnait, J.K., Detection and Estimation with Fixed Lag for Abruptly Changing Systems, *IEEE Trans. Aerospace Electronic System,* Vol. AES-19, 730-739, Sept. 1983.

Matsumoto, K., Sakaguchi, T. and Wake, T., Fault diagnosis of a Power System Based on Description of the Structure and Structure of the Relay System, *Expert Systems*, Vol. 2, 134-138, 1985.*

Maybeck, P.S., *Failure Detection Through Functional Redundancy*, Tech. Report AFFDL-TR-76-93, Wright Patterson Air Force Base, Ohio, Sep. 1976.

Maybeck, P.S., *Stochastic Models, Estimation, and Control*, Vol. 2, Academic Press, 1982.

McFarlane, A.G.J. and Karcanias, N., Poles and Zeros of lincar Multivariable Systems: A Survey of the Algebraic, Geometric and Complex Variable Theory, *Int. J. Control*, Vol. 24, 33-74, 1976.

McLean, Report Prepared by Adaptronics Inc., Virginia, USA, *Feasibility of Using Adaptive Learning Networks for Eddy Current Signal Analysis*, EPRI-NP-723, TP577-723, Final Report, March, 1978.

McMaster, R.C., *Non-Destructive Testing Handbook*, New York: The Ronald Press Company Inc., 1963.

Meditch, J.S., *Stochastic Optimal Linear Estimation and Control*, McGraw-Hill, 1969.

Mehra, R.K. and Peschon, J., An Innovations Approach to Fault Detection and Diagnosis in Dynamic Systems, *Automatica*, Vol. 7, 637-640, 1971.

Meier, L., Ross, D.W. and Glaser, M.B., *Evaluation of the Feasibility of Using Internal Redundancy to Detect and Isolate on Board Control Data Instrumentation Failures*, Tech. Report AFFDL-TR-70172, WPAFB, Ohio, Jan., 1971.

Merrill, W.C. and DeLaat, J.C., A Real-Time Simulation Evaluation of an Advanced Detection, Isolation and Accomodation Algorithm for Sensor Failures in Turbine Engines, *Proc. American Control Conference, Seattle, Wa.*, 162-169, 1986.

Meserole, J.S., *Detection Filters for Fault Tolerant Control of Turbofan Engines*, Ph.D. Thesis, MIT, Cambridge, Ma., 1981.

Milne, R., Strategies for Diagnosis, *IEEE Trans. Systems, Man and Cybernetics*, Vol. SMC-17, 333-339, 1987.

Minsky, M., *Semantics Information Processing*, MIT Press, Cambridge, Ma. USA, 1986.*

Mironowski, L.A., Functional Diagnosis of Linear Dynamic Systems, *Automn. Remote Control*, Vol. 40, 120-128, 1979.

Mita, T., Design of a Zero-Sensitive System, *Int. J. Control*, Vol. 24, 75-81, 1976.

Mita, T., On the Synthesis of an Unknown Input Observer for a Class of Multi-input/output Systems, *Int. J. Control*, Vol. 26, 841-851, 1977.

Mitsutake, T., Tsunoyama, S. and Nanba, H., Application of Autoregressive (AR) Technique to BWR Stability Estimation, *Progr. in Nucl. Energy*, Vol. 9, 675-689, 1982.

Montgomery, R.C., Management of Redundancy in Flight Control System Using Optimal Decision Theory, *Chapter in AGARDOGRAHP No. 251, Theory and Applications of Optimal Control in Aerospace Systems*, Ir. Pieter Kant (ed.), 11-12, July 1981.

Montgomery, R.C., and Caglayan, A.K., Failure Accomodation in Digital Flight Control Systems by Bayesian Decision Theory, *J. Aircraft*, Vol. 13, 1976.

Montgomery, R.C., Horner, G.C., Akpan, I.I. and Vander Velde, W.E., On Incorporating Reliability Considerations into Control System Designs, *Proc. 20th IEEE CDC Conference*, Dec. 16-18, 1981.

Montgomery, R.C. and Price, D.B., Management of Analytical Redundancy in Digital Flight Control Systems for Aircraft, *Proc. AIAA Mechanics and Control of Flight Conference, Anaheim, Calif.*, 1974.

Montgomery, R.C. and Tabak, D., Application of Analytical Redundancy Management to Shuttle Crafts, *Proc. 18th IEEE CDC, San Diego, Calif.*, 1979.

Montgomery, R.C. and Vander Velde, W.E., Reliability Considerations in the Placement of Control System Components, *AIAA Journal of Guidance, Control, and Dynamics*, Vol. 8, 411, 1985.

Montgomery, R.C. and Williams, J.P., Testing of a Failure Accommodation System on a Highly Flexible Grid, *Proc. American Control Conference*, Boston, Ma., 984-989, June 19-21, 1985.

Moore, B.C., On the Flexibility offered by State Feedback in Multivariable Systems beyond Closed-Loop Eigenvalue Assignment, *IEEE Trans. Automatic Control*, Vol. AC-21, 689-692, 1976.

Moraitis, C., *Approximate and Plausible Reasoning using Attribute Grammars*, Research Report, Computer Science Division, National Technical University of Athens (NTUA), Athens, July 1986; Also IEEE Region 8 Student Paper Contest (2nd prize), Paris, 1986.*

Moraitis, C., Papakonstantinou, G. and Tzafestas, S., Attribute Grammars as a Knowledge-Based Diagnosis Tool, *Proc. 1st Europ. Workshop on Fault Diagnostics, Reliability and Related Knowledge-Based Approaches*, S. Tzafestas, M. Singh, C. Schmidt (eds), Vol. 2, pp 53-62, Reidel, 1987.*

Morris, R.A., Quantitative Pulsed Eddy Current Analysis, *Proc. 10th Symposium of Non-destructive Evaluation, San Antonio, Texas*, Apr. 22-25, 1975.

Müller, P.C. and Weber, H.I., Analysis and Optimization of Certain Qualities of Controllability and Observability for Linear Dynamical Systems, *Automatica*, Vol. 8, 237-246, 1972.

Mudge, S.K. and Patton, R.J., Analysis of the Technique of Robust Eigenstructure Assignment with Application to Aircraft Control, *IEE Proc., Part D*, Vol. 135, 275-281, 1988.

Murphy, T.E., Setting up an Expert System, *The Industry and Process Control Magazine (IFCS)*, 54-60, March 1985.

Nakamizo, T., Signal Processing and Failure Detection Method, *System and Control*, Vol. 29, no. 4, 232-242, 1985.

Natesh, R., Nondestructive Testing in the Nuclear Industry, *Proc. of an International Conference, Salt Lake City*, USA, 13-15 Feb. 1978.

Nau, D.S., Expert Computer Systems, *IEEE Computer*, 63-85, Feb. 1983.*

Naylor, C., How to Build an Inference Engine, *Expert Systems*, R. Forsyth (ed.), Ch. 6, pp 63-88, 1984.*

Niccoli, L.G., Wilburn, N.P., Colley, R.W., Alexandro, F., Clark, R.N. and Bouzerdoum, S., Detection of Instrument or Component Failures in a Nuclear Power Plant by Luenberger Observers, *International Topical Meeting on Computer Applications for Nuclear Plant Operation and Control, Pasco, Wa., U.S.A.*, 1985.

Nold, S. and Isermann, R., Identifiability of Process Coefficients for Technical Failure Diagnosis, *Proc. 25th IEEE CDC Conference Athens*, Dec. 10-12, 1986.

O'Reilly, J., Minimal-order Observers for Linear Multivariable Systems with Unmeasureable Disturbances, *Int. J. of Control 28*, 743-751, 1978.

O'Reilly, J., *Observers for Linear Systems*, Vol. 170 in Mathematics in Science and Engineering, Academic Press, 1983.

Ohm, D.Y., Bhattacharya, S.P., and Howze, J.W., Transfer Function Conditions for (C,A,B) Pairs, *IEEE Trans. Automatic Control*, Vol. AC-29, 172-174, 1984. Okita, T., Kobayashi, Y., Tanaka, S. and Hirose, K., Detection of Parameter Changes for Discrete-Time Systems, *Trans. Inst. of Elect. Engineers of Japan*, Series C, Vol. 105, no. 11, 211-218, 1985.

Onken, R. and Stuckenberg, N., Failure Detection in Signal Processing and Sensing in Flight Control Systems, *Proc. 18th IEEE, San Diego, Calif.*, 1979.

Ono, T., Kumamaru, T. and Kumamaru, K., Fault Diagnosis of Sensors Using Vector Gradient Method, *Trans. Society of Instrument and Control Engineers*, Vol. 20, 22-27, 1984.

Panayiotopoulos, T., Papakonstantinou, G. and Stamatopoulos, G., A Theorem-Prover as a Fault Diagnostic Tool, *Proc. 1st Europ. Workshop on Fault Diagnostics, Reliability and Related Knowledge Approaches*, Vol. 2, pp 43-52, Reidel, 1987.*

Papakonstantinou, G. and Kontos, J., Knowledge Representation with Attribute Grammars, *The Computer Journal*, Vol. 29, 241-246, 1986.*

Papakonstantinou, G., Moraitis, C. and Panayiotopoulos, T., An Attribute Grammar Interpreter as a Knowledge Engineering Tool, *Angewandte Informatik* (in press).*

Patton, R.J. and Willcox, S.W., Comparison of Two Techniques of IFD Based on a Nonlinear Stochastic Model of an Aircraft, *Proc. 7th IFAC Symp. on Identification and System Parameter Estimation, York, UK*, Published by Pergamon Press, 1986.

Patton, R.J. and Willcox, S.W., Fault Diagnosis in Dynamic Systems Using a Robust Output Zeroing Design Method, *Proc. First IFAC European Workshop on Failure Diagnosis*, Rhodes, Greece, August 31 - September 3, 1986: in System Fault Diagnostics, Reliability and Related Knowledge-Based Approaches, Vol. 1, Tzafestas, Singh and Schmidt (eds), Reidel Press, 1987..

Patton, R.J., Willcox, S.W. and Winter, S.J., A Parameter Insensitive Technique for Aircraft Sensor Fault Analysis Using Eigenstructure Assignment and Analytical Redundancy, *AIAA Paper 86-2029-CP, Guidance, Navigation and Control Conference, Williamsburg, Va*, Aug. 18-20 1986.

Patton, R.J., Willcox, S.W. and Winter, J.S., A Parameter Insensitive Technique for Aircraft Sensor Fault Analysis, *AIAA J. of Guidance, Control and Dynamics*, 359-367, July/Aug. 1987.

Patton, R.J., Robust Fault Detection Using Eigenstructure Assignment, *Proc. 12th IMACS World Congress on Mathematical Modelling and Scientific Computation, Paris*, Vol. 2, pp 431-434, July 18-22, 1988.

Pau, L.F., *Failure Diagnosis and Performance Monitoring*, published by Marcel Dekker, 1981.

Pau, L.P., Survey of Expert Systems for Fault Detection, Test Generation and Maintenance, *Expert Systems*, Vol. 3, 100-111, 1986.*

Perkins, W.A. and Laffey, T.J., LES: A general expert system and its applications, *SPIE Proc., 485, Applications of Artificial Intelligence*, p. 46, 1984.*

Pipitone, F., The FIS Electronic Troubleshooting System, *IEEE Computer*, 68-76, July, 1986.*

Polanisamy, R. and Lord, W., Finite Element Modelling of Electromagnetic NDT Phenomena, *IEEE Trans. Magnetics*, Vol. MAG-15, 1479-1481, Nov. 1979.

Pople, H., Heuristic Methods for Imposing Structure on Illstructured Problems: The Structuring of Medical Diagnostics, in *Artificial Intelligence in Medicine*, P. Szolovits (ed.), Westview Press, Boulder, Colorado, 1982.*

Prade, H., A Computational Approach to Approximate and Plausible Reasoning with Applications to Expert Systems, *IEEE Trans. PAMI*, Vol. PAMI-7, 260-283, 1985.*

Rault, A., Jaume, D., Verge, M., Industrial Process-Fault Detection and Localisation, *Proc. 9th IFAC World Congress*, Budapest, July 1984, Pergamon Press, Oxford, 1985.

Rault, A., Richalet, A., Barbot, A., Sergenton, J.P., Identification and Modeling of a Jet Engine, *Proc. 1971 IFAC Symposium on Digital Simulation of Continuous Processes*, Gejör, 1971.

Renken, C.J., Theory and Some Applications of Pulsed Eddy Current Fields to the Problems of Eddy Current Testing, *Progress in Applied Materials Research*, Vol. 6, 239-261, Heywood Press, London, 1964.

Richalet, J., Rault, A., Pouliquen, R., *Identification des Processus par la Methode du Modele*, published by Gordon and Breach, 1971.

Rockwell International, *'Space Shuttle OFT Level C FSSR, Part D Redundancy Management'*, Space Div., Downey, USA, Nov. 1977.

Romagnoli, J.A. and Stephanopoulos, G., On the Rectification of Measurement Errors for Complex Chemical Plants, *Chem. Eng. Sci.*, Vol. 35, p. 1067, 1980.*

Rozonoer, L.I., A Variational Approach to the Problem of Invariance of Automatic Control Systems - PART I, *Automation and Remote Control*, Vol. 24, 680-691, Nov. 1963.

Rozonoer, L.I., A Variational Approach to the Problem of Invariance of Automatic Control Systems - PART II, *Automation and Remote Control*, Vol. 24, 793-800, Dec. 1963.

Rubenstein, M., *FLAVOUR: A Zetalisp Based Object-Oriented POPLOG Library File*, Cognitive Studies, University of Sussex, UK, 1985.*

Sandell, H., Bourne, J. and Shiavi, R., GENIE: A Generic Inference Engine for Medical Applications, *Proc. 6th Ann. Conf. IEEE Eng. Med. Biol.*, 66-69, Los Angeles, Sept. 1984.*

Schumacher, J.M., Compensator Synthesis Using (C,A,B)-Pairs, *IEEE Trans. Automatic Control*, Vol. AC-25, 1133-1138, 1980.

Shafer, G., *A Mathematical Theory of Evidence*, Princeton University Press, Princeton, N.J., 1976.*

Shapiro, E.Y. and Decarli, H.E., Analytic Redundancy for Flight Control Sensors on the Lockheed L-1011 Aircraft, *Proc. 18th IEEE CDC Conference, San Diego, Calif.*, 1979.

Shapiro, E. and Takeuchi, A., Object Oriented Programming in Concurrent Prolog, *New Generation Comput.*, Vol. 1, 25-48, 1983.*

Shapiro, S.C., Srihari, S.N., Taie, M.R. and Celler, J., VMES: A Network Versatile Maintenance Expert System, in *Applications of AI in Engineering Problems*, D. Sriram and R. Adey (eds), Vol. II, pp 925-936, Springer-Verlag, 1986.*

Shiozaki, J., Matsuyama, H., O'Shima, E. and Iri, M., An Improved Algorithm for Diagnosis of System Failures in the Chemical Process, *Comput. & Chem. Eng.*, Vol. 9, 285, 1985.*

Shooman, M.L., *Probabilistic Reliability: An Engineering Approach*, McGraw-Hill, 1968.

Shortliffe, E.H. and Buchanan, B.G., A Model of Inexact Reasoning in Medicine, *Mathem. Biosciences*, Vol. 23, 351, 1975.*

Shortliffe, E.H., *Computer-Based Medical Consultations: MYCIN*, American Elsevier/North Holland, 1976.*

Shumway, R.H. and Unger, A.N., Linear Discriminant Function for Stationary Time Series, *J. American Stati. Associ.*, Vol. 69, 948-965, 1974.

Sidhu, G.S. and Biermann, G.J., Integration-Free Interval Doubling for Riccati Equation Solutions, *IEEE Trans. Automatic Control*, Vol. AC-22, 831-834, 1977.

Silverman, L.M. and Payne, H.J., Input-output Structure of Linear Systems with Application to the Decoupling Problem, *SIAM J. Control*, Vol. 9, 199-233, 1971.

Silverthorn, J.T. and Reid, G.J., *Proc. of 9th IEEE Conference on Decision and Control, Albuquerque, New Mexico*, 1206-1207, 1980.

Skattcboe, R., Lihovd, E. and Hystad, R.A., DIAMON: A Knowledge-Based System for Fault Diagnosis and Maintenance Planning for Rotating Machinery, *Proc. 6th Int. Workshop on Expert Systems and Applications*, Avignon, France, Vol. 1, 633-647, Agence de l'Informatique, 1986.*

Sloman, A., NEWOBJ: A POPLOG Object Oriented Library, Cognitive Studies, *Internal Research Report*, University of Sussex, UK, 1985.*

Smestad, T., Orpen Application of Parallel Filters for Malfunction Detection and Alternative Mode Capability, *AGARD Symposium, Sandefjord, Norway*, 1978.

Söderström, T. and Kumamaru, K., On the use of Kullback Discrimination Index for Model Validation and Fault Detection, *Report UPTEC 8520R*, Institute of Technology, Uppsala University, Uppsala, Sweden, 1985(a).

Söderström, T. and Kumamaru, K., Some Model Validation Criteria Based on a Kullback Discrimination Index, *Proc. 24th IEEE CDC Conference, Ft. Lauderdale, Fla*, 219-224, 1985(b).

Söderström, T. and Stoica, P., *System Identification*, Prentice Hall, Hemel Hempstead, 1988.

Sobel, K.M. and Shapiro, E.Y., A Design Methodology for Pitchpointing Flight Control Systems, *AIAA Journal of Guidance, Control and Dynamics*, Vol. 8, 181-187, 1985.

Soeda, T., Yoshimura, T. and Watanabe, K., *A Technique of Plant Failure Diagnosis based on Model Reference Adaptive System*, Technical Report of University of Tokushima, 115-127, 1978.

Solov, E.G. and Thibodeau III, J.R., Failure Detection and Isolation Methods for Redundant Gimballed Inertial Measurement Units, *Proc. AIAA Guidance Navigation & Control Conference*, (paper 73-851), *Key Biscayne, Fla*, Aug. 1973.

Sowizral, H.A., Expert Systems, *Annual Rev. Inform. Sci. and Techn. (ARIST)*, Vol. 20, 179-199, 1985.*

Spal, R. and Kahn, A.H., Eddy Currents in a Conducting Cylinder with a Crack, *Journal of Applied Physics*, Vol. 48, 6135-6138, 1977.

Speyer, J.L., Computation and Transition Requirements for a Decentralized Linear-Quadratic-Gaussian Control Problem, *IEEE Trans. Automatic Control*, Vol. AC-24, 266-269, 1979.

Speyer, J.L. and White, J.E., Shiryayev Sequential Probability Ratio Test for Redundancy Management, *AIAA J. Guidance, Control and Dynamics*, Vol. 7, 588-595, 1984.

Srinathkumar, S., *Spectral Characterisation of Multi-Input Dynamic Systems*, Ph.D. Thesis, Oklahoma State University, U.S.A., 1976.

Srinathkumar, S., Eigenvalue/Eigenvector Assignment using Output Feedback, *IEEE Trans. Automatic Control*, Vol. AC-23, 79-81, 1978.

Stanfill, C., *MACK: A Program which Deduces the Behaviour of Machines from their Forms*, *Sigart Newsletter, no. 93*, 1985.*

Stickel, M., A Prolog Technology Theorem Prover, *Proc. Intl. Symposium on Logic Programming, Atlantic City, N.J.*, 211-217, Feb. 1984.*

Stuckenberg, N., Sensor Failure Detection in Flight Control Systems Using Deterministic Observers, *Proc. 7th IFAC Symp. on Identification and System Parameter Estimation York, UK*, July 1985, H.A. Barker, P. Young (eds), Pergamon Press, 1986.

Stumm, W., Progress in Electromagnetic Tube Testing as an Economization in Tube Production, *Ninth World Conference on Non-Destructive Testing, Melbourne, Australia*, Session 4B, Nov. 1979.

Sundareswaran, K.K, McLane, P.J. and Bayoumi, M.M., Observers for Linear Systems with Arbitrary Plant Disturbances, *IEEE Trans. Automatic Control*, Vol. AC-22, 870-871, 1977.

Sweeney, F.J., Upadhyaya, B.R. and D. J. Shieh, In Core Coolant Flow Monitoring of Pressurized Water Reactors using Temperature and Neutron Noise, *Progr. in Nucl. Energy*, Vol. 9, 201-208, 1985.

Tanaka, S., Diagnosability of Systems and Optimal Sensor Location, *Zeitschrift für Angewandte Mathematik und Mechanik*, Vol.67, no.4, 155-157, 1987.

Tanaka, S. and Müller, P.C., Diagnosability and Failure Detection in Linear Discrete Dynamical Systems, *System Fault Diagnostics, Reliability and Related Knowledge-Based Approaches*, Vol. 1, S. G. Tzafestas, M. Singh and G. Schmidt (eds), Reidel, pp 161-175, 1987.

Tanaka, S., Sagara, M. and Okita, T., On Optimal Sensor Locations for Anomaly Diagnosis, *SICE Trans.*, Vol. 20, 60-67, Japanese, 1984.

Teneketzis, D., Castanon, D.A. and Sandell, N.R., Distributed Estimation for Large-Scale Event-Driven Systems, *Control and Dynamic Systems*, Vol. 22, 1-45, 1985.

Thompson, B.A. and Thompson, W.A., Inside an Expert System, *Byte*, 315-330, Apr. 1985.*

Tomovic, R. and Vukobratovic, M., *General Sensitivity Theory*, Modern Analytic and Computational Methods in Science and Mathematics, Vol. 35, R. Bellman (ed), Elsevier Press, 1972.

Tsuge, Y., Shiozaki, J., Matsuyama, H. and O'Shima, E., Fault Diagnosis Algorithms Based on the Signed Directed Graph and its Modifications, *I. Chem. Eng. Symp. Ser.*, Vol. 92, p. 133, 1985.*

Tugnait, J.K., Adaptive Estimation and Identification for Discrete Systems with Markov Jump Parameters, *IEEE Trans. Automatic Control*, Vol. AC-27, 1054-1065, 1982.

Türkcan, E., Review of Borssele PWR Noise Experiments, Analysis and Instrumentation, *Progr. Nucl. Energy*, Vol. 9, 437-452, 1982.

Tylee, J.L., A Generalised Likelihood Ratio Approach to Detecting and Identifying Failures in Pressuriser Instrumentation, *Nucl. Tech.*, Vol. 56, 484-492, 1982.

Tylee, J.L., On-Line Failure Detection in Nuclear Power Plant Instrumentation, *IEEE Trans. on Automatic Control*, Vol. AC-28, no. 3, 406-415, 1983.

Tylee, J.L. and Purviance, J.E., Providing Nuclear Reactor Control Information in the Presence of Instrument Failures, *Proc. American Control Conferences, Seattle, Wa.*, 170-175, 1986.

Tzafestas, S.G., Knowledge Engineering Approach to System Modeling, Diagnosis, Supervision and Control, *Proc. of the IFAC Symp. on Simulation of Control Systems*, I. Troch, P. Kopacek and F. Breitenecker (eds), 1986. *

Tzafestas, S.G., Artificial Intelligence Techniques in Control: An Overview, *Proc. of the IMACS Symp. on AI, Expert Systems and Languages in Modelling and Simulation, Barcelona*, C. Kulikowski and G. Ferrate (eds), 1987.*

Tzafestas, S.G., Singh, M. and Schmidt, G., *System Fault Diagnostics, Reliability and Related Knowledge-Based Approaches*, Vols. 1,2, Reidel, 1987. *

Tzafestas, S.G., *Knowledge Based Systems Diagnosis, Supervision and Control*, Plenum, New York-London, 1989.*

Umeda, T., Kyriyama, T., O'Shima, E. and Matsuyama, H., A Graphical Approach to Cause and Effect Analysis of Chemical Processing Systems, *Chem. Eng. Sci.*, Vol. 35, p. 2379, 1980.*

Upadhyaya, B.R., Sensor Failure Detection and Estimation, *Nuclear Safety*, Vol. 26, 32-43, 1985.

Upadhyaya, B.R. and Kerlin, T.W., Estimation of Response Time Characteristics of Platimum Resistance Thermometers by the Noise Analysis Technique, *ISA Trans.*, Vol. 17, 21-38, 1978.

Upadhyaya, B.R. and Kitamura, M., Stability Monitoring of Boiling Water Reactors by Time Series Analysis of Neutron Noise, *Nucl. Sci. Eng.*, Vol. 77, 480-492, 1981.

Upadhyaya, B.R., Kitamura, M. and Kerlin, T.W., Multivariate Signal Analysis Algorithms for Process Monitoring and Parameter Estimation in Nuclear Reactors, *Annals. Nucl. Energy*, Vol. 7, 1-10, 1980.

Usoro, P.B., Schick, I.C. and Negahdaripour, S., An Innovation-Based Methodology for HVAC System Fault Detection, *Trans. ASME J. of Dynamic Syst., Meas., and Control*, Vol 107, 284-289, 1985.

Van Dooren, P., Dewilde, P., *The Eigenstructure of an Arbitrary Polynomial Matrix*, Linear Algebra and Applications 50, 545-579, 1983.

Van Dooren, P., The Computation of Kronecker's Canonical Form of a Singular Pencil, *Lin. Algebra and Appl.*, 27, 103-140, 1979.

Van Melle, W., Shortliffe, E.H., Edward, H. and Buchanan B.G., EMYCIN: A Knowledge Engineer's Tool for Constructing Rule-Based Expert Systems, in *Rule-Based Expert Systems*, Buchanan and Shortliffe (eds), Addison Wesley, New York, pp 302-328, 1984.*

Van Trees, H. L., *Detection, Estimation and Modulation Theory, Part I*, Wiley and Sons, Inc., N.Y., 1968.

Viswanadham, N. and Srichander, R., Fault Detection using Unknown Input Observers, *Control-Theory and Advanced Technology*, Vol. 3, 91-101, June 1987.

Wagdi, M.N., An Adaptive Control Approach to Sensor Failure Detection and Isolation, *Journal of Guidance, Control and Dynamics*, Vol. 5, 118-123, 1982.

Waidelich, D.L., Pulsed Eddy Currents, *Research Techniques in Non-destructive Testing*, Sharpe (ed.), Academic Press, New York, 1964.

Wald, A., *Sequential Analysis*, Wiley, 1947 (reprinted by Dover, 1973).

Walker, B.K., *A Semi-Markov Approach to Quantifying Fault-Tolerant System Performance*, Sc.D. Thesis, Dept. of Aero & Astro., MIT, Cambridge, USA, July 1980.

Walker, B.K., Performance Evaluation of Systems that Include Fault Diagnostics, *Proc. Joint Automatic Control Conference* (paper FA-2D), *Charlottesville, Va*, June 1981.

Walker, B.K., Chu, S.K. and Wereley, N.M., Approximate Evaluation of Reliability and Availability via Perturbation Analysis, *Annual Progress Report to Air Force Office of Scientific Research* (AFOSR-84-0160), Dept. of Aero. & Astro., MIT, USA, Dec. 1986.

Walker, B.K. and Gai, E., Fault Detection Threshold Determination Technique Using Markov Theory, *AIAA Journal of Guidance, Control and Dynamics*, Vol.2, 313-319, July-Aug. 1979.

Wang, P.K.C., Invariance, Uncontrollability and Unobservability in Dynamic Systems, *IEEE Trans. Automatic Control*, Vol. AC-10, 366-367, 1965.

Wang, S.H. and Davison, E.J., Observing Partial States for Systems with Unmeasureable Disturbances, *IEEE Trans. Automatic Control*, Vol. AC-23, 481-484, 1978.

Watanabe, K., General Two-Stage Bias Correction Filter and Predictor for Linear Discrete-Time Systems, *Control Theory and Advanced Technology*, Vol. 1, 87-101, 1985.

Watanabe, K., Decentralised Two-Filter Smoothing Algorithms in Discrete-Time Systems, *Int. J. Control*, Vol. 44, 49-63, 1986.

Watanabe, K., Steady-State Covariance Analysis for A Forward-Pass Fixed Interval Smoother, *Trans. ASME Journal of Dynamic System Space, Measurement, and Space Control*, Vol. 108, 136-140, 1986.

Watanabe, K., A Multiple Model Adaptive Filtering Approach to Fault Diagnosis in Stochastic Systems, *Chapter 12 of this book*.

Watanabe, K. and Himmelblau, D.M., Instrument Fault Detection in Systems with Uncertainties, *Int. J. Systems Sci.*, Vol. 13, 137-158, 1982.

Watanabe, K. and Iwasaki, M., A Fast Computational Approach in Optimal Distributed-Parameter State Estimation, *Trans. ASME Journal of Dynamic Systems, Measurement, and Control*, Vol. 105, 1-10, 1983.

Watanabe, K., Yoshimura, T. and Soeda, T., A Diagnosis Design for a Parametric Failure, *Trans. Society of Instrument and Control Engineers*, Vol. 15, 901-906, 1979.

Watanabe, K., Yoshimura, T. and Soeda, T., A Diagnosis Method for Linear Stochastic Systems with Parametric Failures, *Trans. ASME, Journal of Dynamic Systems, Measurement, and Control*, Vol. 103, 28-35, 1981.

Watanabe, K., Yoshimura, T. and Soeda, T., A Discrete-Time Adaptive Filter for Stochastic Distributed Parameter System, *Trans. ASME, Journal of Dynamic Systems, Measurement, and Control*, Vol. 103, 266-278, Sept. 1981.

Weiss, J.L., Pattipati, K.R., Willsky, A.S., Eterno, J.S. and Crawford, J.T., *Robust Detection/Isolation/Accomodation for Sensor Failures*, Research Report no. TR-213, Alphatec Inc., Burlington, Ma., 1985 (see also NASA Contractor Report no. CR-17479, 1985).

Wells, W.R., de Silva, C.W., Failure State Detection of Aircraft Engine Output Sensors, *Proc. Joint Automatic Control Conference, Vol. 2, San Francisco, Ca.*, 1493-1497, 1977.

White, J.E., Speyer, J.L., Detection Filter Design: Spectral Theory and Algorithms, *Proc. ACC, Seattle, Wa.*, 1475-1481, 1986.

Wilkinson, J.H., Linear Differential Equations and Kronecker's Canonical Form; *Recent Advances in Numerical Analysis*, C. de Boor and G.H. Golub (eds), Academic Press, New York, 1978.

Willcox, S.W. and Patton, R.J., *Design and Development of Three Types of Instrument Failure Detection Schemes Based on a Fully Non-Linear Aircraft Model, Department of Electronics*, Research Report, No. YEE1/85, University of York, UK, August 1985.

Williams, J.P. and Montgomery, R.C., Failure Detection and Accommodation in Structural Dynamics Systems Using Analytic Redundancy, *Proc. 24th IEEE CDC Conference*, Dec. 11-13, 1985.

Williams, M.M.R. (ed.), *Progr. in Nucl. Energy*, Vol. 1, 1977.

Williams, M.M.R. (ed.), *Progr. in Nucl. Energy*, Vol. 9, 1982.

Williams, T.L., Orgren, P.J. and Smith, C.L., Diagnosis of Multiple Faults in Nationwide Communications Networks, *Proc. IJCAI-83*, pp. 179-181, 1983.*

Willsky, A.S., A Survey of Design Methods for Failure Detection in Dynamic Systems, *Automatica*, Vol. 12, 601-611, 1976.

Willsky, A.S., Detection of Abrupt Changes in Dynamic Systems, *Lecture Notes in Control and Information Sciences*, Vol. 77, 27-49, 1986.

Willsky, A.S., Chow, A.Y., Gershwin, S.B., Greence, C.S., Houpt, P.K. and Kurkjian, A.L., Dynamic Model-Based Techniques for the Detection of Incidents on Freeways, *IEEE Trans. Automatic Control*, Vol. AC-25, 347-360, 1980.

Willsky, A.S., Deyst, J.J. and Crawford, B.S., Adaptive Filtering and Self-Test Methods for Failure Detection and Compensation, *Proc. 1974 Joint Automatic Control Conference, Austin, Texas*, 637-645, 1974.

Willsky, A.S., Deyst, J.J. and Crawford, B.S., Two Self-Test Methods Applied to an Inertial System Problem, *J. of Spacecraft and Rockets*, Vol. 12, 434-437, 1975.

Willsky, A.S. and Jones, H.L., A Generalised Likelihood Ratio Approach to the Detection and Estimation of Jumps in Linear Systems, *IEEE Trans. Automatic Control*, Vol.AC-21, 108-112, 1976.

Wittig, G., Application of Pulsed Eddy Current Methods in Non-Destructive Testing, *British Steel Corporation Tubes Division*, Translation no. TD1102. British Iron and Steel Industry Translation Service BISI 16234, (Translated from: Materialprüfung Vol. 19, 9, 365-370, 1977), Nov. 1977.

Wonham, W.M., *Linear Multivariable Control: A Geometric Approach*, Springer, New York, 1974.

Woodfield, A.A. and Woods, J.F., Worldwide Experience of Wind Shear during 1981-1982, *AGARD Conference Proc. no. 347*, 1984.

Wright, R.L., *The Microwave Radiometer Spacecraft - A Design Study*, NASA Reference Publication 1079, Dec. 1981.

Wünnenberg, J., Clark, R.N. and Frank, P.M., An Application of Instrument Fault Detection, *Proc. 7th IFAC Symposium on Identification and System Parameter Estimation, York, U.K.*, July 1985, Pergamon Press, 1986.

Wünnenberg, J. and Frank, P.M., Sensor Fault Detection Via Robust Observers, *First European Workshop on Fault Diagnostics, Rhodes*, Greece, Aug. 31-Sept. 3, 1986: in System Fault Diagnostics, Reliability and Related Knowledge-Based Approaches, Vol. 1, Tzafestas, Singh, Schmidt (eds), Reidel Press, 1987.

Wünnenberg, J. and Frank, P.M., Model-Based Residual Generation for Dynamic Systems with Unknown Inputs, *Proc. 12th IMACS World Congress on Scientific Computation, Paris*, Vol. 2, 435-437, July 1988.

Yager, R., On the Relationship of Methods of Aggregating Evidence in Expert Systems, *Cybernetics and Systems*, Vol. 16, 1-21.*

Yager, R., Strong Truth and Rules of Inference in Fuzzy Logic and Approximate Reasoning, *Cybernetics and Systems*, Vol. 16, 23-63.*

Yoshimura, T., Watanabe, K., Konishi, K. and Soeda, T., A Sequential Failure Detection Approach and the Identification of Failure Parameters, *Int. J. Systems Sci.*, Vol. 10, 827-836, 1979.

Young, P. C., Parameter Estimation for Continuous-time models - a Survey, *Automatica*, Vol. 17, 23, 1981.

Yu, B.H., *A Robust Instrument Fault Detection Scheme*, MS Thesis, Department of Electrical Engineering, University of Washington, Seattle, USA, 1985.

Zadeh, L.A., Fuzzy Sets, *Information and Control, Vol. 8*, 338, 1965.*

Zaniolo, C., Object-Oriented Programming in Prolog, *Proc. Int. Symp. Logic Programming*, 1984.*